Proteomics of Microbial Pathogens

Edited by
Peter R. Jungblut and
Michael Hecker

1807–2007 Knowledge for Generations

Each generation has its unique needs and aspirations. When Charles Wiley first opened his small printing shop in lower Manhattan in 1807, it was a generation of boundless potential searching for an identity. And we were there, helping to define a new American literary tradition. Over half a century later, in the midst of the Second Industrial Revolution, it was a generation focused on building the future. Once again, we were there, supplying the critical scientific, technical, and engineering knowledge that helped frame the world. Throughout the 20th Century, and into the new millennium, nations began to reach out beyond their own borders and a new international community was born. Wiley was there, expanding its operations around the world to enable a global exchange of ideas, opinions, and know-how.

For 200 years, Wiley has been an integral part of each generation's journey, enabling the flow of information and understanding necessary to meet their needs and fulfill their aspirations. Today, bold new technologies are changing the way we live and learn. Wiley will be there, providing you the must-have knowledge you need to imagine new worlds, new possibilities, and new opportunities.

Generations come and go, but you can always count on Wiley to provide you the knowledge you need, when and where you need it!

William J. Pesce
President and Chief Executive Officer

Peter Booth Wiley
Chairman of the Board

Proteomics of Microbial Pathogens

Edited by
Peter R. Jungblut and Michael Hecker

WILEY-VCH Verlag GmbH & Co. KGaA

The Editors

Dr. Peter R. Jungblut
Max-Planck Institut für
Infektionsbiologie
Schumannstr. 21–22
10117 Berlin

Prof. Dr. Michael Hecker
Institut für Mikrobiologie
Ernst-Moritz-Arndt-Universität
Friedrich-Ludwig-Jahn-Str. 15
17487 Greifswald

Cover
Drawing of E. coli reproduced with kind permission of David S. Goodsell, The Scripps Research Institute

Cover-Design
A-Plus Design, Achim Bauer, Ludwigshafen

All books published by Wiley-VCH are carefully produced. Nevertheless, authors, editors, and publisher do not warrant the information contained in these books, including this book, to be free of errors. Readers are advised to keep in mind that statements, data, illustrations, procedural details or other items may inadvertently be inaccurate.

Library of Congress Card No.:
applied for

British Library Cataloguing-in-Publication Data
A catalogue record for this book is available from the British Library.

Bibliographic information published by the Deutsche Nationalbibliothek
The Deutsche Nationalbibliothek lists this publication in the Deutsche Nationalbibliografie; detailed bibliographic data is available in the Internet at http://dnb.d-nb.de.

© 2007 WILEY-VCH Verlag GmbH & Co. KGaA, Weinheim

All rights reserved (including those of translation into other languages). No part of this book may be reproduced in any form – by photoprinting, microfilm, or any other means – nor transmitted or translated into a machine language without written permission from the publishers. Registered names, trademarks, etc. used in this book, even when not specifically marked as such, are not to be considered unprotected by law.

Printed in the Federal Republic of Germany
Printed on acid-free paper

Typesetting X Con Media AG, Bonn
Printing Strauss GmbH, Mörlenbach
Binding Litges & Dopf Buchbinderei GmbH, Heppenheim

ISBN: 978-3-527-31759-2

Table of Contents

1	**Genome and proteome analysis of *Chlamydia***	**1**
	Brian B. S. Vandahl, Svend Birkelund and Gunna Christiansen	
1.1	Introduction 1	
1.1.1	*Chlamydia* biology 2	
1.1.1.1	Diseases 2	
1.1.1.2	The developmental cycle 3	
1.2	*Chlamydia* genomes 5	
1.2.1	Sequenced *Chlamydia* genomes 5	
1.2.2	Chlamydial genes 6	
1.2.3	Genome comparison 7	
1.3	Proteome analysis of *Chlamydia* 9	
1.3.1	Early *Chlamydia* proteome studies 10	
1.3.2	*C. trachomatis* proteome studies 10	
1.3.3	*C. pneumoniae* proteome studies 11	
1.3.4	Identification of secreted proteins by comparative proteomics 13	
1.3.5	Proteome studies of comc 15	
1.3.6	Proteome comparison of *S. trachomatis* serovars 15	
1.3.7	Proteome analysis of growth conditions 16	
1.3.8	Considerations in proteomics 16	
1.4	Concluding remarks 17	
2	***Helicobacter pylori* vaccine development based on combined subproteome analysis** **21**	
	Dirk Bumann, Peter R. Jungblut and Thomas F. Meyer	
2.1	Introduction 21	
2.2	Classical whole-cell inactivated *Helicobacter* vaccines 22	
2.3	Subunit *Helicobacter* vaccines: Conventional antigen selection 22	
2.4	Subunit *Helicobacter* vaccines: Global antigen selection based on proteomics 23	
2.4.1	Proteomics as a tool for antigen characterization 23	

Proteomics of Microbial Pathogens. Edited by Peter R. Jungblut and Michael Hecker
Copyright © 2007 WILEY-VCH Verlag GmbH & Co. KGaA, Weinheim
ISBN: 978–3–527–31759–2

2.4.2	The *Helicobacter* proteome 24
2.4.3	Criteria for promising antigen candidates 25
2.4.4	Identification of protective antigens based on multiple criteria 27
2.5	Concluding remarks 28

3 Towards a comprehensive understanding of *Bacillus subtilis* cell physiology by physiological proteomics   31
Michael Hecker and Uwe Völker

3.1	Introduction 32
3.2	Subproteomes *vs.* the total theoretical proteome 33
3.3	The vegetative proteome of growing cells 34
3.4	Proteomes of nongrowing cells – the adaptational network 39
3.5	Proteomic signatures – tools for microbial physiology and their practical application 48
3.6	Transcriptomics *vs.* proteomics – towards a second generation of proteomics 50
3.7	The interactome 52
3.8	The secretome 53
3.9	Post-translational modifications 53
3.10	Protein quality control/protein degradation at a proteomic scale 55
3.11	Gene expression network – horizontal and vertical approach 56
3.12	Concluding remarks 58

4 Web-accessible proteome databases for microbial research   63
Klaus-Peter Pleißner, Till Eifert, Sven Buettner, Frank Schmidt, Martina Boehme, Thomas F. Meyer, Stefan H. E. Kaufmann and Peter R. Jungblut

4.1	Introduction 64
4.2	Materials and methods 64
4.2.1	Data generation and data storage 64
4.2.2	Software tools 65
4.3	Results and discussion 65
4.3.1	Data management, analysis and presentation 65
4.3.2	2D-PAGE database 66
4.3.3	ICAT-LC/MS database 68
4.3.4	FUNC_CLASS database 68
4.3.5	Data analysis and visualization 69
4.4	Concluding remarks 73

5	**A targeted proteomics approach to the rapid identification of bacterial cell mixtures by matrix-assisted laser desorption/ionization mass spectrometry** 75
	Bettina Warscheid and Catherine Fenselau
5.1	Introduction 75
5.2	Materials and methods 76
5.2.1	Chemicals 76
5.2.2	*Bacillus* strains 77
5.2.3	Vegetative cell digestion 77
5.2.4	MALDI-TOF MS and unimolecular decomposition product ion analysis 77
5.2.5	Database searches and identification of *Bacillus* species 78
5.3	Results and discussion 79
5.3.1	On-probe tryptic digestion of bacterial cells 79
5.3.1.1	*Bacillus subtilis* 168 79
5.3.1.2	*Bacillus globigii* and *sphaericus* 14577 79
5.3.1.3	*Bacillus cereus* T and *anthracis* Sterne 81
5.3.2	Partial sequencing for rapid identification of bacterial cells 83
5.3.2.1	Unimolecular decomposition product ion analysis (UDPIA) 83
5.3.2.2	Identification of bacteria 85
5.3.3	Identification of proteins and protein-families in bacterial cell digests 85
5.3.3.1	Flagellin and surface layer protein precursor 85
5.3.3.2	Cold shock and cold shock-like proteins 87
5.3.3.3	Ribosomal proteins 90
5.3.3.4	DNA-binding proteins 90
5.3.3.5	Heat shock proteins 91
5.3.3.6	Other stress related proteins and the prosthetic group of an acyl-carrier protein 92
5.3.4	Analysis of a 1:1 mixture of *B. globigii* and *B. sphaericus* 14577 92
5.4	Concluding remarks 94
6	**Protein identification and tracking in two-dimensional electrophoretic gels by minimal protein identifiers** 97
	Jens Mattow, Frank Schmidt, Wolfgang Höhenwarter, Frank Siejak, Ulrich E. Schaible and Stefan H. E. Kaufmann
6.1	Introduction 98
6.2	Materials and methods 99
6.2.1	Materials 99
6.2.2	Two-dimensional electrophoresis 100
6.2.3	Mass spectrometry 100

6.2.4 Compilation of a theoretical dataset comprising all tryptic peptides of *M. tuberculosis* H37Rv 100
6.2.5 Comparison of MALDI spectra by the program MS-Screener 101
6.2.6 Determination of exogenous contaminant masses 101
6.2.7 Generation of template spectra 102
6.3 Results and discussion 102
6.3.1 Proteome analysis of *M. tuberculosis* H37Rv CSN 102
6.3.2 Determination of exogenous contaminant masses by MS-Screener 102
6.3.3 The MPI approach revealed HspX-specific peptide masses in multiple spectra 106
6.3.4 The MS-Screener analysis revealed truncated variants of Tuf previously not identified by PMF 111
6.3.5 Frequency of tryptic peptides with similar m/z ratios in *M. tuberculosis* H37Rv 113
6.3.6 In mass spectrometry it is important to consider detection probabilities of proteins and peptides 114
6.4 Concluding remarks 117

7 Continued proteomic analysis of *Mycobacterium leprae* subcellular fractions 121
Maria Angela M. Marques, Benjamin J. Espinosa, Erika K. Xavier da Silveira, Maria Cristina V. Pessolani, Alex Chapeaurouge, Jonas Perales, Karen M. Dobos, John T. Belisle, John S. Spencer and Patrick J. Brennan
7.1 Introduction 121
7.2 Materials and methods 125
7.2.1 Preparation and separation of *M. leprae* proteins 125
7.2.2 MS 126
7.2.3 Isolation of basic proteins of *M. leprae* 127
7.3 Results and discussion 127
7.4 Concluding remarks 138

8 CFP10 discriminates between nonacetylated and acetylated ESAT-6 of *Mycobacterium tuberculosis* by differential interaction 141
Limei Meng Okkels, Eva-Christina Müller, Monika Schmid, Ida Rosenkrands, Stefan H. E. Kaufmann, Peter Andersen and Peter R. Jungblut
8.1 Introduction 141
8.2 Materials and methods 142
8.2.1 Protein samples 142
8.2.2 2-DE blot overlay 143
8.2.3 Mass spectrometry 143
8.3 Results 144
8.3.1 High resolution separation of acidic ST-CF proteins 144

8.3.2	Mass analysis of ESAT-6 spots	144
8.3.3	Interaction of recombinant CFP10 with ESAT-6 spots	149
8.4	Discussion	150

9 The cell wall subproteome of *Listeria monocytogenes* 153
Jessica Schaumburg, Oliver Diekmann, Petra Hagendorff, Simone Bergmann, Manfred Rohde, Sven Hammerschmidt, Lothar Jänsch, Jürgen Wehland, and Uwe Kärst

9.1	Introduction	153
9.2	Material and methods	154
9.2.1	Bacterial strain and growth conditions	154
9.2.2	Serial extraction of cell wall proteins	155
9.2.3	Aminopeptidase C assay	155
9.2.4	SDS-PAGE Western blotting, and *N*-terminal sequencing	155
9.2.5	2-D-PAGE	155
9.2.6	Gel staining and protein identification by mass spectrometry	156
9.2.7	Immunoelectron microscopy	156
9.2.7.1	Postembedding labeling studies	156
9.2.7.2	Field emission scanning electron microscopic immunolabeling	157
9.2.8	Overlay blot	157
9.2.9	Ligand fishing	158
9.2.10	Cloning and purification of proteins	158
9.2.11	Kinetic analyses using SPR detection	159
9.2.12	Bioinformatic analysis	159
9.3	Results	160
9.3.1	Prediction and validation of cell wall-associated proteins	160
9.3.1.1	Prediction of exported proteins	160
9.3.1.2	Validation of the cell wall subproteome	160
9.3.2	Analysis of protein processing and localization	167
9.3.3	Analysis of plasminogen-binding proteins	168
9.4	Discussion	173
9.4.1	Analysis of proteins identified from surface extracts	173
9.4.2	Analysis of protein processing and localization in surface extracts	174
9.4.3	Possible function of plasminogen binding in virulence	175
9.5	Concluding remarks	177

10 Low virulent strains of *Candida albicans*: Unravelling the antigens for a future vaccine 181
Elena Fernández-Arenas, Gloria Molero, César Nombela, Rosalía Diez-Orejas and Concha Gil

10.1	Introduction	182
10.2	Materials and methods	183

10.2.1	Microorganism and culture conditions	183
10.2.2	Mice	183
10.2.3	Systemic infection conditions and generation of immune sera	183
10.2.4	T cell purification and passive immunization	184
10.2.5	2-DE	184
10.2.5.1	Protoplast lysate preparation	184
10.2.5.2	Analytical and micropreparative 2-DE	185
10.2.6	Immunoblot analyses	185
10.2.7	MALDI-TOF and MALDI-TOF MS analyses of spots	186
10.2.8	Database search	187
10.3	Results	187
10.3.1	Vaccination assays with different mutant strains and generation of immune sera	187
10.3.2	Importance of cellular immunity in vaccination with *C. albicans* CNC13	188
10.3.2.1	Role of Th1/Th2 cytokines in CNC13 immune response	188
10.3.2.2	Protection induced by passive transfer of sensitized CNC13 lymphocytes	189
10.3.3	Profile of *C. albicans* immunoreactive proteins in the different mutant strains	190
10.3.3.1	Detection and identification of the immunoreactive proteins	190
10.4	Discussion	194
10.4.1	Low virulent *C. albicans* strains as a tool to study the host immune response	194
10.4.2	*C. albicans hog1* mutant induces protection in a vaccination assay	196
10.4.3	*C. albicans* new antigenic proteins	197
10.4.4	Antibody profile linked to successful vaccination against systemic candidiasis	197
10.5	Concluding remarks	198
11	**Proteomic analysis of the sarcosine-insoluble outer membrane fraction of the bacterial pathogen *Bartonella henselae***	**203**
	Thomas A. Rhomberg, Olof Karlberg, Thierry Mini, Ursula Zimny-Arndt, Ulrika Wickenberg, Marlene Röttgen, Peter R. Jungblut, Paul Jenö, Siv G. E. Andersson and Christoph Dehio	
11.1	Introduction	203
11.2	Materials and methods	205
11.2.1	Strains and culture conditions	205
11.2.2	Enrichment of *B. henselae* OMPs	205
11.2.3	Protease exposure	205
11.2.4	1-D SDS-PAGE	206

11.2.5	Protein solubilization and protein quantitation	206
11.2.6	2-D NEPHGE	206
11.2.7	MALDI-TOF-MS	207
11.2.8	Database query	208
11.2.9	*In silico* analysis	208
11.3	Results	208
11.3.1	Enrichment of *B. henselae* OMPs	208
11.3.2	1-D SDS-PAGE of *B. henselae* OMPs and protein assignment by PMF	209
11.3.3	2-D NEPHGE of *B. henselae* OMPs and protein assignment by PMF	211
11.4	Discussion	215

12 **The influence of *agr* and σ^B in growth phase dependent regulation of virulence factors in *Staphylococcus aureus*** 225
Anne-Kathrin Ziebandt, Dörte Becher, Knut Ohlsen, Jörg Hacker, Michael Hecker and Susanne Engelmann

12.1	Introduction	225
12.2	Material and methods	227
12.2.1	Bacterial strains and culture conditions	227
12.2.2	Preparation of the extracellular protein fraction	227
12.2.3	Analytical and preparative PAGE	227
12.2.4	Quantitation of protein spots	228
12.2.5	Transcriptional analyses	228
12.3	Results	228
12.3.1	Growth phase dependent regulation of extracellular proteins	228
12.3.2	The influence of *agr* and σ^B on the extracellular proteome	236
12.3.3	The influence of *agr* and σ^B on the transcription of virulence genes	239
12.3.4	Effect of σ^B on the transcription of *agr* and *sarA*	239
12.4	Discussion	239
12.5	Concluding remarks	245

13 **Comparative proteome analysis of cellular proteins extracted from highly virulent *Francisella tularensis* ssp. *tularensis* and less virulent *F. tularensis* ssp. *holarctica* and *F. tularensis* ssp. *mediaasiatica*** 249
Martin Hubálek, Lenka Hernychová, Martin Brychta, Juraj Lenčo, Jana Zechovská and Jiří Stulík

13.1	Introduction	250
13.2	Material and methods	251
13.2.1	Bacterial cultures and sample preparation for 2-DE	251
13.2.2	Two-dimensional gel electrophoresis	252
13.2.3	Statistical analysis	252

13.2.4	In-gel digestion 253
13.2.5	Mass spectrometric identification 253
13.3	Results and discussion 254
13.4	Concluding remarks 265

14 Proteome comparison of *Vibrio cholerae* cultured in aerobic and anaerobic conditions 267
Biao Kan, Hajar Habibi, Monika Schmid, Weili Liang, Ruibai Wang, Duochun Wang and Peter R. Jungblut

14.1	Introduction 267
14.2	Materials and methods 269
14.2.1	Strains and culture 269
14.2.2	Sample preparation for 2-DE 269
14.2.3	2-DE 269
14.2.4	MS 270
14.2.5	Electron microscope 270
14.3	Results and discussion 270
14.3.1	Identified protein species more abundant in aerobic culture 274
14.3.2	Identified protein species which are more abundant in anaerobic culture 275
14.4	Concluding remarks 276

15 Induction of *Mycobacterium avium* proteins upon infection of human macrophages 279
Lara Brunori, Federico Giannoni, Luca Bini, Sabrina Liberatori, Cristiane Frota, Peter Jenner, Ove Fredrik Thoresen, Graziella Orefici and Lanfranco Fattorini

16 Proteomics-based identification of novel *Candida albicans* antigens for diagnosis of systemic candidiasis in patients with underlying hematological malignancies 289
Aida Pitarch, Joaquín Abian, Montserrat Carrascal, Miguel Sánchez, César Nombela and Concha Gil

16.1	Introduction 290
16.2	Materials and methods 291
16.2.1	Human serum samples 291
16.2.2	Preparation of *C. albicans* protoplast lysates 292
16.2.3	Two-dimensional polyacrylamide gel electrophoresis (2-D PAGE) 292
16.2.4	Immunoblot analysis 294
16.2.5	Peptide sample preparation for MS and MS/MS analyses 294
16.2.5.1	In-gel digestion 294
16.2.5.2	Desalting 294

16.2.6	Mass spectrometric analysis (MS and MS/MS)	295
16.2.6.1	Peptide mapping by MALDI-TOF-MS	295
16.2.6.2	Peptide fragmentation and sequencing by nanoESI-IT-MS	295
16.2.7	Database searching	296
16.3	Results	296
16.3.1	Mapping of C. albicans immunogenic proteins	296
16.3.1.1	Overall 2-D C. albicans antigen recognition pattern	296
16.3.1.2	Identification of C. albicans immunoreactive proteins by peptide mass fingerprinting (PMF)	297
16.3.1.3	Peptide sequencing of C. albicans immunoreactive proteins	298
16.3.1.4	Reference 2-DE C. albicans antigen pattern display on the Net	305
16.3.2	Comparison of 2-D C. albicans antigen recognition patterns obtained with serum samples from patients with and without systemic candidiasis	309
16.3.3	Differences in the 2-D C. albicans antigen recognition profile associated with infection progression	311
16.4	Discussion	312
16.4.1	C. albicans housekeeping enzymes can stimulate the human immune system during systemic candidiasis	316
16.4.1.1	Heat shock proteins (Hsps)	317
16.4.1.2	Metabolic proteins	318
16.4.1.3	Elongation factors and ribosomal proteins	318
16.4.1.4	Miscellaneous proteins	319
16.4.2	Natural anti-Candida antibodies might be correlated with differentiation of the human immune response	319
16.4.3	Serum levels of specific anti-Candida antibodies could be useful for the clinical follow-up of systemic candidiasis	320
	Index	325

Preface
Proteomics of microbial pathogens

Infectious diseases still plague mankind. According to the World Health Report 2004, 19.1% of the deaths estimated in 2002 were caused by infectious diseases. Aids, tuberculosis and malaria each contributed more than 2% to this figure. In June 2006, 387 completely sequenced genomes (http://www.genomesonline.org/) have been published in total, 352 of them from bacteria, an important prerequisite for the analysis of the proteomes of these organisms. In total 940 ongoing bacterial genome projects were reported.

The first successful proteome studies revealed vaccine candidates with promising results in animal models. Immunoproteomics resulted in the detection of antigens which may be used for diagnostics and vaccine candidate prediction. So it can be assumed that proteomics will make a marked contribution to the improvement of worldwide health within the next few years.

Here we look at some of the trends in this field. As there are so many microorganisms currently under investigation, it is not possible to present a comprehensive overview of microbial proteomics. Proteomics technology has been automated within recent years: spot picking, digestion, LC-MS/MS and database searches have increased throughput but produced new bottlenecks in quality control and data evaluation. Microorganisms are ideal models for the application of these new technologies. Bacteria with genomes containing 600 to 7000 predicted genes present a medium-sized complexity which can be used to apply proteomic techniques with a good chance of obtaining an overview of a substantial part of the proteome in combination with prefractionation procedures. Standardization is now an important theme in proteomic technology but the multiple properties of organisms and proteins make standardizing sample preparation nearly impossible. Even related bacteria need different procedures for sample preparation, as outlined in this book in the example of *Mycobacterium leprae*. It may be estimated that in one biological situation more than 50% of the predicted proteins may be identified for genomes such as *Mycoplasma pneumoniae* containing less than 1000 genes, 30% for those containing less than 2500 genes and only 10% for those containing more than 4000 genes. Subfractionation contributes to the number of accessible proteins, but in the future throughput has to be increased further to allow the presentation of the proteome in a kind of film with changing environ-

Proteomics of Microbial Pathogens. Edited by Peter R. Jungblut and Michael Hecker
Copyright © 2007 WILEY-VCH Verlag GmbH & Co. KGaA, Weinheim
ISBN: 978-3-527-31759-2

mental conditions. Only then may more complete proteomes become accessible. Bioinformatics accompanies proteomics through all the technological steps, allowing the data obtained to be stored in a database. A microbial proteomics database system was set up at the Max Planck Institute for Infection Biology (http://www.mpiib-berlin.mpg.de/2D-PAGE/) and by June 2006 it contains 18 bacterial species and 4889 identified spots. Peptide mass fingerprinting data are stored for *Helicobacter pylori* and isotopic labelling results are represented for *Mycobacterium tuberculosis* LC/MS data. Proteomics of microorganisms allow the scientist to start with a hypothesis-free global approach and focus early on the hypotheses elaborated from this first step. In the first few years we learned that posttranslational modifications play a more important role than expected in bacteria, and the resulting protein species composition may be directly visualized by 2-DE/MS but not by LC/MS which has other advantages such as higher throughput and sensitivity potentials. At the moment, for example, the impact of more than 10 ESAT-6 protein species in *Mycobacterium tuberculosis* remains unclear. Proteome analysis at the protein species level is a task for the future.

We wish to thank the authors for their contributions, the referees for their prompt reviewing of the manuscripts and the publishers for their help in producing this book. We also take this opportunity to thank the "Bundesministerium für Bildung und Forschung" in Germany for financing the project "New methods to access the complete proteome of bacteria" and the European Union for support in developing the European Bacterial Proteome Database within the project "Comparative analysis of proteome modulation in human pathogenic bacteria for the identification of new vaccines, diagnostics and antibacterial drugs" (QLRT-1999–31536). Several articles in this book were supported by these two initiatives.

Peter R. Jungblut
Max Planck Institute for Infection Biology
Core Facility Protein Analysis
Berlin

Michael Hecker
Ernst-Moritz-Arndt-Universität Greifswald
Institut für Mikrobiologie
Greifswald

1
Genome and proteome analysis of *Chlamydia**

Brian B. S. Vandahl, Svend Birkelund and Gunna Christiansen

It has been difficult to study the molecular biology of the obligate intracellular bacterium *Chlamydia* due to lack of genetic transformation systems. Therefore, genome sequencing has greatly expanded the information concerning the biology of these pathogens. Comparing the genomes of seven sequenced *Chlamydia* genomes has provided information of the common gene content and gene variation. In addition, the genome sequences have enabled global investigation of both transcript and protein content during the developmental cycle of chlamydiae. During this cycle *Chlamydia* alternates between an infectious extracellular form and an intracellular dividing form surrounded by a phagosome membrane termed the chlamydial inclusion. Proteins secreted from the chlamydial inclusion into the host cell may interact with host cell proteins and modify the host cell's response to infection. However, identification of such proteins has been difficult because the host cell cytoplasm of *Chlamydia* infected cells cannot be purified. This problem has been circumvented by comparative proteomics.

1.1
Introduction

Chlamydia is an obligate intracellular bacterium comprising a number of important animal and human pathogens causing infections with serious sequelae. *Chlamydia trachomatis* is a cause of ocular and genital infections. *Chlamydophila pneumoniae* (previously *Chlamydia pneumoniae*) causes respiratory diseases and has been associated with asthma and atherosclerosis. Sequelae are primarily due to an inflammatory response, which may be sustained by bacteria persisting in the infected organism due to a special intracellular nonreplicative state [1] but delayed-type hypersensitivity may also be involved.

---

* Originally published in Proteomics 2004, 10, 2831–2842

Proteomics of Microbial Pathogens. Edited by Peter R. Jungblut and Michael Hecker
Copyright © 2007 WILEY-VCH Verlag GmbH & Co. KGaA, Weinheim
ISBN: 978–3-527–31759–2

Molecular biological studies of *Chlamydia* have been hampered by the lack of genetic transformation systems. Therefore, sequencing of the genomes of several *Chlamydia* species has been especially important for chlamydial research. The *C. trachomatis* serovar D genome was published in 1998 [2] and in 1999 the first *C. pneumoniae* genome followed [3]. Today *Chlamydia* is one of the most extensively sequenced microorganisms with seven published genomes including four from different isolates of *C. pneumoniae* (http://www.ncbi.nlm.nih.gov:80/PMGifs/Genomes/org.html). Besides direct analysis of genome sequences (genomics), global investigation of transcripts (transcriptomics) and protein content (proteomics) are developed based on the genome sequences.

1.1.1
Chlamydia biology

Traditionally *Chlamydia* was the only genus in the family of *Chlamydiacae* which was the only family in the order of *Chlamydiales*. Since the introduction of *C. pneumoniae* in 1989 [4] there were four species, distinguished mainly by serology: *C. pneumoniae*, *C. trachomatis*, *C. pecorum* and *C. psittaci*. In 1999 a new taxonomy was suggested [5], introducing more genera and species based on phylogenetic relationships with requirement of > 95% 16S rRNA identity within a genus. The suggested taxonomy placed *C. trachomatis* in the genus *Chlamydia* and divided the *C. trachomatis* into three species. The remaining *Chlamydia* species were placed in the new genus *Chlamydophila*, and *C. psittaci* was divided into a number of different species. Similar developmental biology, similar genome size and genome organization of *C. trachomatis* (1.0 Mb) and *C. pneumoniae* (1.2 Mb) [3], representing *Chlamydia* and *Chlamydophila*, respectively, indicate basic similarities but differences are also found [6]. In the present review the new taxonomy will be followed with respect to species names, but *Chlamydia* will be used as a unifying term describing both of the suggested genera *Chlamydophila* and *Chlamydia*.

1.1.1.1 Diseases
The main human pathogenic chlamydiae are *C. trachomatis* and *C. pneumoniae*, but also bird pathogenic *C. psittaci* can cause severe pneumonia, psittacosis, if transferred to humans [7]. *C. trachomatis* is divided into three groups of serovars: (i) serovars A–C are endemic in developing countries and the cause of trachoma, which may lead to blindness by scarring of the cornea [7]; (ii) serovars D–K are sexually transmitted and cause urethritis, cervisitis and salpingitis. It is the most widespread sexually transmitted bacterial disease and infections are often asymptomatic. The infection may cause sterility and increased risk for ectopic pregnancy by scarring of the fallopian tubes if it spreads from the cervix [7]; (iii) serovars L1–L3 are also sexually transmitted but cause lymphogranuloma venereum (LGV). LGV is a more severe infection as it readily spreads to the lymphatic system and becomes systemic [7]. Serovars A–K are known as the trachoma biovar and L1–3 as the LGV biovar.

C. *pneumoniae* is a respiratory pathogen that causes acute and chronic respiratory diseases. Most infections are asymptomatic, but about 30% cause more severe pneumonia, bronchitis or other upper airway illness [8]. About 10% of the cases of community acquired pneumonia in adults and about 5% of the cases of bronchitis and sinusitis are caused by *C. pneumoniae* [9]. Persistent infections have been described [10] and there are indications that treatment may not eliminate the organism [11]. *C. pneumoniae* has been associated with chronic lung diseases [8] and as a possible risk factor for the development of atherosclerosis [12]. *C. pneumoniae* has been detected in atherosclerotic lesions [13] and studies have shown that atheromatous plaques are commonly infected with *C. pneumoniae*. Animal studies suggest that *C. pneumoniae* can accelerate atherosclerosis-like disease [14, 15]. However, other studies fail to detect *C. pneumoniae* in plaques and many studies find no significant association by serology [16, 17]. At present it is not clear whether there is an increased risk of coronary artery disease due to *C. pneumoniae* infection and if there is, the increase may be small.

1.1.1.2 The developmental cycle

Chlamydia is a Gram-negative, obligate intracellular bacterium, characterized by a biphasic developmental cycle. The developmental cycle (Fig. 1) in which the bacteria alternate between an infectious, extracellular form, the elementary body (EB) and a noninfectious intracellular replicating form, the reticulate body (RB) is unique for chlamydiae [18–20]. EBs are small rigid bodies of about 300 nm in diameter that are traditionally described as being metabolically inactive with their DNA packed by histone-like proteins [21, 22]. They are adapted for extracellular survival with a heavily disulfide cross-linked outer membrane, that provides osmotic stability. RBs are about 1 µm in diameter with an outer membrane that is permeable for transport of host cell nutrients and the DNA is unpacked as in other bacteria.

Infectious EBs attach to a susceptible host cell by which they are phagocytosed. The exact mechanism is not known but the uptake is thought to be induced by the bacteria. Inside the phagosome, named the inclusion, the EBs develop into RBs, which divide by binary fission. This includes unpacking of the DNA and reduction of the disulfide bridges of the outer membrane [23], but it is not known what triggers these events. After multiple divisions, the RBs begin conversion into EBs, including packing of the DNA and synthesis of late outer membraneproteins that are disulfide bridged. Ultimately, a new generation of infectious EBs is released upon disruption of the host cell. The bacteria stay inside the inclusion throughout the intracellular stage, which lasts for 72–96 h for *C. pneumoniae* grown in cell culture. The inclusion membrane grows by the acquisition of lipids derived from the host cell [24–26]. It is modified by the insertion of chlamydial proteins, the so-called inclusion membrane proteins (incs), and prevented from fusion with lysosomes [27, 28].

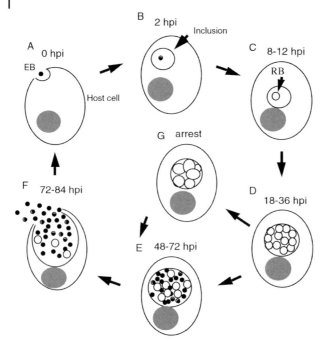

Fig. 1 The developmental cycle of *Chlamydia*. Hours post infection (hpi) are listed for *C. pneumoniae* in cell culture. A, the infectious EB adheres to a host cell and is taken up by endocytosis. B, *Chlamydia* modifies the phagosome, the chlamydial inclusion, to escape the endocytic pathway. C, the EB develops into the metabolically active RB. D, the RBs divide by binary fission and the inclusion grows by incorporation of host cell derived lipids. E, after multiple divisions, the RBs reorganize into EBs. F, ultimately, a new generation of infectious EBs is released by lysis of the host cell. G, low nutrient availability, IFN-γ mediated tryptophan starvation or other stressful conditions can trigger a persistent state with abnormal nondividing RBs. These RBs can be reactivated to enter the developmental cycle when the conditions are again suited for growth. Redrawn from [8].

The developmental cycle of *C. pneumoniae* can be arrested by interferon-gamma (IFN-γ)-induced tryptophan catabolism of the host cell [29]. Tryptophan starvation leads to a nonproductive infection in which enlarged aberrant RBs evolve. These abnormal RBs do not divide and do not mature into EBs, but the developmental cycle can be reactivated [30, 31]. Also *C. trachomatis* can enter a persistent state [32] and in addition to cytokines, limited nutrient availability [33] and treatment with antibiotics that fail to eradicate the infection have been shown to trigger this state [34, 35].

1.2
Chlamydia genomes

1.2.1
Sequenced Chlamydia genomes

The first *Chlamydia* genome sequences of *C. trachomatis* [2] and *C. pneumoniae* [3] are from the *Chlamydia* Genome Project (CGP) (http://chlamydia-www.berkeley.edu:4231/). The sequenced genomes provide insight into genome organization and metabolic pathways of *Chlamydia* and form a basis for further research in gene regulation and protein expression [36]. Genome sequences of *C. muridarum* (previously *C. trachomatis* MoPn) [37], three other isolates of *C. pneumoniae* [37–39] and most recently that of *C. caviae* (previously *C. psittaci* GPIC) [40] have been published. An overview of the sequenced genomes is given in Tab. 1 where the number of predicted protein encoding open reading frames (ORFs) is the number given in the respective references. The number of ORFs is dependent on what sequence length is considered minimum for an expected protein and the cut-off varies slightly between sequencing projects.

C. trachomatis serovar D and *C. muridarum* contain a plasmid, and in *C. caviae* and *C. pneumoniae* AR39 a bacteriophage was found. The genomes of *C. trachomatis* D and *C. muridarum* (human and mouse genital pathogens, respectively), are very similar with an average of about 10% difference between orthologous genes [37]. Most differences between these genomes were found in the replication termination region (RTR) [40] including those in *C. trachomatis* D genes involved in tryptophan synthesis, which are missing in *C. muridarum*.

The *C. pneumoniae* genomes are more than 99.9% identical and the few differences are mainly found in *pmp* [37] and *ppp* genes [41, 42]. A double-stranded circular DNA, the replicative form of a bacteriophage was found upon sequencing the *C. pneumoniae* AR-39 genome [37]. The phage of *C. pneumoniae* AR-39 was suggested as contributing to pathogenicity [43], and a similar phage was identified in *C. abortus* [6].

Tab. 1 Genome size and number of ORFs

Genome	Reference	Base pairs	ORFs	Plasmid/Phage
C. trachomatis D	[2]	1,042,519	894	7493 bp plasmid
C. pneumoniae, CWL029	[3]	1,230,230	1073	–
C. pneumoniae, AR39	[37]	1,229,853	1052	4524 bp phage
C. pneumoniae, J138	[38]	1,226,565	1072	–
C. muridarum	[37]	1,069,412	924	7501 bp plasmid
C. caviae (GPIC)	[40]	1,173,390	1009	7966 bp phage

C. pneumoniae TW-183 has also been sequenced [39] but is not contained in this table as no paper has yet been published on the results

1.2.2
Chlamydial genes

The environment of *Chlamydia* can be considered hostile, since the host cell will attempt to eradicate the bacteria, or friendly, since the bacteria have access to nutrients form the host cell. Analyzing the genome sequences of *Chlamydia* by comparing metabolic pathways and energy systems to those of free-living bacteria reveal many consequences of the availability of nutrients. However, the defense systems implicated by the intracellular nature do not appear as readily from the genome sequences since these may be unique for *Chlamydia*.

No genes are found that encode proteins involved in *de novo* purine and pyrimidine synthesis and the ability to synthesize amino acids is greatly reduced. Correspondingly, a large number of genes encoding different transport proteins have been identified, including many ABC transporters which are primarily involved in transport of smaller peptides and amino acids [2]. Also in good agreement with the intracellular and thus isolated nature of *Chlamydia*, no genes involved in DNA uptake were identified and no insertion sequences were found [2].

Chlamydiae have traditionally been described as energy parasites obtaining ATP from their host cells [19, 44], and the genomes of *C. trachomatis* and *C. pneumoniae* confirmed the presence of two genes *CT065/Cpn0351* and *CT495/Cpn0614* (CT*XXX* and Cpn*XXXX* refer to *C. trachomatis* and *C. pneumoniae* gene numbers, introduced by the CGP [2, 3]), homologous to genes encoding ATP transporting proteins from *Rickettsia prowazekii* [45]. The orthologs from *C. trachomatis* L2 were cloned and used to express functional nucleoside phosphate transporters (npt) in *Escherichia coli*, one (CT065) exchanging ADP for ATP, the other (CT495) transporting all four ribonucleoside triphosphats [46]. Surprisingly, also genes encoding a wide range of ATPases as well as phosphoglycerate kinase, pyruvate kinase, and succinate thiokinase were identified, suggesting the capability of *Chlamydia* to produce ATP itself [2]. This ability may be important in the early and late stages of the developmental cycle where *Chlamydia* supposedly cannot obtain ATP form the host cell [44]. Genes encoding the proteins of an intact glycolytic pathway (although it is questionable whether an enzyme with fructose-1,6-diphosphate aldolase activity is present or this is circumvented), a partial TCA cycle, a complete glycogen synthesis and degradation system, and genes involved in aerobic respiration were also found [2]. Furthermore, proteins encoded by many of these genes were shown to be present in EBs [47] and pyruvate kinase, phosphoglycerate kinase, glyceraldehyde-3-phosphate dehydrogenase and glucose-6-phosphate dehydrogenase were shown to complement *E. coli* mutants when expressed recombinantly [48]. During the intracellular stage, *Chlamydia* may store glycogen that is used to fuel the chlamydiae in the beginning and the end of the developmental cycle together with stored pools of ATP [48].

Four groups of chlamydial proteins have been indicated as especially interesting and important results of the genome project [49]. These groups were (i) peptidoglycan synthesis proteins; (ii) type III secretion proteins; (iii) inclusion membrane proteins (incs); and (iv) polymorphic membrane proteins (pmps).

The presence of a nearly full set of genes involved in peptidoglycan synthesis was unexpected since a peptidoglycan layer is not detected in EBs. However, *Chlamydia* is sensitive to beta-lactam antibiotics and peptidoglycan has been suggested to play a role in the division of RBs [50] supported by the finding of three amidases with probable peptidoglycan degradating activity.

The finding of type III secretion system genes was expected as such genes had earlier been found in *C. caviae* (*C. psittaci* GPIC) [51]. The type III secretion system is known from other Gram-negative bacteria to facilitate the transport of molecules from the bacterial cytosol into a future host cell by penetration of the host cell membrane with a surface protrusion that is thought to function as a channel. Surface projections of both EBs and RBs observed in electron microscopy [52, 53] thought to be involved in nutrient uptake, were speculated to be such type III needles when type III genes in *Chlamydia* were found [54, 55].

Inclusion membrane proteins are chlamydial proteins that are inserted into the inclusion membrane. Such proteins were first identified in *C. caviae* and termed incA, B and C [56, 57]. Homologs of these were found in the genomes of all sequenced chlamydiae but have not been found in any other organism. Several additional incs have since been identified and all of these share a characteristic bilobed hydrophobic region, even though no sequence motif is apparent [58]. Thirty-three genes encoding proteins with this hydrophobicity pattern have been identified in the *C. trachomatis* genome and 93 in the *C. pneumoniae* CWL029 genome [59].

Another group of *Chlamydia* specific proteins found in the genome was the family of polymorphic membrane proteins (pmps). These were initially identified in *C. abortus* (ovine abortion subtype of *C. psittaci*) being immunogenic proteins present in the outer membrane [60]. Nine *pmp* genes were identified in *C. trachomatis*, 17 in *C. caviae* and 21 in *C. pneumoniae*. The pmps are defined by being predicted outer membrane proteins containing repeated sequences of GGAI and FxxN [61] and by protein structure analysis they are predicted to be autotransporters [47, 62]. Incs and pmps are likely to be pivotal for *Chlamydia* biology indicated by the fact that 37.4% of the *Chlamydia* specific coding sequence of *C. pneumoniae* is constituted by *inc* and *pmp* genes (18.9% and 17.5%, respectively) [49].

1.2.3
Genome comparison

Genome sequences are thus available for *C. trachomatis* serovar D, *C. muridarum*, *C. caviae* and four isolates of *C. pneumoniae* (CWL029, AR39, J138 and TW-183), all of these share the unique developmental cycle but they are diverse in tissue tropism; *C. trachomatis* serovar D infects the genital tract of humans, *C. pneumoniae* infects the human respiratory tract; *C. caviae* the conjuctiva of guinea pigs and *C. muridarum* is a mouse pathogen. Hence, genome comparisons may reveal differences that are important for pathogenicity and tissue specificity.

Comparison of the *C. caviae* genome [40] to those of *C. pneumoniae* and *C. muridarum* showed that only 68/1009 *C. caviae* genes were not found in any of the other *Chlamydia* genomes, but differential expression of genes shared by the different organisms may contribute to pathogenicity differences. Seven hundred and ninety-eight genes were found in all genomes and may be the minimum set of genes required for the basic growth and development of *Chlamydia*. Out of the 798 shared genes, 183 could not be found in any other of 70 published microbial genomes in the TIGR database [40]. Investigation of these genes, which include the *inc* and *pmp* genes, may elucidate functions that are specifically related to the intracellular characteristics of *Chlamydia* and its developmental cycle.

The most prominent *C. caviae* specific genes compared to *C. pneumoniae* are the genes required for tryptophan synthesis found in the RTR. *C. caviae* appears to be able to synthesize tryptophan from anthranilate, which is a very early precursor [40]. *C. trachomatis* possess a more limited set of tryptophan synthesis genes [2] and the genital and LGV serovars can produce tryptophan from the intermediate precursor indole, whereas the ocular serovars A and C have a truncated TrpA and serovar B lacks the *trpA* operon [63] similar to that which is found for *C. pneumoniae* [3]. A tox gene similar to cytotoxic genes from enterobacteria has been found in *C. caviae* and *C. muridarum*, the product of which may be secreted by the type III secretion system in order to inhibit actin polymerization [40]. In addition, a gene with homology to an invasin/intimin family protein was identified but the gene is interrupted by two frame shifts [40]. Specific genes found in *C. pneumoniae* that are absent from *C. caviae* include a uridine kinase, two 3-deoxy-D-manno-octulosonic acid (KDO) transferases, and two genes involved in biotin synthesis. In addition, 168 genes with unknown function are present in *C. pneumoniae* but not in any other *Chlamydia* [40].

Comparing *C. pneumoniae* to *C. trachomatis*, 80% of the predicted protein encoding genes have an ortholog in *C. trachomatis* [3]. From the 214 genes found in *C. pneumoniae* but not in *C. trachomatis*, most have no known function, but those that have include genes for purine and pyrimidine salvage pathways and completion of the biotin synthase pathway. A prominent difference is the expansion of the *pmp* gene family from nine members in *C. trachomatis* to 21 members in *C. pneumoniae* [61]. The *C. trachomatis pmp* genes are located in two clusters *pmpA-C* and *pmpE–H* except for one gene, *pmpD*. Most of the difference between *C. pneumoniae* and *C. trachomatis* is accounted for by expansion of *pmpG* to 13 *pmps* (*pmp1–13*) in *C. pneumoniae* [61]. The amino acid identity between pmp1–13 is 34–55%.

The *C. pneumoniae* genomes elucidated that several *pmp* genes contain frame shifts, and these vary between isolates, as listed in Tab. 2. Furthermore, at least pmp10 was shown to be differentially expressed between chlamydiae within the same cell, and this is likely due to a polyG tract that varies in length [64]. Based on the relatively high variability in the *pmp* gene family, considering the otherwise very conserved sequences between isolates, it has been speculated that the pmps may function in surface variation of *Chlamydia* as seen in other pathogenic bacteria [65].

Another gene family in *C. pneumoniae* that shows remarkable variation is the recently identified Cpn1054 family or *C. pneumoniae* polymorphic protein (ppp) family [42]. *Cpn1054* was initially identified as one of eleven paralogous genes located in four hyper-variable regions in *C. pneumoniae* CWL029, one of which is situated between *pmp1* and *pmp2* [66]. The genes were predicted to encode inc

Tab. 2 Variation in *C. pneumoniae* polymorphic membrane protein (*pmp*) genes

pmp	CWL029	AR39	J138
2			frame shift
3	frame shift	frame shift	frame shift
4	frame shift	frame shift	+1 frame shift
5	frame shift	frame shift	frame shift
6		393 bp del.	393 bp del.
10	frame shift		
12	truncation	truncation	truncation
17	frame shift	frame shift	frame shift

proteins by the presence of the characteristic bilobed hydrophobic motif. Many of the genes contain stop mutations that differ between sequenced strains and as in *pmp10*, a poly-G tract was identified in the 5' end of *cpn1054* [66]. Recently, poly-G tracts present in seven of eleven 1054 family members were analyzed by sequencing of a number of clinical isolates [67]. Five out of seven were found to vary in all investigated isolates, and functional analysis of protein products from this gene family will be interesting.

1.3
Proteome analysis of *Chlamydia*

The genome sequence reveals the coding capacity of an organism and thus what proteins it theoretically can produce. The coding capacity is informative, but does not reveal information about when, where and in what quantities the genes are transcribed and whether the possibly resultant proteins are modified or secreted. The direct investigation of proteins in their post-translationally modified and processed form present in a given biological compartment at a specific time and in a defined environment is the task of proteomics.

Proteomics is used to describe any large-scale investigation of proteins and can be approached in many ways but in principle it involves two steps: separation of the proteins in a sample and subsequent identification of these proteins. The perfect proteome study would provide a quantitative measure of every single protein present in the investigated sample. Unfortunately, such a study is so far not possible. Novel quantitative mass spectrometric techniques come close, but these are still in the development phase. Today 2-D gels as a separation tool coupled to mass spectrometry protein identification provides the most comprehensive way of analyzing complex protein mixtures [68].

1.3.1
Early *Chlamydia* proteome studies

In 1985, 2-DE was used to compare the protein content of outer membrane preparations from *C. trachomatis* serovars L2 and F [69]. Chlamydiae were selectively radiolabeled by [^{35}S]methionine incorporation in the presence of the inhibitor of eukaryotic ribosomes, cycloheximide. EBs were purified and *Chlamydia* outer membrane complex (COMC) was prepared by sarkosyl extraction [70]. The COMC was solubilized in a 2-D buffer based on urea with NP-40 as detergent and mercaptoethanol as reducing agent and subjected to 2-D PAGE where the first dimension was carried out in tube gels in which the pH gradient was established during focusing. Three proteins, major outer membrane protein (MOMP), a 60 kDa protein and a 12 kDa protein were observed for *C. trachomatis* F, whereas the 60 kDa protein was missing for *C. trachomatis* L2. However, by NEPHGE it could be concluded that the 60 kDa protein was also present in *C. trachomatis* L2, but migrating more basic than in serovar F [69].

Improvements in 2-D PAGE, including IPG strips for the first dimension, means that today more proteins can be resolved in 2-D gels of COMC [71]. Lambden *et al.* [72] identified the 60 kDa large cysteine-rich outer membrane protein, OmcB (Omp2), and the 12 kDa small cysteine-rich protein, OmcA (Omp3) to be developmentally regulated and transcribed as a polycistronic mRNA late in the developmental cycle. A model of the COMC architecture has been proposed [73] in which Omp2 is localized in the periplasmic space, disulfide cross-linked to Omp3, which is suggested to be anchored in the outer membrane by its lipid moiety. Comparing COMC from different *C. trachomatis* serovar to *C. pneumoniae* and *C. caviae* [74] showed that Omp2 from *C. trachomatis* L2 was resolved in the gels, but migrated one pH unit more basic than Omp2 of *C. trachomatis* F and two pH units more basic than *C. trachomatis* D. No additional proteins were identified for any of the species even though high molecular bands were observed by 1-D SDS gels for *C. trachomatis* serovar D [74].

1.3.2
C. trachomatis proteome studies

The first proteome study on whole *Chlamydia* aimed at identifying early proteins in *C. trachomatis* L2 by pulse labeling with [^{35}S]methionine at 2–4 h post-infection (hpi), 8–10 hpi, 14–16 hpi and 28–30 hpi [75]. Seven proteins were detected earlier than MOMP, four of which were labeled at 2–4 hpi. Three of these were identified by colocalization with proteins detected by immunoblotting with known antibodies. These were the heat shock proteins DnaK and GroEL and the ribosomal protein S1. The remaining four proteins were not identified. Early transcription of the *groEL* gene has recently been confirmed by transcript analysis [76, 77], but *dnaK* was designated a late gene in [78]. However, the designation "late" was based on lower transcription at earlier points in time than 24 hpi and higher transcription at later points in time and this does not exclude early transcription.

A second global study [79] aimed at providing a basis for the development of a protein database of *C. trachomatis* proteins. This was the first *Chlamydia* study to use IPG strips. Approximately 600 spots were separated in the area from pH 4–9 and 10–120 kDa in silver-stained gels. The very good resolution compared to earlier studies can be ascribed to the use of IPGs but also the substitution of mercaptoethanol with dithioerythritol (DTE), and NP-40 with CHAPS may have contributed to the superior results. Mercaptoethanol will more readily migrate out of the first dimension gel than DTE due to its charge, and removal of reducing agent will cause reoxidation and precipitation of proteins. A combination of immunoblotting with known antibodies and *N*-terminal sequencing was used to identify nine known proteins [79]. Seven sequences were obtained from yet uncharacterized proteins distributed in different areas of a 2-DE map and even though the gels showed very good resolution, the study like all other pregenomic proteome studies, suffered from the lack of identification methods for unknown proteins.

In the pregenomic area, 2-DE was most appropriate in studies where antibodies were available for identification of the proteins. One such study demonstrated the superiority of 2-DE in comparison to 1-DE with respect to the resolution of different isoelectric isoforms [80]. A family of high molecular weight *C. abortus* proteins detected by post-abortion sera from sheep were shown to be identical to immunogenic putative outer membrane proteins (POMPs). As the proteins had similar molecular weight, they could not have been distinguished in 1-D gels.

Western blotting of 2-D gels has also been applied to identify immunogenic proteins in *C. trachomatis* using sera from 17 patients suffering from genital inflammatory disease [81]. Fifty-five immunogenic proteins were detected with frequencies varying from 17 to 1. Eight proteins could be identified by colocalization with previously determined proteins. In addition, *N*-terminal sequences were obtained for nine proteins from which six could be identified in the genome sequence. Omp2, GroEL, MOMP and DnaK were the most frequently recognized proteins. These are known antigens, but also previously unknown antigens were detected such as elongation factor TU and ribosomal proteins.

1.3.3
C. pneumoniae proteome studies

The first comprehensive proteome map of *Chlamydia* in the postgenomic area was that of *C. pneumoniae* [47] (Fig. 2). Like Bini *et al.* [79] this study used IPGs in the first dimension and thiourea was incorporated into the 2-DE buffer to obtain the best possible recovery of hydrophobic proteins. Mass spectrometry was used to identify 263 protein spots representing 167 different genes and all identifications were published on the internet at http://www.gram.au.dk in a searchable form. Data for pH 4–7 (Fig. 2) was also included in the bacterial proteome database at the Max Planck Institute for Infection Biology at http://www.mpiib-berlin.mpg.de/2D-PAGE/. The proteome map can thus serve as a reference for 2-D PAGE studies performed in other laboratories. A good agreement between predicted and observed number of proteins was observed in the acidic region, whereas recovery in

12 | 1 Genome and proteome analysis of Chlamydia

Features of the Overview:

1. cursor over cross: protein name appears for identified spots
2. mouse click on sector: zoom into this sector
3. mouse click on cross: show protein information and hyperlinks

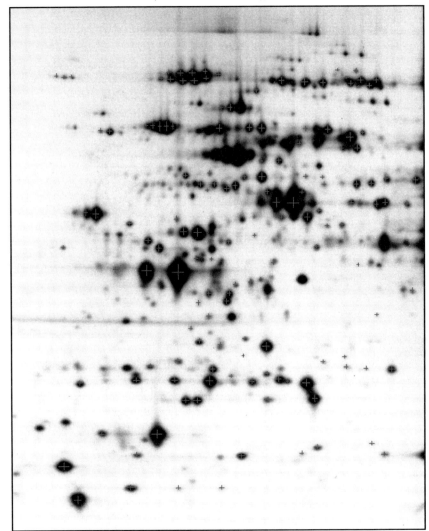

Fig. 2 Screen dump of the clickable IPG4-7 proteome map of C. pneumoniae available at www.mpiib-berlin.mpg.de/2D-PAGE/EBP-PAGE/index.html. Crosses represent identified proteins.

the basic region was poor. The use of basic strips (pH 6–11) did not significantly improve the number of resolved protein spots, but gave a better spatial distribution which is important when proteins are to be excised for further analysis.

To cover the highest possible number of radiolabeled EB proteins in one gel, samples were pooled after labeling with [^{35}S]methionine/cysteine at different points in time during cultivation [47]. Therefore proteins synthesized at different stages of the developmental cycle were labeled, but in consequence the actual protein content of EBs was not reflected by the spot volumes. Radiolabeling was chosen rather than staining to avoid the *Chlamydia* purification step in future experiments. Furthermore, as autoradiography is a more sensitive method for visualization than staining, the protein load can be reduced, which gives better resolution in the gels.

The study provided a reference map [47] but in addition, there were important findings: eight pmps were found to be highly abundant; type III secretion proteins were found in EBs for the first time indicating that this secretion system is present in EBs; the presence of a high number of energy-metabolizing proteins and proteins involved in transcription and translation suggested that EBs are "good to go" when they enter a host cell [47]. Furthermore, a high number of 'hypothetical' proteins were present (31/167) and many of these were abundant. One of the most abundant hypothetical proteins was Cpn0808. This was suggested to be loaded in EBs, ready for secretion by type III secretion upon contact with a future host cell based on its genomic location close to *lcrH1*, which is homologous to a gene encoding a type III secretion chaperone in *Yersinia* [47]. This hypothesis has been further supported by the recent finding of Cpn0809 in the cytoplasm of *C. pneumoniae* infected cells [82].

1.3.4
Identification of secreted proteins by comparative proteomics

Secreted *Chlamydia* proteins may carry out important functions in relation to interaction with the host cell. However, the identification of secreted proteins has been hampered by the fragility of the inclusion and RBs, making it impossible to isolate host cell cytoplasm from infected cells. This problem was circumvented by a 2-D gel comparison approach using the difference between 2-D protein profiles of infected cells and purified chlamydiae as a measure of which *Chlamydia* proteins are found outside the bacteria [83]. The method is outlined in Fig. 3. Proteins present in purified bacteria were subtracted from the protein content of whole lysates of infected cells and the differing proteins identified by MS. Chlamydial proteins were distinguished from eukaryotic by radiolabeling in the presence of cycloheximide. The idea of analyzing chlamydial components present in the infected cell but not in the chlamydiae themselves is somehow parallel to the approach leading to the identification of the first incs by comparing proteins reacting with convalescent sera to those reacting with sera obtained by immunization with inactivated bacteria [56].

The comparative proteomics approach resulted in the identification of CT858 of *C. trachomatis* and Cpn1016 of *C. pneumoniae* as secreted proteins [84] (Fig. 4). These proteins were orthologs and known as a *Chlamydia* protease-like activity factor (CPAF) [85]. CPAF was originally identified by its property to down-regulate host cell transcription factors required for MHC class I and II presentation and subsequently confirmed to be secreted [85]. The expression characteristics of CPAF

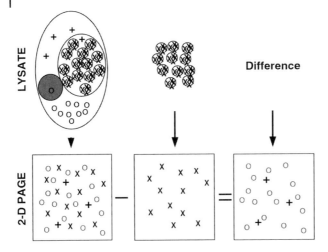

Fig. 3 Schematic drawing of the strategy used in subtractive proteomics. The protein content of whole lysate of infected cells and that of purified bacteria are separated by 2-D PAGE. The infected cells are radioactively labeled and the protein synthesis of the eukaryotic host cell is stopped by cycloheximide. Only chlamydial proteins are labeled. O indicates unlabeled eukaryotic cell proteins; X indicates labeled chlamydial proteins found in EBs; and + indicates labeled, potentially secreted proteins found in the host cell cytoplasm but not in purified bacteria.

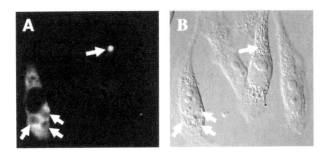

Fig. 4 HEp-2 cells infected with *C. pneumoniae* CWL029 and fixed 54 h postinfection. A, immunofluorescence microscopy with an antibody against the secreted protein Cpn1016 detected by a FITC conjugated secondary antibody. B, Nomarsky image of the same section. Arrows point at *Chlamydia* inclusions. Note that Cpn1016 is detected in the cytoplasm of infected cells but not in the cytoplasm of uninfected cells.

in *C. trachomatis* A, D and L2 as well as in *C. pneumoniae* were further characterized by 2-D PAGE [84]. The study demonstrated how proteome comparison of different biological compartments can lead to the identification of supposedly important molecules, and that genome analysis must be supplemented by further experiments as CPAF was predicted to be an outer membrane protein [86].

1.3.5
Proteome studies of comc

COMC, the sarkosyl-insoluble fraction of EB, is the only separable part of EBs [70]. The proteome analysis of COMC identified several proteins including known outer membrane proteins and the predicted membrane component of the type III secretion apparatus, YscC, indicating that the apparatus is assembled in EBs and that YscC is the membrane component in *Chlamydia* [71]. Other type III secretion proteins identified in the reference map of *C. pneumoniae* [47] were not found in the membrane fraction. Major constituents of the COMC were the pmps, which were characterized by 2-DE with respect to expression in the study by Vandahl *et al.* [87]. Of the 21 pmp genes, 16 were of full length in *C. pneumoniae* CWL029. Proteins encoded by seven of the 16 full length genes were found in COMC. The structure of the pmp proteins has similarities to that of autotransporter proteins [62] with a *C*-terminal part predicted to form a beta-barrel and an *N*-terminal passenger domain. Pmps are heavily up-regulated at the time of conversion of RB to EB, and at least ten pmps are present in EBs. Due to their reaction in formalin fixation it is likely that pmp6, 8, 10, 11 and 21 are surface-exposed [87]. Identified cleavage sites of pmp6 and pmp21 are in agreement with the theory that pmps are autotransporters [87], and this theory has recently been further confirmed by studies by Wehrl *et al.* [88].

1.3.6
Proteome comparison of *S. trachomatis* serovars

2-D reference maps for *C. trachomatis* A, D and L2 EBs were published in 2002 [89]. The general findings for these serovars were very similar to those for *C. pneumoniae*, with many hypothetical proteins (including large amounts of CT579, the ortholog of Cpn0808), type III secretion proteins, highly abundant pmps and many proteins involved in transcription, translation and energy metabolism. Protein products were identified from a total of 134, 133 and 127 different genes in the three serovars, respectively, thereby providing well covered reference maps for further studies. From 144 protein species (including different post-translational variants) identified in all serovars, 55 migrated differently in serovars D and L2, 52 differed between A and L2 whereas only 26 differed between A and D. This reflects the greater similarity between the trachoma serovars A and D than between the LGV serovar L2 and A/D. Most differences are probably caused by substitution of charged amino acid with noncharged (or *vice versa*) and do not have biological implications. Significant differences included a much higher abundance of malate dehydrogenase in L2 than in serovars A and D and the absence of fumarate hydratase (FumC) in L2. The *fumC* gene was confirmed to be truncated in L2 by *in vitro* translation [89]. It was speculated that higher amounts of malate dehydrogenase may be required in L2, if fumarate hydratase is impaired and malate thus must be obtained from the environment.

1.3.7
Proteome analysis of growth conditions

Proteome analysis has also been used to analyze the effect of different growth conditions on protein synthesis. A prominent example is the investigation of the effect of IFN-γ treatment of different *C. trachomatis* serovars [90]. This study reported on up-regulation of tryptophan synthetase in *C. trachomatis* serovar D in response to IFN-γ. Both TrpA and TrpB were found to be up-regulated in serovars A, D and L2, but TrpA was found at a lower molecular weight in serovar A and upon sequencing the gene it was found to be truncated in serovar A. Also, serovar C was found to have a truncated *trpA* gene and in serovar B the gene is missing. The fact that all ocular serovars have impaired *trpB* genes may indicate a role for this process in the development of trachoma [63]. The up-regulation of Trp proteins was confirmed at the transcriptional level by RNA chip analysis [91]. MOMP and other proteins were found to be down-regulated in serovar A but not in serovars D and L2. This is in contrast to the findings of transcript analysis of serovar D, where mRNA encoding MOMP and several other proteins were found to be down-regulated [91]. Also *C. pneumoniae*, which lacks the entire *trp* operon is inhibited by IFN-γ. Several *C. pneumoniae* proteins including MOMP were found up-regulated by IFN-γ and none were found to be down-regulated [92], though the effect was small for all of the proteins.

1.3.8
Considerations in proteomics

Comparison of experimental results is an essential issue in proteome analysis. In transcription analysis, normalization to genome copy numbers can be used to standardize samples [93], and this approach could to some degree be adopted in proteome analysis using parallel samples. As a minimum, gel loading of radio-labeled samples should be adjusted according to scintillation counts and quantified protein levels should be expressed relative to the total amount of a defined set of spots or specific constitutive proteins.

Just as proteome analysis can supplement genomics by elucidating truncations and post-translational modifications it can assist genome annotation by the detection of proteins from unrecognized ORFs. By searching MS/MS sequence tags of small proteins against whole genomes translated in all reading frames, genes may be recognized which were not annotated automatically. A novel *C. trachomatis* D protein of 7 kDa specific for RBs was identified by proteome analysis as a product from a previously unrecognized gene of 204 base pairs, located between *ct804* and *ct805* [94]. Proteome analysis can be specifically designed to identify small genes by casting gels with high resolution in the low molecular weight area and analyzing the protein products by MS/MS.

Compared to proteomics, RNA analysis has the advantage that amplification can be performed and that every gene can be studied at a given time. Thus, transcript analysis is well suited for studying global regulations upon environmental changes.

In such studies, transcript measurements relative to standard amounts of genomic DNA [93] seem better than measurements relative to RNA levels at standard conditions [78]. To minimize RNA degradation total RNA should be extracted from infected cells and bacterial RNA then purified, rather than purifying the bacteria first [95]. Although *Chlamydia* has been the subject of both global transcriptome studies [93, 81] and large-scale proteome studies [47, 89] no comparative studies have been made. By transcript analysis, information can be obtained on the regulation of genes encoding proteins that are too low in abundance to be quantified by proteomic approaches. Proteomics suffers from the demand of protein solubility and low abundant proteins may be hard to detect. Still, small regulations of abundant proteins are better studied and quantified by proteomics, and post-translational modifications can only be studied by proteomics.

1.4
Concluding remarks

Methods for generation of genome, transcriptome and proteome data are now available and such data will be valuable for further investigations. In recent years, many chlamydial proteins have been identified that are at the interface of interaction with the host cell. These include novel surface exposed proteins, inclusion membrane proteins and secreted proteins. Many of the proteins were identified by approaches made possible only by the availability of the *Chlamydia* genomes. In addition, the structure and function of the individual proteins must be further investigated. Proteome analysis has identified proteins of COMC that should be analyzed for surface exposure. This may provide evidence for surface variation due to changes in the expression of interchangeable surface proteins [64].

As more chlamydial proteins that interact with the host cell are identified and characterized, we will learn more about *Chlamydia* developmental biology, the infection and ultimately even more about the eukaryotic cell. The humane genome sequence will aid the identification of interaction partners for chlamydial proteins.

1.5
References

[1] Stephens, R. S., *Trends Microbiol.* 2003, *11*, 44–51.

[2] Stephens, R. S., Kalman, S., Lammel, C., Fan, J. *et al.*, *Science* 1998, *282*, 754–759.

[3] Kalman, S., Mitchell, W., Marathe, R., Lammel, C. *et al.*, *Nat. Genet.* 1999, *4*, 385–389.

[4] Grayston, J. T., Wang, S. P., Kuo, C. C., Campbell, L. A., *Eur. J. Clin. Microbiol. Infect. Dis.* 1989, *3*, 191–202.

[5] Everett, K. D., Bush, R. M., Andersen, A. A., *Int. J. Syst. Bacteriol.* 1999, *49*, 415–440.

[6] Everson, J. S., Garner, S. A., Lambden, P. R., Fane, B. A., Clarke, I. N., *J. Bacteriol.* 2003, *185*, 6490–6492.

[7] Schachter, J., Caldwell, H. D., *Annu. Rev. Microbiol.* 1980, *34*, 285–309.

[8] Hahn, D. L., Azenabor, A. A., Beatty, W. L., Byrne, G. I., *Front. Biosci.* 2002, *7*, e66–76.

[9] Grayston, J. T., *J. Infect. Dis.* 2000, *181*, Suppl. 3, 402–410.

[10] Falck, G., Gnarpe, J., Gnarpe, H., *Scand. J. Infect. Dis.* 1996, *28*, 271–273.

[11] Kutlin, A., Roblin, P. M., Hammerschlag, M. R., *Antimicrob. Agents Chemother.* 2002, *46*, 409–412.

[12] Belland, R. J., Ouellette, S. P., Gieffers, J., Byrne, G. I., *Cell Microbiol.* 2004, *6*, 117–127.

[13] Shor, A., Kuo, C. C., Patton, D. L., *S. Afr. Med. J.* 1992, *82*, 158–161.

[14] Campbell, L. A., Kuo, C. C., *J. Med. Microbiol.* 2002, *51*, 623–625.

[15] Leinonen, M., Saikku, P., *Lancet Infect. Dis.* 2002, *2*, 11–17.

[16] Wong, Y. K., Gallagher, P. J., *Heart* 1999, *81*, 232–238.

[17] Danesh, J., Whincup, P., Lewington, S., Walker, M. et al., *Eur. Heart J.* 2002, *23*, 371–375.

[18] Schachter, J., *Curr. Top. Microbiol. Immunol.* 1988, *138*, 109–139.

[19] Moulder, J. W., *Microbiol. Rev.* 1991, *55*, 143–190.

[20] Wyrick, P. B., *Cell Microbiol.* 2000, *2*, 275–282.

[21] Costerton, J. W., Poffenroth, L., Wilt, J. C., Kordova, N., *Can. J. Microbiol.* 1976, *22*, 16–28.

[22] Christiansen, G., Pedersen, L. B., Koehler, J. E., Lundemose, A. G., Birkelund, S., *J. Bacteriol.* 1993, *175*, 1785–1795.

[23] Hatch, T. P., Miceli, M., Sublett, J. E., *J. Bacteriol.* 1986, *165*, 379–385.

[24] Hackstadt, T., Scidmore, M. A., Rockey, D. D., *Proc. Natl. Acad. Sci. USA* 1995, *92*, 4877–4881.

[25] Hackstadt, T., Rockey, D. D., Heinzen, R. A., Scidmore, M. A., *EMBO J.* 1996, *15*, 964–977.

[26] Wylie, J. L., Hatch, G. M., *J. Bacteriol.* 1997, *179*, 7233–7242.

[27] Friis, R. R., *J. Bacteriol.* 1972, *110*, 706–721.

[28] Al-Younes, H. M., Rudel, T., Meyer, T. F., *Cell Microbiol.* 1999, *1*, 237–247.

[29] Summersgill, J. T., Sahney, N. N., Gaydos, C. A., Quinn, T. C., Ramirez, J. A., *Infect. Immun.* 1995, *63*, 2801–2803.

[30] Mehta, S. J., Miller, R. D., Ramirez, J. A., Summersgill, J. T., *J. Infect. Dis.* 1998, *177*, 1326–1331.

[31] Pantoja, L. G., Miller, R. D., Ramirez, J. A., Molestina, R. E., Summersgill, J. T., *Infect. Immun.* 2001, *69*, 7927–7932.

[32] Beatty, W. L., Morrison, R. P., Byrne, G. I., *Microbiol. Rev.* 1994, *58*, 686–699.

[33] Coles, A. M., Reynolds, D. J., Harper, A., Devitt, A., Pearce, J. H., *FEMS Microbiol. Lett.* 1993, *106*, 193–200.

[34] Clark, R. B., Schatzki, P. F., Dalton, H. P., *Arch. Microbiol.* 1982, *133*, 278–282.

[35] Clark, R. B., Schatzki, P. F., Dalton, H. P., *Med. Microbiol. Immunol.* 1982, *171*, 151–159.

[36] Stephens, R. S. (Ed.), *Introduction to Chlamydia Intracellular Biology Pathogenesis, and Immunity*, American Society for Microbiology, Washington DC, 1999.

[37] Read, T. D., Brunham, R. C., Shen, C., Gill, S. R. et al., *Nucleic Acids Res.* 2000, *28*, 1397–1406.

[38] Shirai, M., Hirakawa, H., Kimoto, M., Tabuchi, M. et al., *Nucleic Acids Res.* 2000, *28*, 2311–2314.

[39] http://www.ncbi.nlm.nih.gov/entrez/viewer.fcgi?val=AE009 440.

[40] Read, T. D., Myers, G. S., Brunham, R. C., Nelson, W. C. et al., *Nucleic Acids Res.* 2003, *31*, 2134–2147.

[41] Daugaard, L., Christiansen, G., Birkelund, S., *FEMS Microbiol. Lett.* 2001, *203*, 241–248.

[42] Rocha, E. P., Pradillon, O., Bui, H., Sayada, C., Denamur, E., *Nucleic Acids Res.* 2002, *30*, 4351–4360.

[43] Karunakaran, K. P., Blanchard, J. F., Raudonikiene, A., Shen, C. et al., *J. Clin. Microbiol.* 2002, *40*, 4010–4014.

[44] Hatch, T. P., Al-Hossainy, E., Silverman, J. A., *J. Bacteriol.* 1982, *150*, 662–670.

[45] Andersson, S. G., *Biochim. Biophys. Acta* 1998, *1365*, 105–111.

[46] Tjaden, J., Winkler, H. H., Schwoppe, C., Van Der Laan, M. et al., *J. Bacteriol.* 1999, *181*, 1196–1202.

[47] Vandahl, B. B., Birkelund, S., Demol, H., Hoorelbeke, B. et al., *Electrophoresis* 2001, *22*, 1204–1223.

[48] Iliffe-Lee, E. R., McClarty, G., *Mol. Microbiol.* 1999, *33*, 177–187.

[49] Rockey, D. D., Lenart, J., Stephens, R. S., *Infect. Immun.* 2000, *68*, 5473–5479.

[50] Chopra, I., Storey, C., Falla, T. J., Pearce, J. H., *Microbiology* 1998, *144*, 2673–2678.

[51] Hsia, R. C., Pannekoek, Y., Ingerowski, E., Bavoil, P. M., *Mol. Microbiol.* 1997, *25*, 351–359.

[52] Matsumoto, A., *J. Bacteriol.* 1982, *150*, 358–364.

[53] Matsumoto, A., *J. Bacteriol.* 1982, *151*, 1040–1042.

[54] Bavoil, P. M., Hsia, R. C., *Mol. Microbiol.* 1998, *28*, 860–862.

[55] Hatch, T., *Science* 1998, *282*, 638–639.

[56] Rockey, D. D., Heinzen, R. A., Hackstadt, T., *Mol. Microbiol.* 1995, *15*, 617–626.

[57] Bannantine, J. P., Rockey, D. D., Hackstadt, T., *Mol. Microbiol.* 1998, *28*, 1017–1026.

[58] Bannantine, J. P., Griffiths, R. S., Viratyosin, W., Brown, W. J., Rockey, D. D., *Cell Microbiol.* 2000, *2*, 35–47.

[59] Toh, H., Miura, K., Shirai, M., Hattori, M., *DNA Res.* 2003, *10*, 9–17.

[60] Longbottom, D., Russell, M., Jones, G. E., Lainson, F. A., Herring, A. J., *FEMS Microbiol. Lett.* 1996, *142*, 277–281.

[61] Grimwood, J., Stephens, R. S., *Microb. Comp. Genomics* 1999, *4*, 187–201.

[62] Henderson, I. R., Lam, A. C., *Trends Microbiol.* 2001, *9*, 573–578.

[63] Shaw, A. C., Christiansen, G., Roepstorff, P., Birkelund, S., *Microbes Infect.* 2000, *2*, 581–592.

[64] Pedersen, A. S., Christiansen, G., Birkelund, S., *FEMS Microbiol. Lett.* 2001, *203*, 153–159.

[65] Brunham, R. C., Plummer, F. A., Stephens, R. S., *Infect. Immun.* 1993, *61*, 2273–2276.

[66] Daugaard, L., Christiansen, G., Birkelund, S., *FEMS Microbiol. Lett.* 2001, *203*, 241–248.

[67] Viratyosin, W., Campbell, L. A., Kuo, C. C., Rockey, D. D., *BMC Microbiol.* 2002, *2*, 38.

[68] Fey, S. J., Larsen, P. M., *Curr. Opin. Chem. Biol.* 2001, *5*, 26–33.

[69] Batteiger, B. E., Newhall, W. J. 5th, Jones, R. B., *Infect. Immun.* 1985, *50*, 488–494.

[70] Caldwell, H. D., Kromhout, J., Schachter, J., *Infect. Immun.* 1981, *31*, 1161–1176.

[71] Vandahl, B., Christiansen, C., Birkelund, S., *Proc. 10th Intl. Symp. Human Chlamydial Infections* 2002, ISBN 0966438310, pp. 547–550.

[72] Lambden, P. R., Everson, J. S., Ward, M. E., Clarke, I. N., *Gene* 1990, *87*, 105–112.

[73] Everett, K. D., Hatch, T. P., *J. Bacteriol.* 1995, *177*, 877–882.

[74] Moroni, A., Pavan, G., Donati, M., *Arch. Microbiol.* 1996, *165*, 164–168.

[75] Lundemose, A. G., Birkelund, S., Larsen, P. M., Fey, S. J., Christiansen, G., *Infect. Immun.* 1990, *58*, 2478–2486.

[76] Shaw, E. I., Dooley, C. A., Fischer, E. R., Scidmore, M. A. et al., *Mol. Microbiol.* 2000, *37*, 913–925.

[77] Slepenkin, A., Motin, V., de la Maza, L. M., Peterson, E. M., *Infect Immun.* 2003, *71*, 2555–2562.

[78] Nicholson, T. L., Olinger, L., Chong, K., Schoolnik, G., Stephens, R. S., *J. Bacteriol.* 2003, *185*, 3179–3189.

[79] Bini, L., Sanchez-Campillo, M., Santucci, A., Magi, B. et al., *Electrophoresis* 1996, *17*, 185–190.

[80] Giannikopoulou, P., Bini, L., Simitsek, P. D., Pallini, V., Vretou, E., *Electrophoresis* 1997, *18*, 2104–2108.

[81] Sanchez-Campillo, M., Bini, L., Comanducci, M., Raggiaschi, R. et al., *Electrophoresis* 1999, *20*, 2269–2279.

[82] Lugert, R., Kuhns, M., Polch, T., Gross, U., *Med. Microbiol. Immunol. (Berl.)* 2003, epub 31. 10. 2003.

[83] Vandahl, B. B., Birkelund, S., Christiansen, G., *Methods Enzymol.* 2002, *358*, 277–288.

[84] Shaw, A. C., Vandahl, B. B., Larsen, M. R., Roepstorff, P. et al., *Cell Microbiol.* 2002, *4*, 411–424.

[85] Zhong, G., Fan, P., Ji, H., Dong, F., Huang, Y., *J. Exp. Med.* 2001, *193*, 935–942.

[86] Stephens, R. S., Lammel, C. J., *Curr. Opin. Microbiol.* 2001, *4*, 16–20.

[87] Vandahl, B. B., Pedersen, A. S., Gevaert, K., Holm, A. et al., *BMC Microbiol.* 2002, *2*, 36.

[88] Wehrl, W., Brinkmann, V., Jungblut, P. R., Meyer, T. F., Szczepek, A. J., *Mol. Microbiol.* 2004, *51*, 319–334.

[89] Shaw, A. C., Gevaert, K., Demol, H., Hoorelbeke, B. et al., *Proteomics* 2002, *2*, 164–186.

[90] Shaw, A. C., Christiansen, G., Birkelund, S., *Electrophoresis* 1999, *20*, 775–880.
[91] Belland, R. J., Nelson, D. E., Virok, D., Crane, D. D. et al., *Proc. Natl. Acad. Sci. USA* 2003, *23*, 15971–15976.
[92] Molestina, R. E., Klein, J. B., Miller, R. D., Pierce, W. H. et al., *Infect. Immun.* 2002, *70*, 2976–2981.
[93] Belland, R. J., Zhong, G., Crane, D. D., Hogan, D., *Proc. Natl. Acad. Sci. USA* 2003, *100*, 8478–8483.
[94] Shaw, A. C., Larsen, M. R., Roepstorff, P., Christiansen, G., Birkelund, S., *FEMS Microbiol. Lett.* 2002, *212*, 193–202.
[95] Byrne, G. I., *Proc. Natl. Acad. Sci. USA* 2003, *100*, 8040–8042.

2
Helicobacter pylori vaccine development based on combined subproteome analysis*

Dirk Bumann, Peter R. Jungblut and Thomas F. Meyer

Effective vaccines could provide long-term solutions to many important infectious diseases, however, vaccine development has been hampered by the slow identification of protective antigens. Proteomics provides global information about relevant antigen properties and thus might be ideally suited for identifying promising vaccine antigen subsets. *Helicobacter pylori* proteomics data are stored in a proteomics database (http://www.mpiib-berlin.mpg.de/2D-PAGE/). In this review, we describe how a combined *Helicobacter* subproteome analysis resulted in the rapid identification of novel, highly protective antigens. This illustrates the great potential of pathogen proteomics for vaccine development.

2.1
Introduction

Helicobacter pylori chronically infects roughly half of the world's human population [1]. In a subset of patients, *Helicobacter* infections eventually result in gastric or duodenal ulcers, or even gastric cancer. It has been estimated that about one million people die each year because of *Helicobacter* infections. Various combinatorial chemotherapies have proven to be effective for *Helicobacter* eradication in most patients, yet high costs prevent routine therapy in developing countries, and rapidly rising antibiotic resistance, and substantial adverse effects prevent routine therapy in developing countries suggesting that chemotherapy alone will be insufficient to control this important pathogen. To achieve the goal of an improved protection and therapy, effective vaccines are considered the most promising strategy and hence extensive preclinical and clinical efforts are put into vaccine development [2]. However, an effective well-tolerated vaccine still appears to be far from being accomplished indicating that further basic research is needed.

---

* Originally published in Proteomics 2004, 10, 2843–2848

Proteomics of Microbial Pathogens. Edited by Peter R. Jungblut and Michael Hecker
Copyright © 2007 WILEY-VCH Verlag GmbH & Co. KGaA, Weinheim
ISBN: 978-3-527-31759-2

2.2
Classical whole-cell inactivated *Helicobacter* vaccines

Initial attempts have used the classical approach of whole-cell inactivated vaccines. Inactivation can be achieved by lysing the bacteria using physical or genetic approaches, or by formalin treatment. Such vaccines have shown some efficacy in rodent *Helicobacter* infection models but one clinical trial failed to show any effect on the *Helicobacter* load of infected patients [3]. Whole cell vaccines contain a complex mixture of many antigens and thus might induce broad protective immune responses. On the other hand, the large number of components increases the likelihood of adverse effects [4], and also represents a difficult task for quality control. For these reasons, there is a general trend to replace whole-cell inactivated vaccines by so called subunit vaccines containing only one or a mixture of a few defined antigens.

2.3
Subunit *Helicobacter* vaccines: Conventional antigen selection

In the past few years, the complete genome sequences of most important pathogens including *H. pylori* have been elucidated. These data sets provide a comprehensive list of all possible protein antigens (in the case of *Helicobacter* around 1700 predicted proteins [5]) that could be used for vaccination. A recent microarray study revealed that clinical isolates of *H. pylori* considerably differ in gene content with a commonly present core set of only 1281 genes [6]. Since a *Helicobacter* vaccine should be broadly protective against a wide variety of strains, antigens for a subunit vaccine should preferentially be selected from this core set. Usually only very few of all encoded antigens are actually protective, resulting in a rather cost-intensive identification process [7]. For *H. pylori*, a large scale screening identified only ten protective *Helicobacter* antigens out of 400 different candidates that were tested in preclinical animal immunization experiments [8]. Unfortunately, the identity of these antigens and their relative protective efficacy has not yet been published. However, the available summary data clearly indicate that either large-scale empirical testing or powerful selection criteria are needed to identify protective *Helicobacter* antigens.

In the light of the low frequency of protective antigens, it is surprising that simple testing of suspected antigens successfully identified a number of potent antigens some of which have already entered clinical trials [2]. In particular, the enzymes urease and catalase that were initially recognized through standard microbiological characterization of *Helicobacter*, the main seroreactive antigen CagA, and the vacuolating toxin VacA all turned out to be highly protective in preclinical infection models. However, clinical testing of urease revealed poor efficacy in humans [9]. Clinical studies on the protective ability of CagA and VacA have recently been finished but not yet published [2]. A large number of strains do not produce CagA or functional VacA and thus infection with such strains could not be prevented with the respective antigens. Extensive evidence seems to suggest that

Tab. 1 Properties of protective *H. pylori* antigens.

Name	Detection by proteomics[a]	Abundant[b]	Seroreactive[b]	Surface exposed[c]
Urease A	+	+	+	+
Urease B	+	+	+	+/−
Catalase	+	+	+	+
HspA	+	+	−	+/−
HspB	+	+	+	+/−
VacA	+	+	−	+
Lipoprotein Lpp20	−	n.d.	−	+
L7/L12 ribosomal protein	+	+	+	n.d.
Hypothetical secreted protein HP1488	−	n.d.	−	poss.
Hypothetical secreted protein HP1117	−	n.d.	−	poss.
Hemolysin secretion protein precursor	+	+	+	poss.
Citrate synthase	+	+	−	n.d.
NapA	+	+	+	+
CagA	+	+	+	+
HP0231	+	+	+	+
HP0410	+	+	+	+

a) combined data from [12], [13]
b) data from [30]; n.d., not detected
c) data from [21], [22], [23]; +/− data not conclusive; poss., not yet experimentally confirmed as surface exposed, but possible based on sequence properties; n.d., not detected

CagA/VacA negative strains are less pathogenic but this view has been repeatedly challenged and it is clear that at least some gastric cancer patients had been infected with such seemingly non-virulent strains [10, 11]. Other virulence factors and enzymes have also been identified but only a few of them have been tested in preclinical immunization trials (Tab. 1). Additional protective antigen candidates are thus still needed.

2.4
Subunit *Helicobacter* vaccines: Global antigen selection based on proteomics

2.4.1
Proteomics as a tool for antigen characterization

In contrast to one-by-one testing of individual antigen candidates, global techniques such as DNA microarray analysis and proteomics allow a large fraction of all antigens that are encoded by the pathogen's genome to be rapidly evaluaded in parallel. Based on the resulting data for various relevant antigen properties, it should be possible to select a small number of attractive candidate antigens for subsequent immunization studies. Sequence analysis, transcriptomics, and proteomics all offer valuable, complementary information for such an approach. Proteomics, however, is particularly useful as it directly reveals actual antigen properties such as abundance, localization, and seroreactivity, instead of indirect data as provided by the other techniques. There are also critical limitations to proteomics, such as the difficulty in detecting low-abundance proteins and poor resolution of highly hydrophobic integral membrane proteins on 2-DE gels, but these limitations might not be critical for *Helicobacter* vaccine development. Low abundance proteins that are difficult to detect with current proteome techniques may also be poorly recognized by the host immune system because antigen recognition is generally dose-dependent. On the other hand, surface exposed membrane proteins might be particularly immunogenic and at least in Gram-negative bacteria, the typical outer membrane proteins containing amphiphilic β-sheets are well resolved on published *Helicobacter* 2-DE gels [12]. In contrast, the highly hydrophobic α-helical integral membrane proteins of the inner membrane are largely missing in *H. pylori* 2-DE studies in agreement with the analysis of other samples. Since this protein class is typically not surface exposed in *H. pylori* it may be of minor relevance for vaccine development.

2.4.2
The *Helicobacter* proteome

The proteome of *H. pylori* has been extensively analyzed by several groups using both 2-DE as well as LC-MS and large data sets of identified proteins are available making the *Helicobacter* proteome one of the best characterized microbial proteomes [13–17]. In particular, different subproteomes including the immunoproteomes, the secretome, and surface exposed proteins have been characterized (http://www.mpiib-berlin.mpg.de/2D-PAGE/) and these subproteomes offer especially useful information for vaccine development as will be discussed below. A comprehensive analysis of total lysates of *H. pylori* strain 26695 resolved on large 2-DE gels yielded 384 identified protein species that correspond to 290 different genes [12]. Some of these proteins have differential PTMs resulting in multiple spots. Interestingly, more than 95% of these identified proteins belong to the core set of 1281 genes [6] that are conserved in diverse clinical isolates (Bumann *et al.* manuscript in preparation). Only twelve identified proteins are absent in some of the characterized 15 different isolates [6] including members of the CAG pathogenicity island and some outer membrane proteins that all play a role in virulence of a subset of *Helicobacter* isolates. None of the eight strain 26695-specific putative gene products [6] have been detected. The proteomics data set is thus biased towards conserved proteins that are of primary interest for vaccine development.

Helicobacter can exist in two morphologically distinct forms, spiral rods and coccoid bacteria, and these forms have been observed both in human stomach biopsies and in *in vitro* cultures. *In vitro*, the spiral rods represents the actively growing form, while the coccoid form is mostly observed in late stationary cultures. The biological function of the coccoid form is controversial. Our recent proteome analysis indicates that it contains a number of stage-specific protein species, but all eleven identified species were merely minor variants of proteins that are abundant in both rod-shaped and coccoid *Helicobacter* cells (Bumann *et al.*, manuscript in preparation) suggesting that there are few if any additional antigens in coccoid cells. This result is another example for the minor changes in *H. pylori* protein composition under diverse environmental conditions. This finding is consistent with the presence of rather fiew regulatory systems compared to other bacteria [18]. However, a recent study using selective capture of transcribed sequences suggested that *H. pylori* expresses fourteen genes exclusively in infected stomachs but not in exponentially growing *in vitro* cultures [19]. Indeed, none of the corresponding gene products have yet been detected in *in vitro* cultures. Seven of these infection-specific fourteen genes are commonly present in diverse *H. pylori* isolates and thus might represent attractive vaccine antigens. At present, *H. pylori* proteomics is solely based on *in vitro* cultures and, consequently, the said seven and potentially other *in vivo* induced antigens might be missing in existing proteome data sets. However, several highly protective antigens have been clearly detected in published proteomics studies indicating that many relevant antigens are indeed accessible using this *in vitro* approach.

2.4.3
Criteria for promising antigen candidates

The current data set of about 300 *H. pylori* proteins thus represents a suitable starting point for vaccine antigen selection but it still remains a challenge to identify a small subset of particularly promising antigens for rapid testing in preclinical immunization trials. Based on previous experience, protective antigens represent only a rather small percentage (2.5%) of all antigens [8] suggesting that identifying a single protective antigen would require on average testing of about 40 randomly selected candidates. To improve this low hit rate, relevant antigen parameters are needed for preselecting promising candidates. Unfortunately, such parameters are still poorly defined owing to a lack of experimental evidence. Circumstantial evidence suggests that highly abundant surface exposed antigens that are recognized by the immune system during infection might represent a promising set of candidates. Abundant antigens might be more immunogenic compared to low abundance proteins since immune responses are generally dose-dependent. Surface exposed proteins of *H. pylori* diffuse more easily into the stomach mucosa compared to cytosolic antigens [20]. In the mucosa, such antigens might be efficiently taken up by resident antigen presenting cells resulting in potent immune responses. Seroreactivity may be a useful parameter since serorecognition of an antigen suggests that it is expressed *in vivo* and has become accessible to the host's immune system. While all these hints are tentative, a subset of antigens that fulfills all these requirements might be more likely to induce protective immune responses compared to the average of randomly chosen antigens. This is supported by the available evidence for known protective *Helicobacter* antigens (Tab. 1).

Proteomics offers rapid solutions to assess a large number of antigens for these three parameters, *i.e.* abundance, surface exposure and seroreactivity. In particular, the staining intensity of spots on 2-DE gels allows a semi-quantitative estimate for the abundance of the respective protein species to be obtained. Surface exposed antigens can be either detected in culture supernatants or after surface labelling. In addition, immunoblotting of 2-DE gels reveals global information on the immunoproteome. The most abundant *H. pylori* protein species have been reported in several independent proteome studies and the large overlap between these data sets supports the utility of the results [13–17]. However, as mentioned above, about 4–7% of all *H. pylori* genes are highly expressed in infected stomachs but not in *in vitro* cultures [19]. Thus, the available *in vitro* proteomes are likely to lack some potentially interesting vaccine antigen candidates.

Two proteomics studies analyzed the secretome of *H. pylori* during broth culture by analysis of precipitated cell culture supernatants [21, 22]. *H. pylori* has a pronounced tendency of spontaneous autolysis resulting in a potential contamination of culture supernatants with nonspecifically released cytoplasmic proteins. In one study, optimized culture conditions resulted in minimal autolysis which allowed 23 specifically secreted proteins to be identified [21]. In the other study, radioactive labeling was used to determine the relative abundance of distinct protein species in total extracts and culture supernatants [22]. Sixteen proteins which were overrepresented in the supernatants were considered to be secreted. The large overlap between both studies supports the utility of these complementary approaches. Besides antigens that are released to the external environment, surface attached antigens are also potentially attractive vaccine antigens. Early attempts to purify the outer membrane of *H. pylori* largely failed because of technical difficulties in removing inner membrane contaminations. We used selective labeling of surface exposed proteins with a highly hydrophilic biotinylation reagent that poorly permeates the membranes [23]. Biotinylated proteins were affinity purified and separated on large 2-DE gels resulting in the identification of 18 proteins of which only one had been previously predicted to be surface exposed based on typical sequence properties. Interestingly, many of the secreted proteins were found to be also partially surface attached suggesting that readsorption of released antigens might modify the surface of *H. pylori* [21].

To determine which *H. pylori* antigens are recognized by the immune system of infected patients, several immunoproteomics studies were performed [24–29]. In each study, there was a considerable inter-individual variation which might partially be related to specific antigen patterns of the various *H. pylori* strains which infected the different patients [30]. In addition, methodological differences resulted in somewhat complementary data sets. A combination of these various data yields a comprehensive representation of the *H. pylori* immunoproteome. However, T cells instead of antibodies might play a dominant role for protective immunity against *Helicobacter* [31]. At present, suitable methods for global identification of *H. pylori* antigens that are recognized by T cells from infected patients are lacking, but new developments including the tetramer array technology [32] might soon improve this situation.

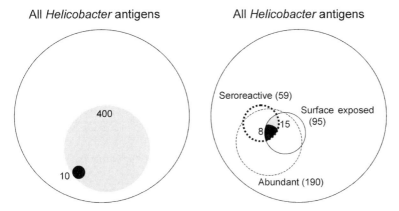

Fig. 1 Schematic comparison of conventional random antigen selection (left) and subproteome guided antigen selection (right). The large circles represent the entire proteome of *Helicobacter*, the gray areas represent the individually tested antigens, and the black areas represent the identified protective antigens. The number of identified antigens that belong to each of three subproteomes in the right panel is given in parentheses. The intersection of all three subproteomes contains 15 antigens (gray), at least eight of which are highly protective (black). A large number of randomly chosen candidates have to be tested to identify a few protective antigens (left), while candidate preselection using multiple subproteomes results in a rather high proportion of protective antigens (right).

2.4.4
Identification of protective antigens based on multiple criteria

As discussed above, the potentially relevant antigen parameters, *i.e.* abundance, surface exposure, and seroreactivity, by themselves are still rather vague. It thus appeared logical, to combine these soft criteria in order to enhance the hit rate for antigen selection. Following this rationale, we prioritized a subset of fifteen *H. pylori* antigens that were present in all three discussed subproteomes (high abundance, surface exposition, seroreactivity) [33]. Interestingly, this set of 15 antigens contained six well-known protective antigens. Two of the newly identified antigens are particularly attractive because they appear to be *Helicobacter* specific. In fact, they have proven to be highly protective in the murine *Helicobacter* infection model. Subproteome guided antigen selection thus achieved a very high hit rate (at least eight out of 15 predicted candidates) that is substantially superior to previous hit rates (10 protective out of 400 tested candidates [8]) (Fig. 1).

2.5 Concluding remarks

The proteomes of many important microbial pathogens have been recently characterized. Data mining using a subproteomics approach similar to that summarized here for *Helicobacter* might rapidly yield small sets of highly attractive antigen candidates for vaccine development. It will be interesting to see if such a strategy can indeed achieve similarly high success rates with regard to other pathogens.

Our work was supported in part by grants from the Deutsche Forschungsgemeinschaft (Bu 971/2-1; Bu 971/4-2, SFB621-A9) to D.B. and T.F.M. and the Bundesministerium für Bildung und Forschung (031U107C-031U207) to P.J. and T.F.M.

2.6 References

[1] Walker, M. M., Crabtree, J. E., *Ann. N. Y. Acad. Sci.* 1998, *859*, 96–111.

[2] Ruggiero, P., Peppoloni, S., Rappuoli, R., Del Giudice, G., *Microbes. Infect.* 2003, *5*, 749–756.

[3] Kotloff, K. L., Sztein, M. B., Wasserman, S. S., Losonsky, G. A. et al., *Infect. Immun.* 2001, *69*, 3581–3590.

[4] Moran, A. P., Knirel, Y. A., Senchenkova, S. N., Widmalm, G. et al., *J. Biol. Chem.* 2002, *277*, 5785–5795.

[5] Alm, R. A., Ling, L. S., Moir, D. T., King, B. L. et al., *Nature* 1999, *397*, 176–180.

[6] Salama, N., Guillemin, K., McDaniel, T. K., Sherlock, G. et al., *Proc. Natl. Acad. Sci. USA* 2000, *97*, 14668–14673.

[7] Adu-Bobie, J., Capecchi, B., Serruto, D., Rappuoli, R., Pizza, M., *Vaccine* 2003, *21*, 605–610.

[8] Ferrero, R. L., Labigne, A., *Scand. J. Immunol.* 2001, *53*, 443–448.

[9] Sougioultzis, S., Lee, C. K., Alsahli, M., Banerjee, S. et al., *Vaccine* 2002, *21*, 194–201.

[10] Rahman, M., Mukhopadhyay, A. K., Nahar, S., Datta, S. et al., *J. Clin. Microbiol.* 2003, *41*, 2008–2014.

[11] Hsu, P. I., Hwang, I. R., Cittelly, D., Lai, K. H. et al., *Am. J. Gastroenterol.* 2002, *97*, 2231–2238.

[12] Krah, A., Schmidt, F., Becher, D., Schmid, M. et al., *Mol. Cell Proteomics* 2003, *2*, 1271–1283.

[13] Jungblut, P. R., Bumann, D., Haas, G., Zimny-Arndt, U. et al., *Mol. Microbiol.* 2000, *36*, 710–725.

[14] Bumann, D., Meyer, T. F., Jungblut, P. R., *Proteomics* 2001, *1*, 473–479.

[15] Lock, R. A., Cordwell, S. J., Coombs, G. W., Walsh, B. J., Forbes, G. M., *Pathology* 2001, *33*, 365–374.

[16] Cho, M. J., Jeon, B. S., Park, J. W., Jung, T. S. et al., *Electrophoresis* 2002, *23*, 1161–1173.

[17] Govorun, V. M., Moshkovskii, S. A., Tikhonova, O. V., Goufman, E. I. et al., *Biochemistry* 2003, *68*, 42–49.

[18] Kelly, D. J., *Adv. Microb. Physiol.* 1998, *40*, 137–189.

[19] Graham, J. E., Peek, R. M., Jr., Krishna, U., Cover, T. L., *Gastroenterology* 2002, *123*, 1637–1648.

[20] Mai, U. E., Perez-Perez, G. I., Allen, J. B., Wahl, S. M. et al., *J. Exp. Med.* 1992, *175*, 517–525.

[21] Bumann, D., Aksu, S., Wendland, M., Janek, K. et al., *Infect. Immun.* 2002, *70*, 3396–3403.

[22] Kim, N., Weeks, D. L., Shin, J. M., Scott, D. R. et al., *J. Bacteriol.* 2002, *184*, 6155–6162.

[23] Sabarth, N., Lamer, S., Zimny-Arndt, U., Jungblut, P. R. et al., *J. Biol. Chem.* 2002, *277*, 27896–27902.

[24] McAtee, C. P., Fry, K. E., Berg, D. E., *Helicobacter* 1998, *3*, 163–169.

[25] McAtee, C. P., Lim, M. Y., Fung, K., Velligan, M. *et al.*, *Clin. Diagn. Lab Immunol.* 1998, *5*, 537–542.

[26] Hocking, D., Webb, E., Radcliff, F., Rothel, L. *et al.*, *Infect. Immun.* 1999, *67*, 4713–4719.

[27] Kimmel, B., Bosserhoff, A., Frank, R., Gross, R. *et al.*, *Infect. Immun.* 2000, *68*, 915–920.

[28] Haas, G., Karaali, G., Ebermayer, K., Metzger, W. G. *et al.*, *Proteomics* 2002, *2*, 313–324.

[29] Krah, A., Miehlke, S., Pleissner, K. P., Zimny-Arndt, U. *et al.*, *Int. J. Cancer* 2004, *108*, 456–463.

[30] Bumann, D., Holland, P., Siejak, F., Koesling, J. *et al.*, *Infect. Immun.* 2002, *70*, 6494–6498.

[31] Blanchard, T. G., Czinn, S. J., Redline, R. W., Sigmund, N. *et al.*, *Cell Immunol.* 1999, *191*, 74–80.

[32] Soen, Y., Chen, D. S., Kraft, D. L., Davis, M. M., Brown, P. O., *PLoS. Biol.* 2003, *1*, E65.

[33] Sabarth, N., Hurwitz, R., Meyer, T. F., Bumann, D., *Infect. Immun.* 2002, *70*, 6499–6503.

3
Towards a comprehensive understanding of *Bacillus subtilis* cell physiology by physiological proteomics*

Michael Hecker and Uwe Völker

Using *Bacillus subtilis* as a model system for functional genomics, this review will provide insights how proteomics can be used to bring the virtual life of genes to the real life of proteins. Physiological proteomics will generate a new and broad understanding of cellular physiology because the majority of proteins synthesized in the cell can be visualized. From a physiological point of view two major proteome fractions can be distinguished: proteomes of growing cells and proteomes of non-growing cells. In the main analytical window almost 50% of the vegetative proteome expressed in growing cells of *B. subtilis* were identified. This proteomic view of growing cells can be employed for analyzing the regulation of entire metabolic pathways and thus opens the chance for a comprehensive understanding of metabolism and growth processes of bacteria. Proteomics, on the other hand, is also a useful tool for analyzing the adaptational network of nongrowing cells that consists of several partially overlapping regulation groups induced by stress/starvation stimuli. Furthermore, proteomic signatures for environmental stimuli can not only be applied to predict the physiological state of cells, but also offer various industrial applications from fermentation monitoring up to the analysis of the mode of action of drugs. Even if DNA array technologies currently provide a better overview of the gene expression profile than proteome approaches, the latter address biological problems in which they can not be replaced by mRNA profiling procedures. This proteomics of the second generation is a powerful tool for analyzing global control of protein stability, the protein interaction network, protein secretion or post-translational modifications of proteins on the way towards the elucidation of the mystery of life.

---

* Originally published in Proteomics 2004, 12, 3727–3750

Proteomics of Microbial Pathogens. Edited by Peter R. Jungblut and Michael Hecker
Copyright © 2007 WILEY-VCH Verlag GmbH & Co. KGaA, Weinheim
ISBN: 978-3-527-31759-2

3.1
Introduction

With the publication of the first genome sequence of a living organism in 1995 [1] a new era in biology was opened. This era of functional genomics provides for the first time a global and comprehensive view of life in general towards the elucidation of the mystery of life. During the past 10 years we have learned a lot from these genomics data. Comparative genomics combined with bioinformatics, for instance, provides the key to a new understanding of the evolution of bacteria. The genomic sequence, however, presents only the blue-print of life, not life itself. Now functional genomics is required to bring this static genome information to the dynamics of life. The proteome is highly dynamic and flexible and thus the definite protein pattern of bacteria is determined by the environmental stimuli imposed to the cell. This review summarizes some new data that illustrate how the proteomic information can provide clues to a comprehensive understanding of bacterial cellular physiology. It is mainly focused on *Bacillus subtilis*, the model organism of Gram-positive bacteria. Interest in *Bacillus'* sporulation as a valuable model system for analyzing cell differentiation at a molecular level as well as its industrial potential for the production of many extracellular enzymes have both contributed to its attractiveness and its development into a model organism. Therefore, we now have an extensive knowledge of the genetics, biochemistry, and physiology of *B. subtilis*. The combination of the sophisticated molecular tools, an extensive database and the toolbox of functional genomics makes functional genomics approaches in Bacilli particularly rewarding.

Sequencing of the *B. subtilis* genome revealed about 4100 genes including 1700 genes with still unknown functions [2] thus indicating that many chapters of the "Bible of *Bacillus*", one of the most intensively studied organisms at all, are still empty [3]. The elucidation of the function of this surprisingly high number of unknown genes is a big challenge for future research and a main goal of functional genomics. Joint research programs in Japan and Europe aimed at the construction of a mutant library containing mutants in each single gene of unknown function and an accompanying comprehensive phenotypic screening program which should help to get first information on the physiological roles of the corresponding genes [4].

Proteomics of *B. subtilis* is even older than the publication of the genome sequence, first proteomic pictures date back to the mid-eighties [5, 6] relying on the two-dimensional polyacrylamide gel electrophoresis (2-DE), a highly sensitive technique introduced by O'Farrell, Klose, and others almost 30 years ago [7–10]. Already at that time it was possible to look for changes in protein patterns in response to stress or starvation stimuli at a "proteomic scale", a field mainly pioneered by Fred Neidhardt and Ruth Van Bogelen for *Escherichia coli* [11, 12]. However, it was almost impossible or at least very difficult to identify the interesting proteins. Later, N-terminal sequencing and the growing DNA data bases opened the chance to identify some of the proteins under study already before the entire genome sequence became available. Finally, the publication of the genome sequence was a real break-through in proteomics, because it allowed routine identification of proteins on 2-D protein gels by means of mass spectrometry (MS), mainly by MALDI-TOF-MS.

Even if non-gel based alternatives to 2-DE are coming more and more into the focus [13, 14], 2-DE is still state-of-the art and will continue to be particularly valuable for bacterial physiology and comparative physiological proteomics involving multiple samples. The high reproducibility of 2-DE is particularly valuable for multisample comparisons such as kinetic studies. Non-gel based technologies on the other hand make groups of proteins accessible to proteome analysis that have not been covered thus far, such as membrane or low-abundance proteins and proteins with extreme pI and/or molecular weight. If combined with stable isotope-labeling techniques like isotope-coded affinity tagging (ICAT) [15], multidimensional liquid chromatography or multidimensional protein identification technology (MudPIT) [16] in combination with on-line ESI-MS/MS provide the potential for very reliable relative or even absolute quantification [17]. However, non-gel based techniques thus far fail to correlate changes observed at the peptide level to individual protein isoforms, which is accomplished in 2-DE-based separations. Therefore, with the improvement of the reproducibility of non-gel based techniques, traditional 2-D gel approaches will probably be more and more supplemented but not replaced by new technologies.

Proteomic studies are well advanced for diverse bacteria, such as the model bacteria *Escherichia coli* and *Bacillus subtilis* as well as many pathogenic bacteria including *Mycoplasma pneumoniae*, *Haemophilus influenzae*, *Helicobacter pylori*, *Staphylococcus aureus*, and *Mycobacterium tuberculosis*. The highest proteome coverage of the genome has been accomplished for *Mycoplasma pneumoniae* where over 81% of the genomically predicted ORFs were detected by proteomics [18]. Because of its small genome *Mycoplasma pneumoniae* is without any doubt a suitable model organism for the visualization of the entire proteome. However, it is not very attractive for physiological proteomics because the genomic data imply a strong limitation in the regulation of gene expression with no alternative sigma factor or two-component regulatory system available so far. A global non-gel based proteome analysis of the radiation-resistant bacterium *Deinococcus radiodurans* identified 61% of its predicted proteome (1910 proteins) with high confidence, the largest protein number identified for any organism to date [19]. In model organisms, such as *E. coli* and *B. subtilis*, recent studies identified approximately 1480 [20] and 900 proteins [21], covering an essential portion of the actual proteome of both species.

3.2
Subproteomes *vs.* the total theoretical proteome

Preferentially, one would like to access almost all proteins synthesized in a cell in order to unleash the full potential of proteomics and to gain a comprehensive view of cellular events. Unfortunately, not all cellular proteins can be visualized on one single gel. In contrast to DNA arrays that may cover the entire genome, currently only parts of the cellular proteins are accessible to proteome techniques. However, if one combines the main subproteomic fractions, such as neutral/weakly acidic/alkaline cytosolic proteins, cell-wall/surface associated, and extracellular proteins, the majority of proteins synthesized in a bacterial cell can be visualized by the proteome approach. The greatest subproteomic fractions not covered by the routine approaches are proteins with multiple membrane-spanning domains. Special procedures, such as SDS-PAGE combined with multidimensional chromatography and MS/MS-techniques, allow visualization of at least a portion of this membrane subproteomic fraction [22]. In addition, low-abundance proteins and very alkaline or acidic as well as extremely large or small proteins are not covered by the standard experimental separation protocols.

The analysis of the entire proteome is also limited by physiological constraints. From a physiological point of view, only a subpopulation of proteins will be expressed because only a part of the genome is active at any time. Based on the physiological state of bacteria two major classes of proteomes should be defined: proteomes of growing cells and proteomes of nongrowing cells. The proteomes of growing cells (vegetative proteome) will be dominated by house-keeping proteins mainly required for growth and reproduction forming the core portion of the vegetative proteome. This more or less stable core fraction has to be supplemented by proteins used for degradation of carbon/energy sources or for synthesis of amino acids and nucleotides, a fraction mainly depending on the composition of the growth medium. Furthermore, the rate of expression of the translational apparatus, of DNA replication and RNA synthesis machines, and of many anabolic reactions mainly depends on the growth rate of bacteria [23]. For nongrowing cells proteomic signatures for various stress or starvation stimuli will indicate if the nongrowing cell population suffered from heat or oxidative stress or from glucose or phosphate starvation. The assembly of various subproteomic fractions not only from an analytical but also from a physiological point of view is the main approach on the way towards the entire proteome of bacteria.

Physiological proteomics of the first generation (see below) mainly depends on quantitative comparative protein expression profiling of cells grown under different conditions or of wild-type with mutant cells. Obviously, such a comparative approach has to rely on only a minimal number of subfractions/gels because only a small number of standard gels can be managed within studies involving multiple time points/samples. Therefore, it is essential to identify a maximal number of proteins on a minimal number of gels. However, if one intends to gain access to low-abundance or membrane proteins one can apply different prefractionation protocols (*e.g.*, free-flow electrophoresis, *etc*.) in combination with very tight pH gradients or analyze protein mixtures by a combination of a multidimensional LC and MS/MS techniques.

3.3
The vegetative proteome of growing cells

According to DNA array data, growing cells of *B. subtilis* express about 2500 genes. From these 2500 genes, 1550 proteins should be visualized in the standard window of pI 4 – 7 (the remaining are mostly membrane proteins with an alkaline pI). With the identification of about 700 proteins more than 40% of the predicted proteins in the pI

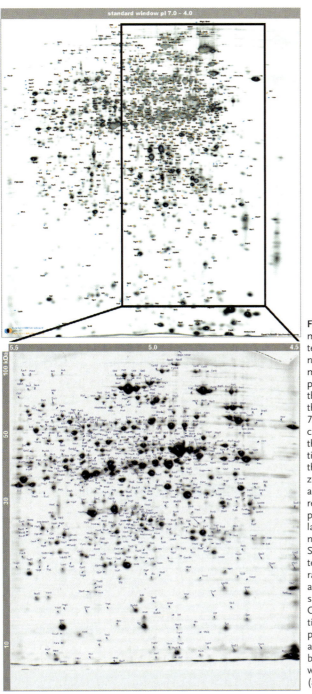

Fig. 1 Reference 2-DE maps of cytosolic proteins of B. subtilis exponentially growing in minimal medium. The upper part of the figure displays the 2-D reference map of the standard p*I* range 4–7. Resolution of overcrowded gel regions and thus protein representation can be improved by the inclusion of ultrazoom gels into the study as exemplified for the p*I* region 4.5–5.5 (lower part). Protein spots are labelled with protein names according to the SubtiList database. Protein extracts were separated on 18 cm IPG strips and 2-DE gels were stained with colloidal CBB. Detailed information onto each individual protein identified is available at http://microbio2.biologie.uni-greifswald.de:8880/sub2d.htm (adapted from [21]).

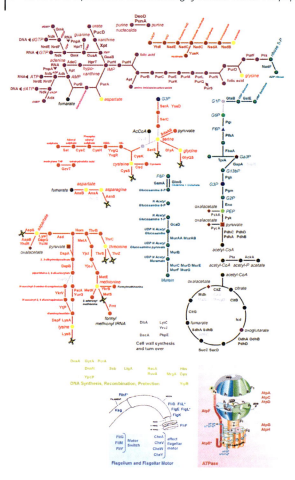

Fig. 2 Assignment of proteins identified to biochemical pathways. Proteins that have not been identified in the 2-D gel images thus far are shaded light grey. Proteins exclusively identified from the membrane protein fraction are indicated with asterisks. Components of the carbohydrate metabolism, such as glycolysis (right side of the left page), pentose phosphate shunt (left side of the right page), and amino sugar synthesis for murein synthesis (center of the left page), are indicated in dark cyan. Citric acid cycle (lower right side of the left page), biotin and fatty acid metabolism (upper center of the right page) are indicated by dark grey. Amino acid metabolism is colored red. Purine (upper left corner) and pyrimidine (lower right) metabolism are encoded in purple and the nicotinate metabolism (upper left page) in brown. Components not directly involved in metabolic pathways but in essential cell structures are presented in boxes. DNA related functions in yellowish green (bottom left page), flagellum and chemotaxis related components in azure (bottom left page), and ATPase components in pink (bottom left page). Transporters are highlighted in dark blue, components of the transcriptional machinery in green and the ribosome and other components of the translational apparatus are encoded in ochre (adapted from [21]).

3.3 The vegetative proteome of growing cells | 37

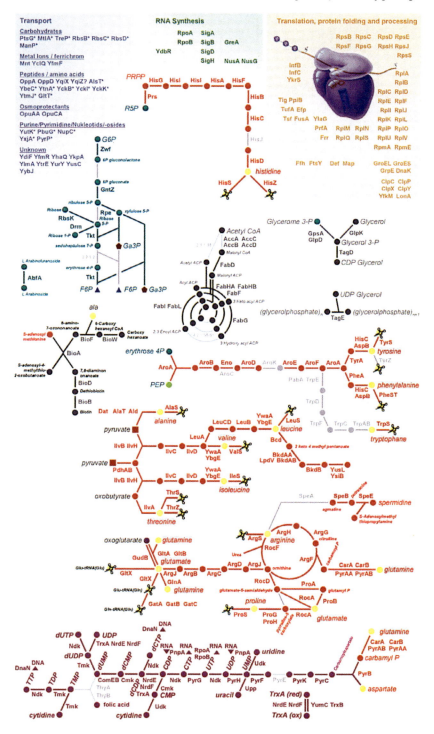

range 4 – 7 are currently accessible to proteomic approaches [21] (Fig. 1). Insertion of these mostly house-keeping proteins into a metabolic map revealed that the majority of basic metabolic pathways is already represented in most recent proteomic studies (Fig. 2). Besides the components of the translation and protein folding machine, ATPase, the flagellum/flagellar motor and many proteins involved in transport processes the pathways covered include glycolysis, citric acid cycle, fatty acid synthesis, amino acid or purine/pyrimidine biosynthesis. The coverage is most comprehensive for the parts of the translational apparatus (*e.g.*, 89% of the aminoacyl-tRNA synthetases, 83% of the translation elongation factors), the enzymes of main glycolytic pathways (81%), the citric acid cycle (53%), the metabolism of nucleotides and nucleic acids (51%), and the enzymes involved in metabolism of amino acids and related molecules (45%) [21]. The coverage seems to be low since all paralogs were considered. For the TCA cycle, for instance, all biochemical steps are covered with at least one enzyme, but only 53% of all paralogs have been detected in total. A quantitative evaluation assigned many of the proteins involved to the 400 most abundant protein spots and allowed an estimation of the fraction of overall translation invested for sustaining these metabolic functions. The experimentally derived ranking of protein abundance also gains support from some in silico predictions of presumably highly expressed genes that are based on the genome sequence and codon usage [24, 25]. Furthermore, 91 of the 271 essential genes of *B. subtilis* [26] code for proteins covered by the 400 most abundant protein spots. These top 400 proteins also include 88 proteins of thus far undefined function, some which are highly abundant proteins. Analysis of their function might provide new insights into essential, but still unknown physiological processes. Looking for proteomic signatures in the corresponding mutants would be a proteomic approach to predict their functions (see Section 5).

The vegetative proteome of *B. subtilis* with 693 cytosolic proteins in the pI 4–7 region, 53 alkaline and 168 intrinsic membrane proteins [21] is now ready for physiological application. Despite the fact that the main metabolic routes of *B. subtilis* are probably already known there is only limited information available on its regulation. Proteomic approaches now offer the chance to analyze the regulation of entire metabolic pathways. Even for the main catabolic routes, such as glycolysis or TCA cycle, new information became available by this global proteomic view [27], leading to a new understanding of its regulation which is exemplarily illustrated below.

Glycolysis is strongly stimulated by glucose and the citric acid cycle is repressed if glutamate is available even in the presence of oxygen (Fig. 3). Therefore, during glucose excess ATP might be mainly produced *via* substrate phosphorylation. The excess of metabolic intermediates can not enter the citric acid cycle because it is repressed and therefore need to be secreted into the extracellular medium (overflow metabolism) (Fig. 3). This phenomenon, known as the crabtree effect, seemed to depend on CcpA, the global regulator of carbon catabolite repression. However, subsequent detailed studies did not support a direct role of CcpA in mediating the crabtree effect. Mutational inactivation of *ccpA* rather triggered hyperphosphorylation of HPr at the serine residue 46 thus causing reduced glucose uptake in the mutant and that diminished the level of glycolytic intermediates below a level required for the regulation [27–29]. This example shows that proteomics is an attractive approach to initiate new physio-

3.4 Proteomes of nongrowing cells – the adaptational network

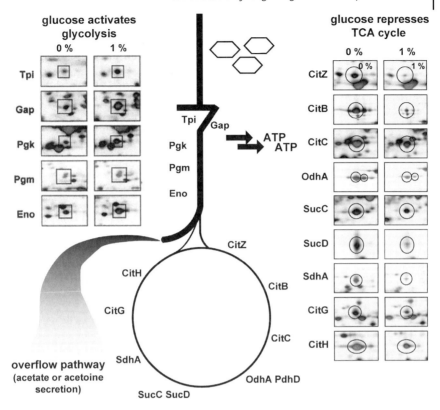

Fig. 3 Schematic presentation of the regulation of glycolysis and the TCA cycle in B. subtilis grown with an excess of glucose. Glucose activates glycolysis (left part) and represses the TCA cycle (right part). Therefore, excess glycolytic intermediates that can not enter the TCA cycle have to be secreted into the extracellular medium as a result of an overflow mechanism. ATP production is mainly accomplished via substrate level phosphorylation (adapted from [27, 61]).

logical studies because of its global view of the processes. The vegetative proteome map with almost 900 entries in total covering many metabolic pathways should promote related studies to gain a more comprehensive picture of metabolism and its regulation in B. subtilis showing that even in the best studied model system of Gram-positive bacteria many interesting phenomena still wait for their elucidation.

3.4
Proteomes of nongrowing cells – the adaptational network

From a physiological point of view, proteins produced in response to stress or starvation are of crucial importance because stress or starvation are the rule and not the exception in natural ecosystems. The proteomes of nongrowing cells are probably more heterogeneous than proteomes of growing cells because the single stress/starvation stimuli that trigger the nongrowing state normally induce a great number of stress and starvation specific proteins [30, 31]. These stress/starvation-induced proteins are organized in a complex and highly sophisticated adaptational network which consists of stimulons, regulons, and modulons. Proteomics supported by DNA array techniques is an excellent tool for analyzing this network, for its dissection into the single modules and for studying their adaptive functions. Four main steps are necessary for exploring the network: (i) The definition of stimulons (entire set of proteins induced or repressed by one stimulus, [32]). (ii) The dissection of stimulons into single regulons which form the basic modules of global gene expression. (iii) The analysis of overlapping regions in different regulons and of a hierarchical ordered regulon interaction network (modulons). (iv) The analysis of regulons that activate each other in a sequential manner towards the elucidation of complex gene regulation programmes.

The first step in analyzing stress adaptation is to look for all proteins induced (or repressed) by one stimulus at a proteomic scale because all newly induced proteins together accomplish stress adaptation. The dual channel imaging technique is particularly suited for the search for proteins belonging to stimulons or regulons [33]. This technique allows a rapid allocation of proteins to functional or regulation groups (stimulons, regulons) simply by a comparative analysis of protein content *vs.* protein synthesis. Two digitized images of 2-D gels have to be generated and combined in alternate additive dual-color channels (Fig. 4). The first one (densogram) displays protein content (accumulated proteins) visualized by Sypro Ruby staining or some other staining techniques in intensity-dependent green shades. The second image (autoradiograph) showing the proteins synthesized during a 5-min pulse-labelling with ^{35}S-L-methionine is false-colored in shades of red. When the two images are combined, proteins accumulated and synthesized in growing cells are colored yellow. However, looking for red-colored or green-colored proteins will rapidly define induced and repressed proteins, respectively. Proteins synthesized predominantly after the imposition of a stress/starvation stimulus that have not yet accumulated in the cell are colored red because radiolabeling is more sensitive than the staining. Identifying red proteins is a simple technique for finding all proteins induced by a single stimulus and thus defining the stimulon structure. Heat stress, for instance, induces more than 100 proteins, some of which perform known functions while the function of others has not been explored yet. For these unknown proteins a preliminary prediction of their function may be feasible: because of their classification as members of the heat stress stimulon they are likely involved in adaptation to heat stress. A similar technique can be used for a preliminary functional prediction of unknown proteins involved in adaptation to osmotic, acid, oxidative stress, *etc.* Proteins repressed by heat stress can be visualized by their green color, which indicates that they are present in the cell and probably still active, but no longer synthesized.

In most cases stimulons consist of more than one regulon. The next step in exploring the network is the dissection of stimulons into regulons, which are better defined from a genetical point of view (Fig. 5). Different induction kinetics of individual proteins of the stimulon frequently indicate such heterogeneity within this

3.4 Proteomes of nongrowing cells – the adaptational network

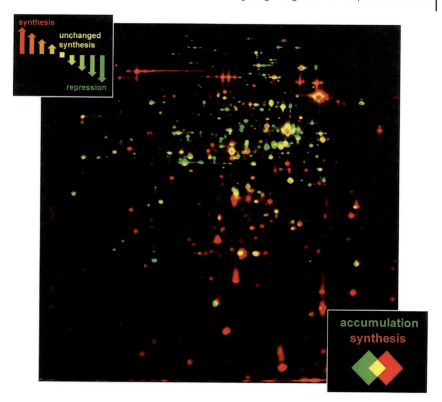

Fig. 4 Comparative dual channel imaging of protein synthesis and accumulation. During exposure to heat shock (48°C) *B. subtilis* cells were pulse labeled with ^{35}S-L-methionine. Subsequent to the separation of the crude protein extract by 2-DE, the protein content was visualized by Sypro Ruby and the protein synthesis rate was determined by phosphorimaging. Both images were overlaid with the aid of the Delta-2D software package (Decodon, Greifswald, Germany). Protein content and synthesis are displayed in green and red, respectively. Members of the heat shock stimulon appear in red because their synthesis is strongly induced following heat shock (for details on the dual-channel imaging technique see [33]).

protein group. The procedure for this dissection of stimulons into regulons is to look for proteins that follow the same induction pattern and kinetics, as dictated by the global regulator that controls the regulon, and then to analyze a mutant in this regulator: proteins no longer induced or repressed in the mutant will probably belong to the regulon. Using specific software packages (gel-warping programs, DIGE technology based programs, *etc.*) proteins only induced in the wild type and not in the mutant can easily be detected. Applying this approach, two main groups of regulons were distinguished: proteins induced by a particular stress or starvation stimulus with a specific adaptive function against that stimulus, and more general stress or starvation proteins induced by a diverse set of environmental stimuli.

Biochemical and physiological studies have shown that the specific adaptive functions of the first group, the stress-specific proteins include: (i) neutralization of the stress factor, (ii) adaptation to its presence, and/or (iii) repair of damage caused by the stress stimulus. Starvation-specific proteins, on the other hand, may allow: (i) uptake of the limiting substrate with very high affinity, (ii) a search for alternative substrates not used in the presence of the preferred one (catabolite repression), (iii) replacement of the limiting substrate by others, or (iv) moving to new nutrients by chemotaxis. In addition to these stress- or starvation-specific proteins induced by a single stimulus or a few very similar stimuli, another set of proteins induced by wide range of nonrelated stress or starvation stimuli was discovered [34]. This pattern of induction by stress or starvation indicates that such proteins may have a relatively nonspecific, but nevertheless essential, protective function under stress, regardless of the specific growth-restricting signal. Therefore, these proteins have been called general stress proteins [34].

Two major general responses can be classified for vegetative *B. subtilis* cells: the σ^B-dependent general stress response protects the nongrowing cell against "future stress", an essential feature of a dormancy of vegetative nonsporulating cells as an alternative to sporulation [31, 35–37] and the nonspecific stringent response that prevents nutrient wasting during long-term stationary phases [38–41]. This stringent response in turn is intertwined with a number of early stationary phase responses mainly *via* the ppGpp-mediated decrease in the GTP-level [42–44]. Sensing of the GTP-level, as a general signal of nutrient availability, is accomplished by the GTP-activated global repressor CodY [45] that primarily functions to adjust a number of diverse stationary phase responses including antibiotic production, competence, and sporulation [46–48] to the overall nutritional status of the cell.

The allocation of proteins to regulons can also be used for analyzing the size and function of new, still unknown regulons provided that the physiological conditions that activate the regulon are known. Under inducing conditions the protein pattern of the wild type and a mutant lacking the global regulatory protein will be compared. In the case of global repressors the approach is even simpler, since regulon members will be derepressed in the mutant even during exponential growth. After the identification of the regulon members by mass spectrometry the putative function of the entire regulon may be predicted based on the biochemical/physiological assignment of a group of members of the regulon. However, this functional prediction should be confirmed by detailed physiological studies of mutants. Relying on such functional genomics approaches (either proteomics, transcriptomics or a combination of both) the functions of some originally unknown regulons were predicted [49–53]. There are, however, still many unknown regulons encoded in the *B. subtilis* that wait for detailed analyses. Currently, we can only speculate how many of the 4100 genes are organized in regulons and how many regulons exist in *B. subtilis*, probably more than 250. Similar estimations have been made for *E. coli* where 300 of the 4405 probable ORFs corresponding to 8% of the coding capacity encode transcription factors [54].

The dissection of the entire genome into its basic modules of global gene regulation, and the assignment of the thus far uncharacterized proteins into these functional boxes is a good and simple approach for an overall, but still preliminary

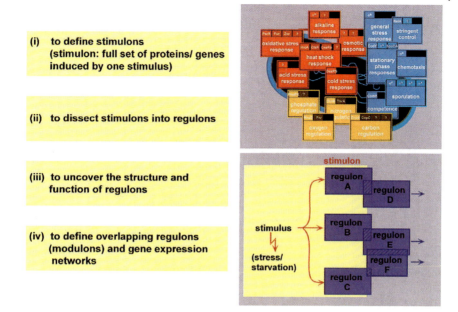

Fig. 5 Exploration of gene expression networks: from stimulons *via* regulons and modulons (overlapping regulons) to the network (adapted from [61]).

prediction of the function of all the unknowns which in combination with *in silico* predictions (Blast searcher, *e.g.*, neighbourhood arrangements, *etc.*) [55, 56] provided a good starting point for a more detailed functional analysis.

Finally, all functional genomics data can be assembled into an adaptational network consisting of a great number of stress/starvation stimulons and regulons (Fig. 5). A prominent feature of this adaptational network which is probably of considerable physiological significance is the interplay between stress- or starvation-specific regulons and general stress/starvation regulons [57]. The individual specific and general regulons do not exist independently from each other, but instead, are tightly connected, forming the adaptational network. The individual regulons are connected by genes that are controlled by more than one global regulator. Proteomics again is a good strategy for finding such regulon-overlapping areas. Protein members of regulons displaying a somehow atypical induction pattern are good candidates for such a double control. The σ^B-dependent ClpC or ClpP proteins of *B. subtilis*, for instance, are strongly induced by diamide stress which is not a typical inducer of the σ^B-response. This atypical induction is the result of the double control of both genes by σ^B and the CtsR-repressor that is inactivated following the imposition of diamide stress [58]. *Via* the σ^B-dependent proteins YvyD or YtxH there is a direct link between the σ^B regulon and the σ^H regulon because both genes are under σ^B and σ^H double control [59, 60]. The σ^H regulon in turn is connected with the RelA regulon and with the sporulation regulons. Employing

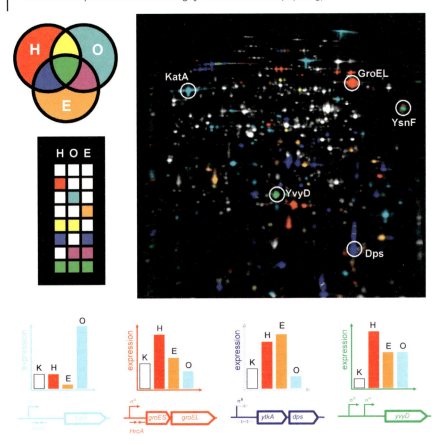

Fig. 6 Multicolor imaging of expression patterns. The Delta-2D software (Decodon) was utilized to visualize complex expression patterns on the 2-D gel image in the standard pH range 4 – 7. The color code is presented in the top left corner. Proteins only induced by single stresses are colored red (heat), light-blue (oxidative stress), and orange (ethanol), respectively. Proteins induced by oxidative stress as well as by heat stress are colored yellow, proteins induced by ethanol stress and heat stress are colored dark-blue, proteins induced by oxidative and ethanol stress can be recognized as purple spots and finally, proteins induced by all three stimuli are displayed in green. Expression patterns of selected examples and a schematic presentation of the corresponding gene regulation are displayed at the bottom of the figure (Bernhardt et al., in preparation).

such a "regulon walking" the complex adaptational network can be explored step by step [61]. Recently, the exploration of such complex expression patterns was greatly simplified with the introduction of a new software package (Decodon, Greifswald, Germany) that is particularly suited for the visualization of more complex proteome profiling studies. Expression profiling involving large numbers of experiments essentially depends on perfect matching of proteins spots across all gel images. Usually, the information content of larger experimental sets is seriously limited by accumulating matching errors. The new concept of introducing standard positions, derived from the average position in all gels, is more accurate and allows creation of expression profiles that include the whole data collection. Using this concept of average-derived standard positions, an image fusion containing all observed spots is generated and utilized for multicolor visualization [62], an example of which is illustrated in Fig. 6.

Proteins only induced by oxidative stress or heat are colored light-blue and red, respectively. Proteins induced by oxidative stress as well as by heat stress are colored yellow, proteins induced by ethanol stress (orange) and heat stress (red) are colored dark-blue and finally, proteins induced by all three stimuli are displayed in green. The PerR-dependent KatA protein only induced by oxidative stress is colored light-blue and GroEL only induced by heat stress in HrcA-dependent manner is colored red. The typical σ^B-dependent Dps protein is colored blue because σ^B-dependent proteins are induced in response to heat as well as ethanol stress. However, YvyD, that has also been shown to belong to the σ^B-regulon is colored green. The reason for this additional induction by oxidative stress seems to be a second, σ^H-dependent promoter [59], that has already been mentioned above and that is perhaps induced by oxidative stress (which has to be shown experimentally). For other proteins such as YsnF the mechanisms permitting induction by multiple factors is less clear but can now be studied in the future. These selected examples should demonstrate that the combination of image fusion and color-coding is a useful tool for finding proteins that are characterized by a complex regulation pattern at a proteomic scale (Bernhardt et al., in preparation). This approach can of course also be applied to the identification of proteins that are only induced either by phosphate, amino acid or by glucose starvation or by a combination of two or all three starvation stimuli, and will help in general to uncover new overlapping regions between regulation groups in the complex adaptational network.

The final goal is to establish a model of an adaptational network collecting all the information from functional genomics, including transcriptomics, proteomics, and metabolomics. This database should contain all information on the single stimulons and regulons: (i) gene regulation mechanisms, including their regulon-specific DNA target sequences, (ii) the collection of genes and proteins involved, (iii) the stress/starvation-specific induction/repression ratios both at the mRNA and protein level as well as (iv) data on protein stability and post-translational modifications at a proteome scale. The elucidation of the function of all proteins arranged in the stress/starvation boxes in stress/starvation adaptation is one of the very ambitious objectives of such an approach. Such a "functional genomics cell model" that connects genome sequence information with cell physiology and biochemistry will provide a most comprehensive understanding of cell adaptation to stress and starvation. It will constitute an excellent example of functional genomics which shows how to proceed from genome sequence information via transcriptomics and proteomics to real life in a systems biology approach. Naturally, such a systems biology approach will also have to incorporate protein interaction networks as well as metabolomics and require a strong bioinformatics/modeling platform that extends the individual snapshots of the adaptational network to a time-resolved view.

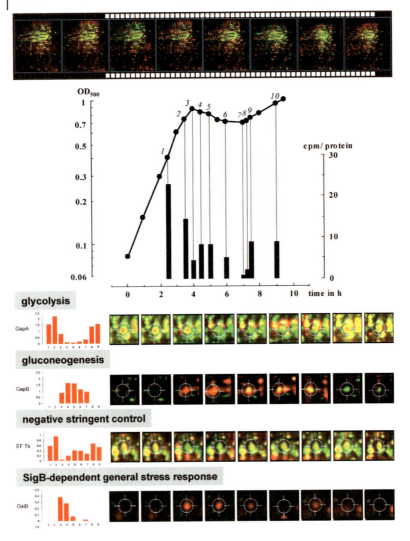

Fig. 7 Dynamics of protein synthesis profiles of growing and glucose-starved cells of B. subtilis. Individual dual channel 2-DE patterns of protein synthesis and accumulation recorded during the different phases of the growth curve are assembled into a movie of life (top part). Growth curve (optical density at 500 nm) and ^{35}S-L-methionine incorporation (million cpm per 60 μg protein) are displayed in the middle part. Patterns of selected examples representing different branches of cellular physiology are displayed in the lower part of the figure. Sample points correspond to the following growth phases depicted in the middle part: 1,2, exponential growth, 3–7, glucose starvation; 8,9 recovery of growth after the re-addition of glucose. The bar graphs on the left display normalized relative synthesis rates of the individual proteins at the different time points. This glucose starvation movie is available on the internet at: http://microbio2.biologie.uni-greifswald.de:8880/sub2d.htm (adapted from [61, 63]).

The network we have presented so far has depicted single moments in the life of a B. subtilis cell. Because the network is highly dynamic, with sequential gene expression programs as an essential feature it is necessary to assemble "snapshot by snapshot" through time to visualize growth and developmental processes at a proteomic scale as in a "movie of life". Such an sequential adaptation has already been analyzed for growing cells entering a glucose-starvation-induced stationary phase [63], a typical environmental situation for B. subtilis in nature (Fig. 7).

The transition from growing cells to glucose-starved cells is accompanied by a very comprehensive reorganization of the gene expression program [63]. Comparative dual channel analysis of protein content (protein staining; green) vs. protein synthesis (pulse labeling with ^{35}S-L-methionine; red) revealed at least 400 proteins that change in color from yellow to green during entry into stationary phase because their synthesis has been switched off in the stationary-growth phase. This expression kinetics is typical of proteins with housekeeping functions necessary for rapid growth and cell division. On the other hand, more than 150 proteins have been seen to be induced in a sequential order as indicated by the appearance of new protein spots colored red. These 150 proteins have been assigned to different general and specific regulation groups with the aid of mutants lacking global regulators. The powerful dual channel imaging technique not only allows the description of stimulons or regulons activated in a time-dependent manner but also permits the detailed kinetic analysis of individual proteins with the three colors: red standing for newly induced but not yet accumulated, yellow representing balance between synthesis and accumulation, and green indicating that the protein is present and probably still active but no longer synthesized. This expression profile is shown for AcoB, a subunit of the acetoin dehydrogenase, required for the utilization of the carbon overflow product acetoine as a secondary carbon source (Fig. 8). AcoB is induced after glucose exhaustion (red) followed by a balance between synthesis and accumulation and finally is switched off again, but this transient induction seems to be sufficient for the accumulation of the protein needed during glucose starvation (Fig. 8). This pattern is typical of most of the glucose-starved inducible proteins considered in this study (e.g., MalA, BglH, RbsK).

The extensive reprogramming of the gene expression network was confirmed by the application of DNA macroarray techniques (data not shown). The data clearly show that about the half of all B. subtilis genes are involved in the process, changing their expression pattern. Almost 1000 (vegetative) genes were switched off in cells entering glucose starvation, and the same number was induced during the course of the experiment. The integration of all the data particularly the combination of proteomics/transcriptomics data with basic cell physiology provides a comprehensive understanding of the processes occurring in cells entering a glucose-starvation-induced stationary phase. The induction and the kinetics of specific (repression of glycolysis, induction of gluconeogenesis, induction of alternate carbon degradation pathways, etc.) and general responses (σ^B-dependent general stress regulon, stringent response, Spo0A- or CodY-dependent responses) can be followed at a transcriptomic and proteomic scale. The final outcome of integrative ap-

protein accumulation and synthesis (AcoB)

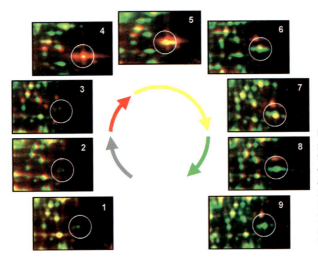

Fig. 8 Kinetics of synthesis (red) and accumulation (green) of acetoine dehydrogenase subunit AcoB during growth (1, 2), glucose starvation (3–7) and recovery of growth following re-addition of glucose (8, 9) (adapted from [61, 63]).

proaches involving transcriptomics, proteomics, and metabolomics of wild-type and mutant cells should be a dynamic model of cellular adaptation along the growth curve from the growing into the nongrowing state.

3.5
Proteomic signatures – tools for microbial physiology and their practical application

The collection of data on the proteins induced by environmental stimuli and their integration into the adaptational network may also provide information of practical relevance. Proteomics signatures [64] which have become a valuable tool for microbial physiology in general can be utilized for the prediction of the physiological state of cells grown in a bioreactor or in a biofilm. This tool for molecular diagnosis of cells does not only apply for growing cells where proteomic signatures can reveal which amino acids or carbon sources were available in the medium, *etc*. Relying on proteomic signatures one can predict if nongrowing cells suffered from oxidative stress, phosphate starvation, or from other stimuli. Examples of typical indicator proteins include KatA or AhpC for oxidative stress or GroEL/ClpC/ClpE for heat/protein stress, *etc*. (see Fig. 9).

Proteomic signatures were also applied to the characterization of the growth behavior of *B. licheniformis*, an enzyme producer important at an industrial scale. In stationary phase cells grown in a complex medium there was a strong proteomic signature for oxidative stress, followed by a signature for protein stress (ClpC) probably as a consequence of protein oxidation. This protein damage induced by oxidative stress was indicated by the fragmentation of EF-Tu. This example illustrates that proteomic signatures can be used to monitor fermentation processes in large bioreactors. Since *B. licheniformis* is a main producer of extracellular enzymes at an industrial scale, proteomic signatures can be valuable tools for the diagnosis of the

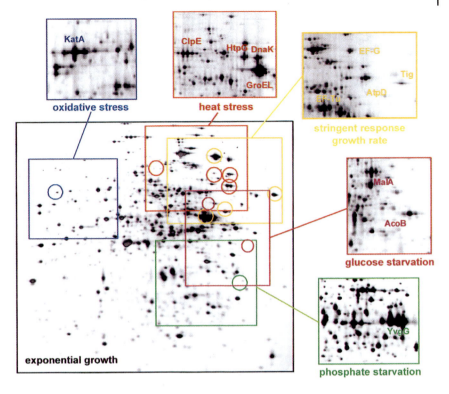

Fig. 9 Proteomics signatures of different physiological stress conditions. Comparisons of the protein profile of exponentially growing as well as stressed B. subtilis cultures reveal signature-like changes that are particular for certain stress stimuli (e.g., induction of the catalase KatA by oxidative stress). The individual section of the 2-DE gels display typical parts of the proteomic signatures of oxidative or heat stress, the stringent response or limitation of glucose or phosphate.

cell physiology of the producer strain under industrial fermentation conditions. The indication of severe oxidative stress occurring in a bioreactor might also have significant consequences for the quality of the proteins to be produced [65, 66].

Such a comprehensive proteome signature library for stress/starvation stimuli can also be exploited for the prediction of the mode of action (MOA) of drugs. Nitrofurantoin, for instance, shows the same proteomic signature as diamide does, a drug that induces oxidative damage *via* non-native disulfide bonds [58]. According to Guay [67], nitrofurantoin, still used for the treatment of urinary tract infections, should primarily affect DNA or RNA synthesis or the activity of metabolic enzymes. The proteomic signature rather suggests that nitrofurantoin induces oxidative stress presumably *via* non-native disulfide formation.

The stress/starvation proteomic signature library was combined with an antibiotic proteomic signature library that proved to be an excellent tool for the evaluation of unknown drugs. A pyrimidinone derivative developed by Bayer (BAY 50–2369)

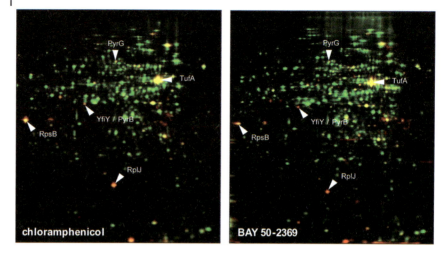

Fig. 10. Profiling of the mode of action of new antimicrobial substances. B. subtilis cultures were exposed to chloramphenicol or BAY 50–2369 during exponential growth and crude protein extracts of treated as a wells as untreated control cultures were separated by 2-DE. Subsequent to the staining of the gels, drug induced changes in the protein profile were identified by the dual channel techniques relying on the overlay of exponentially growing controls and treated samples.

showed a very similar proteomic signature as chloramphenicol strongly suggesting that the peptidyl transferase is the target protein for this substance [68] (Fig. 10). Phenyl-thiazolylurea sulfonamides seems to inhibit the activity of aminoacyl-tRNA synthetase (probably Phe-RS) because the drug showed a clear proteomic signature for a stringent response normally induced by an excess of uncharged tRNA molecules [69]. This approach was introduced almost 15 years ago when Ruth Van Bogelen and Frederick C. Neidhardt [70] found that antibiotics acting on the ribosome can be allocated to two main groups, drugs leading to an increase of mistranslation or premature translation termination and thereby triggering the protein stress response similar to a heat shock (puromycin, streptomycin, kanamycin) and drugs inhibiting the peptidyl transferase reaction (chloramphenicol) and thereby slowing the translational elongation simulating a cold shock response.

3.6
Transcriptomics *vs.* proteomics – towards a second generation of proteomics

The value of physiological proteomics very much depends on the degree of coverage of the cellular proteome. Whereas whole genome arrays cover all genes of an organism per definition, due to the technical as well as physiological constraints already discussed above particular 2-DE gels can only display a selection of the cellular protein profile. Furthermore, the cellular mRNA levels do not display such a wide dynamic range as the encoded proteins and thus whole genome arrays are believed to provide a much more comprehensive overview of the actual gene expression pattern than proteome studies. There are indeed a number of global gene expression studies employing both transcriptomics as well as proteome approaches that support the general view that the DNA array technologies record changes in gene expression more completely than proteomics [38, 58, 71–77]. Accordingly, gene expression profiling or the definition of regulons or stimulons in mind it is generally recommended to complement the proteomic approach by transcriptomics data. Nevertheless, proteomics will maintain its central position in functional genomics when it focuses on fields which cannot be replaced by DNA arrays or which substantially extend transcriptome data. The proteins and not the mRNAs are the main players of life. mRNA molecules are only highly unstable transmitters on the path from the genes to the ribosome, but the proteome depicts the final level of gene expression. The protein profiling, however, only provides information on the concentration of the individual proteins in the cell. However, this is not enough to make an organism viable. The elucidation of the mystery of life also needs to comprehend all the information related to the fate and destination of each single protein within and outside the cell that can be addressed by the proteomics of the second generation.

In our own proteomic studies we rely on the following approach. Using protein expression profiling, which can also be described as first generation proteomics, the most essential cellular events that have previously not been accessible by conventional approaches can be visualized by the panorama view of proteomics. However, the comprehensive proteomics information should only be regarded as a starting point to come to a deep understanding of the events visualized by protein expression profiling. These more detailed studies should combine classical biochemistry and molecular genetics with proteomics of the second generation. This new type of proteomics studies can once again be directed towards a global analysis of particular protein modifications, such as protein phosphorylations/dephosphorylations or protein oxidations triggered by stress or starvation. This sophisticated global protein analysis of course requires elaborate analytical techniques such as the combination of high-throughput peptide separations with MS techniques that allow identification and localization of the encountered post-translational modifications. Analysis of protein complexes is another promising and rapidly developing field that requires a combination of sophisticated biochemical tagging methods, prefractionation of samples by chromatographic techniques, and again identification of protein complex components by MS/MS analysis. All these approaches will profit from well-advanced molecular techniques because detailed characterization will require the comparative analysis of specific mutants (*e.g.*, lacking kinase or phosphatase activities). Furthermore, in the future proteomics techniques will also be necessary to correlate changes in the activity of individual biochemically well-defined enzymes with cellular reactions to environmental changes, such as the oxidative inactivation of glyceraldehyde-3-phosphate dehydrogenase that is observed following exposure of *S. aureus* to hydrogen peroxide [78].

In the future, proteomics should address very crucial questions of cellular life such as (i) how are proteins transferred to their final destination (sorting/targeting/secretion); (ii) how do proteins form the protein interaction network; (iii) how are proteins post-translationaly modified: (iv) how are proteins damaged, repaired and in hopelessly cases degraded (protein quality control); (v) how is the stability of the proteins regulated; and (vi) how is the activity of proteins regulated [79, 80] at a proteomic scale.

In the last part of the review the application of the second generation proteomics for addressing physiological questions will be illustrated in the light of a few selected examples.

3.7
The interactome

Genome-wide mRNA profiling does not provide any information about the activity, arrangement or final destination of the gene products, the proteins. Proteomics, on the other hand, if applied in a combination of the first and second level can provide the appropriate answers. Only in rare cases do proteins act completely independent of other proteins. Metabolism does not function in a bag of enzymes, but instead seems to be highly organized in complex higher order structures. Cellular life in general is organized as a complex protein interaction network, with many proteins taking part in multicomponent protein aggregates. A *B. subtilis* protein interaction database of DNA-replication relying on yeast two hybrid systems data has recently been published [81], but a description of the interactome based upon affinity tag/pull down/MS/MS approaches at a proteome wide scale (see [82]) is not yet available.

The majority of proteins act in the cytoplasm, but many proteins have specific sorting signals that lead them to their final destination outside the cytoplasm, *e.g.*, to the cytoplasmic membrane, to the membrane cell wall interspace, to the cell wall, or even to the extracellular space. Two main approaches to visualize these final destinations may be considered. One is to look for localization of proteins by sophisticated microscopic techniques after labelling the proteins with fluorescent dyes. In practice, some proteins may have a fixed position in the cell, while others move, *e.g.*, from cell pole to cell pole as shown by time lapse microscopy. The application of these modern fluorescence microscopic techniques has revolutionized our view of the subcellular organization of bacteria, leading to the discovery of highly specialized structures that are placed in time at or near the cell poles or at midcell (for reviews see [83, 84]). These structures are accomplished by complex and dynamic changes in the subcellular localization of signal transduction and cytoskeleton proteins, that even drive chromosome segregation and cell division [85–91]. It is now clear that the cytoskeleton is not unique to eukaryotes and that proteins that are structural and functional homologous to tubulin (FtsZ) and actin (MreB) are present in bacteria (for review see [87]). A "molecular topology" of the cell may be the intriguing outcome of these novel studies. On the other hand, the more "classical" proteomic approach analyzes the protein composition of subcellular fractions, the membrane fraction, the periplasmic fraction, and the "outer membrane" proteins for Gram-negative bacteria, the cell-wall-associated proteins, or the extracellular proteins.

3.8
The secretome

Collaboration between a group with a long-standing experience in the genetics of protein secretion with a proteomics laboratory offered the opportunity of gaining new information on protein secretion at a proteome-wide scale. At first this proteomic view of protein secretion opened the chance to bring the genome-based signal peptide predictions to the real life of more than 100 secreted proteins detected in the extracellular space of *B. subtilis* [92, 93]. Furthermore, cell-wall associated and membrane proteins were identified among them not only lipid-anchored proteins but even intrinsic membrane proteins with more than one membrane-spanning domain (see [2, 21, 94, 95]). Recently, Bunai *et al.* [22] identified many membrane-anchored proteins, among them 30 of the 37 predicted high-affinity solute-binding proteins from various ABC transporters, relying on cell membrane treatment by mixtures of detergents. This proteome-wide view of protein secretion provided much new information on protein secretion, such as (i) the major role of SecA and the prokaryotic signal recognition particles (*e.g.*, Ffh) in protein secretion; (ii) the low specificity of the five signal peptidases in protein secretion; (iii) the role of signal peptidase II and lgt in lipid anchoring and processing; (iv) the minor role of the twin-arginine translocation machinery in protein secretion of *B. subtilis*; (v) the major role of the lipoprotein PrsA as a extracellular folding catalyst in protein folding at the surface of the membranes which is also a peptidase-prolyl *cis/trans*-isomerase; (vi) the HtrA protein as an indicator for protein secretion stress (see [22, 95, 96]).

3.9
Post-translational modifications

Proteins that show more than one position on the 2-D gel are candidates for post-translational modification reactions. However, the detection of post-translational protein modifications either *via* migration differences in 2-DE or by MS/MS does not necessarily indicate *in vivo* modifications but can be also caused by *in vitro* preparation artefacts. Nevertheless, proteomics is a reasonable technique to get a first impression on protein modifications at a proteome-wide scale. A shift of almost all proteins to a more acidic position was found when cells were treated with actinonin, an antibiotic that inhibits the deformylase activity [97]. Immediately after pulse-labeling many newly synthesized proteins still contain formyl-methionine at the first position that is later cleaved off by peptide deformylase. Prevention of this removal of the formylated (negatively charged) start amino acid by actinonin results in a more acidic protein spot. In other cases the more acidic protein spot is the result of protein phosphorylation exemplarily shown for the anti-anti sigma factor RsbV (Fig. 11) or DegU. The staining with ProQ Diamond Phosphoprotein Stain provides an overview of the cellular phosphoproteome (Ser, Thr, Tyr) and is an excellent tool for analyzing protein phosphorylation at a proteome-wide scale. Using this approach the role of still unknown Tyr/Ser/Thr protein kinases encoded

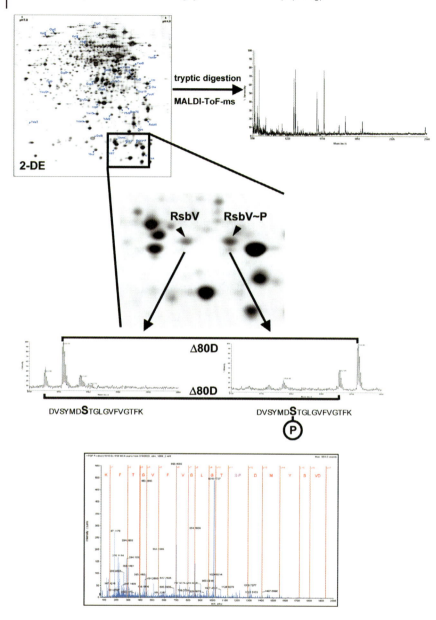

Fig. 11. Analysis of post-translational modifications – the RsbV example. PMF revealed at least two different isoforms of the RsbV protein. Detailed comparison of the spectra identified a peptide carrying a potential serine phosphorylation in the more acidic form. The site of phosphorylation was subsequently confirmed by ESI-MS/MS-analysis and shown to be Ser56 that had been postulated to be phosphorylated before (see also Fig. 12).

in the *B. subtilis* genome in signal transduction might be elucidated, provided that the experiments are permitted under conditions allowing kinase activity [98, 99]. The challenge is to find the target proteins for the unknown kinases and to identify the phosphorylated amino acid residues.

A shift to a more acidic spot position might also occur in response to oxidative stress when cysteine residues are oxidized to sulfonic acid [78, 100–102]. This example illustrates that the proteome approach also enables a proteome-wide view of protein damage and protein quality control systems. Protein carbonylation as a particular product of an oxidative stress induced reaction can be visualized by the oxo-blot technique [103–105]. The oxidation of sulfhydryl groups to S-S-bridges can also be monitored at a proteomic scale by specifically analyzing the disulfide proteome (Hochgräfe *et al.*, in preparation).

3.10
Protein quality control/protein degradation at a proteomic scale

It is the primary goal of protein quality control systems to repair the damaged protein providing the native folding state again. The induction of the GroEL/DnaK machine (HrcA regulon) and the Clp machine (CtsR regulon) is a good indicator for protein stress in the cell. Diamide treatment, for instance, that induces non-native S-S-bridges is a strong inducer of the CtsR-regulon and a weak inducer of the HrcA-regulon [58] suggesting that both regulons respond in a different way to protein stress. The main function of the Clp proteins is to refold the damaged proteins and to degrade the hopelessly destroyed proteins because protein waste is toxic for the cell. In addition to malfolded proteins or specific regulators no longer active [106], such as CtsR itself [107, 108], vegetative proteins no longer required in nongrowing cells can also form targets for Clp-dependent protein degradation. This was revealed by a 2-DE approach as a new tool for visualizing protein degradation at a proteomic scale. Pulse-chase labeling with ^{35}S-L-methionine was done in growing cells leading to radiolabelling of vegetative proteins. Then the stability of these labelled house-keeping proteins was followed in glucose-starved stationary phase cells of the wild type and a *clpP* mutant providing a proteome-wide view of protein stability by comparing the intensity during exponential growth with different stages in stationary phase. Many vegetative proteins were degraded in the wild type in a time-dependent scale but remained stable in the *clpP* mutant. This observation suggests that the degradation of vegetative proteins no longer required in nongrowing cells (at least not at the high level required in growing cells) occurs at a considerable rate presumably providing alternative energy sources in the energy-starved cell. Among the vegetative proteins degraded there are key enzymes of anabolic reactions, such as MurAA, a key enzyme in peptidoglycan biosynthesis [109], or PurF a key enzyme in purine biosynthesis (Gerth *et al.*, unpublished data).

These data show that in nongrowing cells the synthesis of vegetative proteins is not only switched off [38, 63] but the proteins still present and probably active are at least partially degraded by a controlled proteolysis adjusting for each protein the new level required under the changed physiological conditions. This comprehensive proteolytic attack on vegetative proteins mainly depends on the Clp machine. Future studies will now have to address the questions if there is a specific degradation signal (*e.g.*, adapter protein) that marks the vegetative proteins for degradation in cells no longer able to grow.

3.11
Gene expression network – horizontal and vertical approach

First-generation proteomics (expression proteomics) is predominantly concerned with the analysis of the regulatory network by defining regulon by regulon (horizontal approach). However, the dissection of the genome into regulons and the elucidation of the single regulon members is only the first step in understanding functions of the regulon within the adaptational network. Accordingly, in the second step the individual regulons should be selected for a detailed analysis (vertical approach) as illustrated for the σ^B-dependent general stress regulon as a model (Fig. 12). This regulon was selected because the induction of the σ^B-dependent general stress proteins proved to be the most obvious change at the proteome level.

In cells exposed to severe stress up to 20% or even more of the residual translational capacity were invested in the expression of σ^B-dependent general stress proteins [110]. Accordingly, the proteomic view of cellular events allowed the selection of one of the most essential stress/starvation responses. Using a combination of traditional genetic approaches as well as proteomics and transcriptomics about 150 genes were defined as σ^B-dependent as a first step [60, 110–119]. In parallel a regulatory network that activates the transcription factor σ^B *via* at least three signal transduction pathways in response to (i) environmental stress (heat, osmotic, acid, ethanol); (ii) energy starvation (limitation of glucose, phosphate or oxygen), and (iii) growth at low temperature has been elucidated, but the actual perception of the stimuli by the regulatory cascade remains to be discovered [35, 37]. In a next step, the overall function of the regulon was predicted from regulon members with biochemically defined function and subsequently confirmed by comparative phenotypic studies of wild-type strains and *sigB* mutants. σ^B-dependent general stress proteins are now known to provide the nongrowing cell with a nonspecific, multiple, and preventive stress resistance [120, 121]. Furthermore, the regulon provides protection to cells growing at low temperature [122]. Currently, we are analyzing the multiple stress resistance phenotypes of individual mutants in the majority of σ^B-dependent genes in order to understand the protective function of each single member of the regulon. These mutant screening studies will be supplemented by related techniques of functional genomics (*e.g.*, comprehensive post-translational modification analysis of the σ^B-dependent proteins, protein interaction network at a regulon scale – in order to establish a "guilty by association" network, *etc.*) and combined with traditional biochemical or enzymatic techniques to accomplish a comprehensive understanding of this crucial component of the stress/starvation

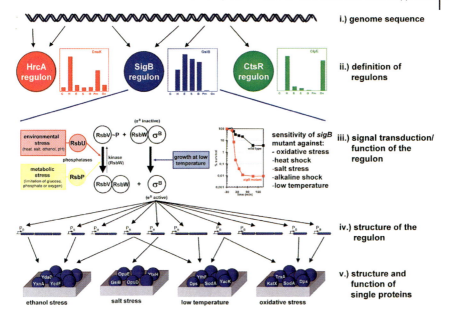

Fig. 12. Comprehensive analysis of structure, signal transduction pathways, and function of the SigB-dependent general stress regulon. The SigB-dependent general stress regulon is one of the cellular responses that are mounted when cells experience a heat shock. However, the SigB-regulon is not only induced by heat shock but a whole collection of environmental insults. Activation of the transcription factor SigB can be accomplished *via* at least three different signal transduction pathways. The environmental- and metabolic-stress-sensing branches of the signal transduction cascade convey their output *via* two PP2C-type phosphatases, RsbU and RsbP, to the antagonist protein RsbV [31, 35]. Dephosphorylation of RsbV~P activates RsbV that can then free SigB from the inhibitory RsbW/SigB complex resulting in the active transcription of the entire general stress regulon. Low-temperature growth [122] triggers SigB activation *via* a thus unresolved RsbV-independent pathway, that also functions in an *rsbVUP* triple mutant. Mutational inactivation of SigB prevents induction of the general stress genes and dramatically increases sensitivity of *B. subtilis* against a whole collection of stress factors. Finally, comparative phenotypical analysis of mutants lacking individual stress proteins and a corresponding wild-type strain will reveal the physiological importance and function of each of the regulon members.

response within the network (Fig. 12). This example should illustrate the true potential of proteomics when it is applied to obvious (and interesting) physiological problems and supplemented with more detailed experiments using traditional biochemical or genetic approaches.

3.12
Concluding remarks

Functional genomics techniques, such as transcriptomics as well as proteomics, provide a new understanding of cellular events because they enable an unbiased view of cellular adaptation in a new and wider context. Proteomics is particularly rewarding for microorganisms because not just a small percentage but a major fraction of cellular proteins is accessible. When combined with appropriate bioinformatics tools, such expression studies constitute perfect screening tools that generate a large list of interesting candidate genes and proteins. However, one has to escape the temptation of the descriptive level and has to supplement the global view by detailed physiological studies employing traditional biochemical and genetic approaches in order to gain a comprehensive understanding of the phenomena studied.

This is the reason why proteomics has to be developed from expression profiling to a new second level addressing important physiological questions, such as a cellular perspective of post-translational modifications, protein stability, protein localization within particular compartments, and finally protein-protein interaction networks. The ambitious goal of systems biology approaches – modeling of cellular life – will only be accomplished if the toolboxes of functional genomics, bioinformatics as well as traditional biochemistry and genetics are applied in concert. Within the ensemble of functional genomics techniques proteomics will keep its central position because it deals with the main players of life, with the proteins.

We thank the members of the Hecker and Völker Laboratories for their contributions to the work presented in this article. We are grateful to J. Bernhardt D. Becher and K. Büttner for providing some of the figures for this article. We are indebted to DECODON GmbH for the close cooperation and the pre-release access to new software tools. M.H. and U.V. were supported by grants from the Bundesministerium für Bildung und Forschung, the Deutsche Forschungsgemeinschaft, the Max-Planck-Society, and the Fonds der Chemischen Industrie.

3.13
References

[1] Fleischmann, R. D., Adams, M. D., White, O., Clayton, R. A. et al., *Science* 1995, *269*, 496–512.

[2] Kunst, F., Ogasawara, N., Moszer, I., Albertini, A. M. et al., *Nature* 1997, *390*, 249–256.

[3] Sonenshein, A. L., Hoch, J. A., Losick, R. (Eds.), *Bacillus subtilis and its Closest Relatives. From Genes to Cells*, ASM Press, Washington, DC 2002.

[4] Schumann, W., Ehrlich, D., Ogasawara, N. (Eds.), *Functional Analysis of Bacterial Genes: A Practical Manual*, John Wiley & Sons, Chichester 2001.

[5] Streips, U. N., Polio, F. W., *J. Bacteriol.* 1985, *162*, 434–437.

[6] Richter, A., Hecker, M., *FEMS Microbiol. Lett.* 1986, *36*, 69–71.

[7] O'Farrell, P. H., *J. Biol. Chem.* 1975, *250*, 4007–4021.

3.13 References

[8] Klose, J., *Humangenetik* 1975, *26*, 231–243.

[9] Scheele, G. A., *J. Biol. Chem.* 1975, *250*, 5375–5385.

[10] MacGillivray, A. J., Rickwood, D., *Eur. J. Biochem.* 1974, *41*, 181–190.

[11] Herendeen, S. L., VanBogelen, R. A., Neidhardt, F. C., *J. Bacteriol.* 1979, *139*, 185–194.

[12] VanBogelen, R. A., Kelley, P. M., Neidhardt, F. C., *J. Bacteriol.* 1987, *169*, 26–32.

[13] Link, A. J., *Trends Biotechnol.* 2002, *20*, S8–13.

[14] Gevaert, K., Van Damme, J., Goethals, M., Thomas, G. R. et al., *Mol Cell Proteomics* 2002, *1*, 896–903.

[15] Gygi, S. P., Rist, B., Gerber, S. A., Turecek, F. et al., *Nat Biotechnol* 1999, *17*, 994–999.

[16] Washburn, M. P., Wolters, D., Yates III, J. R., *Nat. Biotechnol.* 2001, *19*, 242–247.

[17] Lu, Y., Bottari, P., Turecek, F., Aebersold, R., Gelb, M. H., *Anal Chem* 2004, *76*, 4104–4111.

[18] Jaffe, J. D., Berg, H. C., Church, G. M., *Proteomics* 2004, *4*, 59–77.

[19] Lipton, M. S., Pasa-Tolic, L., Anderson, G. A., Anderson, D. J. et al., *Proc. Natl. Acad. Sci. USA* 2002, *99*, 11049–11054.

[20] Taoka, M., Yamauchi, Y., Shinkawa, T., Kaji, H. et al., *Mol Cell Proteomics* 2004, *3*, 780–787.

[21] Eymann, C., Dreisbach, A., Albrecht, D., Bernhardt, J. et al., *Proteomics* 2004, *4*, 2849–2876.

[22] Bunai, K., Ariga, M., Inoue, T., Nozaki, M. et al., *Electrophoresis* 2004, *25*, 141–155.

[23] Bremer, H., Dennis, P., in: Neidhardt, F., Curtiss, R., Ingraham, J. L., Lin, E. C. C. et al. (Eds.), *Escherichia coli and Salmonella: Cellular and Molecular Biology*. ASM Press, Washington, DC 1996, pp. 1553–1569.

[24] McHardy, A. C., Pühler, A., Kalinowski, J., Meyer, F., *Proteomics* 2004, *4*, 46–58.

[25] Rocha, E. P., Danchin, A., *Mol. Biol. Evol.* 2004, *21*, 108–116.

[26] Kobayashi, K., Ehrlich, S. D., Albertini, A., Amati, G. et al., *Proc. Natl. Acad. Sci. USA* 2003, *100*, 4678–4683.

[27] Tobisch, S., Zühlke, D., Bernhardt, J., Stülke, J., Hecker, M., *J. Bacteriol.* 1999, *181*, 6996–7004.

[28] Ludwig, H., Homuth, G., Schmalisch, M., Dyka, F. M. et al., *Mol. Microbiol.* 2001, *41*, 409–422.

[29] Ludwig, H., Rebhan, N., Blencke, H. M., Merzbacher, M., Stulke, J., *Mol. Microbiol.* 2002, *45*, 543–553.

[30] Hecker, M., Schumann, W., Völker, U., *Mol. Microbiol.* 1996, *19*, 417–428.

[31] Hecker, M., Völker, U., *Adv. Microbial. Physiol.* 2001, *44*, 35–91.

[32] Smith, M. W., Neidhardt, F. C., *J. Bacteriol.* 1983, *154*, 344–350.

[33] Bernhardt, J., Büttner, K., Scharf, C., Hecker, M., *Electrophoresis* 1999, *20*, 2225–2240.

[34] Hecker, M., Völker, U., *FEMS Microbiol. Ecol.* 1990, *74*, 197–213.

[35] Price, C. W., in: Storz, G., Hengge-Aronis, R. (Eds.), *Bacterial Stress Responses*, ASM Press, Washington, DC 2000, pp. 179–197.

[36] Hengge-Aronis, R., in: Storz, G., Hengge-Aronis, R. (Eds.), *Bacterial Stress Responses*, ASM, Washington, DC 2000, pp. 161–178.

[37] Price, C. W., in: Sonenshein, A. L., Hoch, J. A., Losick, R. (Eds.), *Bacillus subtilis and its Closest Relatives. From Genes to Cells*, ASM Press, Washington, DC 2002, pp. 369–384.

[38] Eymann, C., Homuth, G., Scharf, C., Hecker, M., *J. Bacteriol.* 2002, *184*, 2500–2520.

[39] Wendrich, T. M., Marahiel, M. A., *Mol. Microbiol.* 1997, *26*, 65–79.

[40] Rhaese, H. J., Dichtelmuller, H., Grade, R., *Eur. J. Biochem.* 1975, *56*, 385–392.

[41] Cashel, M., Gentry, D. R., Hernandez, V. J., Vinella, D., in: Neidhardt, F. C., Curtiss III, R., Ingraham, J. L., Lin, E. C. C. et al. (Eds.), *Escherichia coli and Salmonella: Cellular and Molecular Biology*, ASM Press, Washington, DC 1996, pp. 1458–1496.

[42] Ochi, K., Kandala, J. C., Freese, E., *J. Biol. Chem.* 1981, *256*, 6866–6875.

[43] Lopez, J. M., Marks, C. L., Freese, E., *Biochim. Biophys. Acta* 1979, *587*, 238–252.

[44] Lopez, J. M., Dromerick, A., Freese, E., *J. Bacteriol.* 1981, *146*, 605–613.

[45] Ratnayake-Lecamwasam, M., Serror, P., Wong, K. W., Sonenshein, A. L., *Genes. Dev.* 2001, *15*, 1093–1103.

[46] Molle, V., Nakaura, Y., Shivers, R. P., Yamaguchi, H. et al., *J. Bacteriol.* 2003, *185*, 1911–1922.

[47] Inaoka, T., Takahashi, K., Ohnishi-Kameyama, M., Yoshida, M., Ochi, K., *J. Biol. Chem.* 2003, *278*, 2169–2176.

[48] Inaoka, T., Ochi, K., *J. Bacteriol.* 2002, *184*, 3923–3930.

[49] Mascher, T., Margulis, N. G., Wang, T., Ye, R. W., Helmann, J. D., *Mol. Microbiol.* 2003, *50*, 1591–1604.

[50] Hecker, M., Völker, U., *Mol. Microbiol.* 1998, *29*, 1129–1136.

[51] Cao, M., Wang, T., Ye, R., Helmann, J. D., *Mol. Microbiol.* 2002, *45*, 1267–1276.

[52] Cao, M., Helmann, J. D., *J. Bacteriol.* 2004, *186*, 1136–1146.

[53] Wiegert, T., Homuth, G., Versteeg, S., Schumann, W., *Mol. Microbiol.* 2001, *41*, 59–71.

[54] Martinez-Antonio, A., Collado-Vides, J., *Curr. Opin. Microbiol* 2003, *6*, 482–489.

[55] von Mering, C., Zdobnov, E. M., Tsoka, S., Ciccarelli, F. D. et al., *Proc. Natl. Acad. Sci. USA* 2003, *100*, 15428–15433.

[56] Bork, P., Orengo, C. A., *Curr. Opin. Struct. Biol.* 2004, *14*, 261–263.

[57] Pragai, Z., Harwood, C. R., *Microbiology* 2002, *148*, 1593–1602.

[58] Leichert, L. I., Scharf, C., Hecker, M., *J. Bacteriol.* 2003, *185*, 1967–1975.

[59] Drzewiecki, K., Eymann, C., Mittenhuber, G., Hecker, M., *J. Bacteriol.* 1998, *180*, 6674–6680.

[60] Varon, D., Brody, M. S., Price, C. W., *Mol. Microbiol.* 1996, *20*, 339–350.

[61] Hecker, M., *Adv. Biochem. Eng. Biotechnol.* 2003, *83*, 57–92.

[62] Luhn, S., Berth, M., Hecker, M., Bernhardt, J., *Proteomics* 2003, *3*, 1117–1127.

[63] Bernhardt, J., Weibezahn, J., Scharf, C., Hecker, M., *Genome Res.* 2003, *13*, 224–237.

[64] VanBogelen, R. A., Schiller, E. E., Thomas, J. D., Neidhardt, F. C., *Electrophoresis* 1999, *20*, 2149–2159.

[65] Voigt, B., Schweder, T., Becher, D., Ehrenreich, A. et al., *Proteomics* 2004, *4*, 1465–1490.

[66] Antelmann, H., Darmon, E., Noone, D., Veening, J. W. et al., *Mol. Microbiol.* 2003, *49*, 143–156.

[67] Guay, D. R., *Drugs* 2001, *61*, 353–364.

[68] Bandow, J. E., Brötz, H., Leichert, L. I., Labischinski, H., Hecker, M., *Antimicrob. Agents Chemother.* 2003, *47*, 948–955.

[69] Beyer, D., Kroll, H. P., Endermann, R., Schiffer, G. et al., *Antimicrob. Agents Chemother.* 2004, *48*, 525–532.

[70] VanBogelen, R. A., Neidhardt, F. C., *Proc. Natl. Acad. Sci. USA* 1990, *87*, 5589–5593.

[71] Yoshida, K., Kobayashi, K., Miwa, Y., Kang, C. M. et al., *Nucleic Acids Res.* 2001, *29*, 683–692.

[72] Jürgen, B., Hanschke, R., Sarvas, M., Hecker, M., Schweder, T., *Appl. Microbiol. Biotechnol.* 2001, *55*, 326–332.

[73] Mäder, U., Antelmann, H., Buder, T., Dahl, M. K. et al., *Mol. Genet. Genomics* 2002, *268*, 455–467.

[74] Mäder, U., Homuth, G., Scharf, C., Büttner, K. et al., *J. Bacteriol.* 2002, *184*, 4288–4295.

[75] Mostertz, J., Scharf, C., Hecker, M., Homuth, G., *Microbiology* 2004, *150*, 497–512.

[76] Graumann, P., Schröder, K., Schmid, R., Marahiel, M. A., *J. Bacteriol.* 1996, *178*, 4611–4619.

[77] Beckering, C. L., Steil, L., Weber, M. H., Völker, U., Marahiel, M. A., *J. Bacteriol.* 2002, *184*, 6395–6402.

[78] Weber, H., Engelmann, S., Becher, D., Hecker, M., *Mol. Microbiol.* 2004, *52*, 133–140.

[79] Saghatelian, A., Jessani, N., Joseph, A., Humphrey, M., Cravatt, B. F., *Proc. Natl. Acad. Sci. USA* 2004, *101*, 10000–10005.

[80] Jessani, N., Cravatt, B. F., *Curr. Opin. Chem. Biol.* 2004, *8*, 54–59.

[81] Noirot-Gros, M. F., Dervyn, E., Wu, L. J., Mervelet, P. et al., *Proc. Natl. Acad. Sci. USA* 2002, *99*, 8342–8347.

[82] Gavin, A. C., Bosche, M., Krause, R., Grandi, P. et al., *Nature* 2002, *415*, 141–147.

[83] Shapiro, L., McAdams, H. H., Losick, R., *Science* 2002, *298*, 1942–1946.

[84] Shapiro, L., Losick, R., *Science* 1997, *276*, 712–718.

[85] Ben-Yehuda, S., Rudner, D. Z., Losick, R., *Science* 2003, *299*, 532–536.

[86] Errington, J., Daniel, R. A., Scheffers, D. J., *Microbiol. Mol. Biol. Rev.* 2003, *67*, 52–65.

[87] Errington, J., *Nat. Cell Biol.* 2003, *5*, 175–178.

[88] Wu, L. J., Errington, J., *Cell* 2004, *117*, 915–925.

[89] Mascarenhas, J., Soppa, J., Strunnikov, A. V., Graumann, P. L., *EMBO J.* 2002, *21*, 3108–3118.

[90] Ben-Yehuda, S., Losick, R., *Cell* 2002, *109*, 257–266.

[91] Gueiros-Filho, F. J., Losick, R., *Genes Dev.* 2002, *16*, 2544–2556.

[92] Antelmann, H., Tjalsma, H., Voigt, B., Ohlmeier, S. et al., *Genome Res.* 2001, *11*, 1484–1502.

[93] van Dijl, J. M., Braun, P. G., Robinson, C., Quax, W. J. et al., *J Biotechnol.* 2002, *98*, 243–254.

[94] Antelmann, H., Yamamoto, H., Sekiguchi, J., Hecker, M., *Proteomics* 2002, *2*, 591–602.

[95] Hirose, I., Sano, K., Shioda, I., Kumano, M. et al., *Microbiology* 2000, *146*, 65–75.

[96] Tjalsma, H., Antelmann, H., Jongbloed, J. D., Braun, P. G. et al., *Microbiol. Mol. Biol. Rev.* 2004, *68*, 207–233.

[97] Bandow, J. E., Becher, D., Büttner, K., Hochgrafe, F. et al., *Proteomics* 2003, *3*, 299–306.

[98] Madec, E., Stensballe, A., Kjellstrom, S., Cladiere, L. et al., *J. Mol. Biol.* 2003, *330*, 459–472.

[99] Madec, E., Laszkiewicz, A., Iwanicki, A., Obuchowski, M., Seror, S., *Mol. Microbiol.* 2002, *46*, 571–586.

[100] Wagner, E., Luche, S., Penna, L., Chevallet, M. et al., *Biochem J* 2002, *366*, 777–785.

[101] Yang, K. S., Kang, S. W., Woo, H. A., Hwang, S. C. et al., *J. Biol. Chem.* 2002, *277*, 38029–38036.

[102] Rabilloud, T., Heller, M., Gasnier, F., Luche, S. et al., *J. Biol. Chem.* 2002, *277*, 19396–19401.

[103] Dukan, S., Nystrom, T., *Genes Dev.* 1998, *12*, 3431–3441.

[104] Dukan, S., Farewell, A., Ballesteros, M., Taddei, F. et al., *Proc. Natl. Acad. Sci. USA* 2000, *97*, 5746–5749.

[105] Mostertz, J., Hecker, M., *Mol. Genet. Genomics* 2003, *269*, 640–648.

[106] Jenal, U., Hengge-Aronis, R., *Curr. Opin. Microbiol.* 2003, *6*, 163–172.

[107] Krüger, E., Witt, E., Ohlmeier, S., Hanschke, R., Hecker, M., *J. Bacteriol.* 2000, *182*, 3259–3265.

[108] Krüger, E., Zühlke, D., Witt, E., Ludwig, H., Hecker, M., *EMBO J.* 2001, *20*, 852–863.

[109] Kock, H., Gerth, U., Hecker, M., *Mol. Microbiol.* 2004, *51*, 1087–1102.

[110] Bernhardt, J., Völker, U., Völker, A., Antelmann, H. et al., *Microbiology* 1997, *143*, 999–1017.

[111] Boylan, S. A., Redfield, A. R., Price, C. W., *J. Bacteriol.* 1993, *175*, 3957–3963.

[112] Boylan, S. A., Thomas, M. D., Price, C. W., *J. Bacteriol.* 1991, *173*, 7856–7866.

[113] Price, C. W., Fawcett, P., Ceremonie, H., Su, N. et al., *Mol. Microbiol.* 2001, *41*, 757–774.

[114] Igo, M., Losick, R., *J. Mol. Biol.* 1986, *191*, 615–624.

[115] Helmann, J. D., Wu, M. F., Kobel, P. A., Gamo, F. J. et al., *J. Bacteriol.* 2001, *183*, 7318–7328.

[116] Völker, U., Engelmann, S., Maul, B., Riethdorf, S. et al., *Microbiology* 1994, *140*, 741–752.

[117] Antelmann, H., Bernhardt, J., Schmid, R., Mach, H. et al., *Electrophoresis* 1997, *18*, 1451–1463.

[118] Petersohn, A., Brigulla, M., Haas, S., Hoheisel, J. D. et al., *J. Bacteriol.* 2001, *183*, 5617–5631.

[119] Petersohn, A., Bernhardt, J., Gerth, U., Höper, D. et al., *J. Bacteriol.* 1999, *181*, 5718–5724.

[120] Völker, U., Maul, B., Hecker, M., *J. Bacteriol.* 1999, *181*, 3942–3948.

[121] Gaidenko, T. A., Price, C. W., *J. Bacteriol.* 1998, *180*, 3730–3733.

[122] Brigulla, M., Hoffmann, T., Krisp, A., Völker, A. et al., *J. Bacteriol.* 2003, *185*, 4305–4314.

4
Web-accessible proteome databases for microbial research*

Klaus-Peter Pleißner, Till Eifert, Sven Buettner, Frank Schmidt, Martina Boehme,, Thomas F. Meyer, Stefan H. E. Kaufmann and Peter R. Jungblut

The analysis of proteomes of biological organisms represents a major challenge of the post-genome era. Classical proteomics combines two-dimensional electrophoresis (2-DE) and mass spectrometry (MS) for the identification of proteins. Novel technologies such as isotope coded affinity tag (ICAT)-liquid chromatography/mass spectrometry (LC/MS) open new insights into protein alterations. The vast amount and diverse types of proteomic data require adequate web-accessible computational and database technologies for storage, integration, dissemination, analysis and visualization. A proteome database system (http://www.mpiib-berlin.mpg.de/2D-PAGE) for microbial research has been constructed which integrates 2-DE/MS, ICAT-LC/MS and functional classification data of proteins with genomic, metabolic and other biological knowledge sources. The two-dimensional polyacrylamide gel electrophoresis database delivers experimental data on microbial proteins including mass spectra for the validation of protein identification. The ICAT-LC/MS database comprises experimental data for protein alterations of mycobacterial strains BCG *vs.* H37Rv. By formulating complex queries within a functional protein classification database "FUNC_CLASS" for *Mycobacterium tuberculosis* and *Helicobacter pylori* the researcher can gather precise information on genes, proteins, protein classes and metabolic pathways. The use of the R language in the database architecture allows high-level data analysis and visualization to be performed "on-the-fly". The database system is centrally administrated, and investigators without specific bioinformatic competence in database construction can submit their data. The database system also serves as a template for a prototype of a European Proteome Database of Pathogenic Bacteria. Currently, the database system includes proteome information for six strains of microorganisms.

---

* Originally published in Proteomics 2004, 5, 1305–1313

Proteomics of Microbial Pathogens. Edited by Peter R. Jungblut and Michael Hecker
Copyright © 2007 WILEY-VCH Verlag GmbH & Co. KGaA, Weinheim
ISBN: 978-3-527-31759-2

4.1
Introduction

The analysis of proteomes of diverse biological organisms represents one of the challenges in the post-genome era and is a rich source of biological information [1]. In contrast to sequence data of more than 90 bacterial genomes [2, 3] accessible *via* public databases, proteome data are characterized by diverse data types and are stored in proprietary databases located worldwide. The diversity of data is due to the various methods applied for proteome research such as 2-DE/MS, isotope coded affinity tag (ICAT)-LC/MS, protein sequencing, and other methods. Proteome databases were established by different research groups and in different ways. Databases such as SWISS-2DPAGE, 2D-PAGE, HSC-2DPAGE referred in the WORLD-2DPAGE index (http://www.expasy.org/ch2d/2d-index.html) or ProteomeWeb [4] and YPRC-PDB [5] can serve as examples. Additionally, efforts have been made to unite such heterogeneous databases by defining a set of rules for federation [6] or to create standards for modeling, capturing and disseminating proteome experimental data [7]. Further immunologic/bacterial proteome databases deal with human primary T helper cells [8], *Streptomyces coelicolor* (http://proteom.biomed.cas.cz/strepto/cc1_strep.php) and *Bacillus subtilis* (http://microbio2.biologie.uni-greifswald.de:8880/sub2d), for instance.

For immunologic research, we published our proprietary microbial proteome 2-DE database "2D-PAGE" [9, 10]. This database comprises data for *Mycobacterium tuberculosis, Helicobacter pylori, Chlamydophila pneumoniae, Borrelia garinii, Francisella tularensis* and *Mycoplasma pneumoniae*. Additionally, proteome data of Jurkat-T cells, mammary gland (mouse) and rat heart may be obtained. The 2D-PAGE database is mainly characterized by the consequent application of the relational data model for database construction and the application of open source software tools. Here we describe our web-accessible proteome database system for microbial research which reflects an effort to integrate 2-DE/MS, ICAT-LC/MS and functional classification data of proteins with genomic, metabolic and other knowledge sources in molecular biology, such as 3D-structure or protein-protein interaction databases. By formulating complex biological queries the researcher can gather information on genes, proteins, functional protein classes of *M. tuberculosis* and *H. pylori* and metabolic pathways at a glance. Thus, the storage, analysis and visualization of proteomic data contribute to a more holistic view on microorganisms.

4.2
Materials and methods

4.2.1
Data generation and data storage

The proteome database contains diverse data types generated by 2-DE (gel images), MS (spectra), ICAT-LC/MS data (spectra), MS-database search results and textual information describing experimental protocols (for example, sample preparation) or results of protein identification. The sample preparation procedures and protocols of 2-DE are documented for each proteome [11, 12]. The 2-DE gels are scanned and analyzed either by our own developed gel image analysis software TopSpot [13] which can be downloaded from the "Download area" of the 2D-PAGE website free of charge or by PDQuest (Bio-Rad Laboratories, Hercules, CA, USA). The scanned images as well as the processed sets of gels (matchsets) containing the differentially regulated protein spots as markers for comparative 2-DE experiments are additionally stored in the laboratory information management system (SQL*LIMS; Applied Biosystems, Foster City, CA, USA) based on Oracle. Mass spectra used for protein identification are obtained using the mass spectrometer Voyager Elite (Perspective Biosystems, Framingham, MS, USA) and also stored as proprietary binary files or attachments into the laboratory information management system (LIMS). MS-database search results from MASCOT (Matrix Science, London, UK; http://www.matrixscience.com) or MS-Fit (UCSF Mass Spectrometry Facility; http://prospector.ucsf.edu) and their descriptive information can be stored in the LIMS using a parser written in Python. Because only part of the complex data within the LIMS should be published worldwide, a Java-GUI enabling user-controlled data transfer from the LIMS into the corresponding tables of the 2D-PAGE database was developed. The usage of a LIMS assures a given quality of experimental data. Thus, the LIMS represents our central laboratory data repository. A detailed description of LIMS and its adaptation to our specific needs will be provided in the future.

4.2.2
Software tools

For the software development of our database system, only open source software tools were applied. Specifically, we are using the relational database management system MySQL (http://www.mysql.com), PERL, PHP, HTML, JavaScript, Java, the GD graphic library (http://www.boutell.com/gd/) and the language for data analysis and graphics R (http://www.r-project.org/) [14]. To integrate proteome research with genomic or metabolic data dynamically created HTML hyperlinks are essential. Therefore, the accession methods and URLs to public genomic and metabolic databases must also be known. Further, specific proteomic tools such as MASCOT for MS database searches and PDQuest or TopSpot for gel image analysis are applied.

4.3
Results and discussion

4.3.1
Data management, analysis and presentation

A schematic overview of our proteome informatics approach for microbial research is illustrated in Fig. 1. It mainly consists of three parts: data acquisition, data analysis, and web-accessible data presentation. For the acquisition of heterogeneous

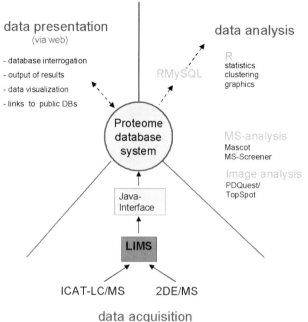

Fig. 1 Schematic overview of our proteome informatics approach for microbial research.

proteome data, a LIMS is applied. Results of gel image analysis performed by PDQuest can also be transmitted into the LIMS. A Java-GUI enables transfer of selected data from the LIMS into the database system consisting of 2D-PAGE, FUNC_CLASS (functional classification) and ICAT-LC/MS databases, whereby the 2D-PAGE plays the most important role. The database system is interconnected with public genomic, metabolic and other knowledge bases. To accomplish data analyses, specific proteomic tools such as MASCOT, PDQuest or others are used. Furthermore, continuative explorative data analysis, for example calculation of statistical tests or determination of percent distribution of protein classes, is carried out using R, a language for data analysis and graphics. Web-accessible data presentation includes the possibilities of complex interrogations and the presentation of results (tables, graphics, download of tab-separated files) *via* a web browser enabling access to public databases by hyperlinks.

4.3.2
2D-PAGE database

The 2D-PAGE proteome database (http://www.mpiib-berlin.mpg.de/2D-PAGE/), first published *via* the internet in 1999 [9], is a multispecies database containing proteome information on diverse strains of microorganisms, tissues and cells. The number of identified spots, the methods of protein identification, the number of antigens and other information on each strain of microorganism are summarized on the statistics page of the database. A part of the database structure is shown in

4.3 Results and discussion

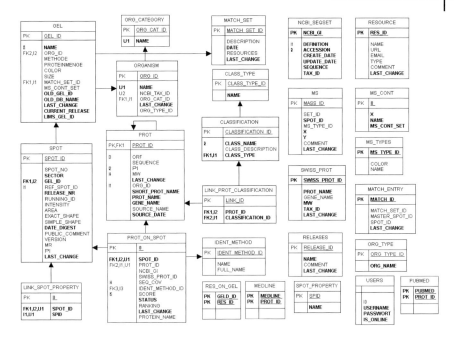

Fig. 2 Part of the structure of the 2D-PAGE database.

Fig. 2 where the main tables and their relations are schematically depicted. The self-explanatory headings of the tables show how the information on gels, spots, proteins, identification methods, MS, public sequence, literature databases, functional protein classes, and organisms are stored. The fact that one spot may contain several proteins or one protein may be found in different spots is also taken into consideration in the development of the database schema.

Preparing 2-D gel images for database construction is usually done by the Top-Spot gel image processing system. If a spot list was generated by other gel image analysis systems (for example, PDQuest or MELANIE) this data may also be accepted. The structure of a spot list is described in the "Technical Description of Data Submission" at our website. By using TopSpot, for example, the spot detection procedure yields both the spot positions and also describes the spot contours by polygons. These polygon approximations of protein spots serve as sensitive clickable areas within web-accessible gel images and provide the links to the annotated information which is stored in corresponding database tables.

Beside the specific interactions between descriptive information and gel image manipulations our database includes information on mass spectrometric data (peptide mass fingerprints) to comprehend the process of protein identification. Thus, synthetic MS spectra are generated from peak lists of peptide mass fingerprint data. 2D-PAGE is a centrally administrated database. The preparation of gel images and acquisition of descriptive data should be carried out according to data submission rules. The image and data files must be transferred to the database administrator by email or ftp. We are then able to establish a user-specific 2-DE database without further user interactions at the administration level. Thus, we offer a service to build up 2-DE databases for investigators without bioinformatic background. 2D-PAGE also serves as a template for a prototype of a European Proteome Database of Pathogenic Bacteria (EBP). Currently, the database system contains proteomic data on six strains of microorganisms and four eukaryotic proteomes.

4.3.3
ICAT-LC/MS database

The ICAT-LC/MS technology is a powerful tool to complement classic proteomic methods [15, 16]. For ICAT-LC/MS we have developed a relational database which contains information on mass error, coverage of predicted ions, gene name, protein name, sequence and the ratio of intensity for light/heavy isotope tags for mycobacterial strains BCG *vs.* H37Rv. The light/heavy isotope ratio represents the main information source in the ICAT-LC/MS technology. Because of the huge amount of generated spectra produced by ICAT-LC/MS it is advantageous to store and query data *via* a database that is more comfortable in comparison to an EXCEL approach. Furthermore, the formulation of complex queries can be carried out easily by a web-accessible interface. Additionally, the output of resulting records matching the search criteria can be limited to relevant data field descriptions. Currently, the ICAT-LC/MS database (http://www.mpiib-berlin.mpg.de/bioinfo/ICAT/) contains 1894 entries, of which 183 were manually evaluated.

4.3.4
FUNC_CLASS database

After sequencing the genomes of *M. tuberculosis* and *H. pylori,* protein coding genes were automatically predicted and functionally categorized or classified. The information on protein coding genes with their functional categories is available for *M. tuberculosis* at (http://www.sanger.ac.uk/Projects/Mtuberculosis/Genelist/) and for *H. pylori* at the comprehensive microbial resource of TIGR (http://www.tigr.org/tigr-scripts/CMR2/gene_attribute_form.dbi). We used these information sources to establish a functional classification database (FUNC_CLASS), consisting of CLASS_Mtb and CLASS_HP26695 for the functional classification of *M. tuberculosis* and *H. pylori,* respectively. The functional classification databases were also linked with our local databases 2D-PAGE and ICAT-LC/MS for mycobacterial strain BCG *vs.* H37RV and KEGG (http://www.genome.ad.jp/kegg/) for metabolic pathways and with the Protein Extraction, Description, and Analysis Tool (PEDANT) (http://pedant.gsf.de/) [17]. Using the capabilities of PEDANT, a comprehensive analysis of complete genomic sequences can be provided. For example, the sequence positions of protein coding genes or the 3-D presentation of protein structures can be shown. The CLASS_Mtb contains information on ORFs, short names of genes, protein names, class_ID, class name (category), M_r and pI. In total, 3924 genes are stored.

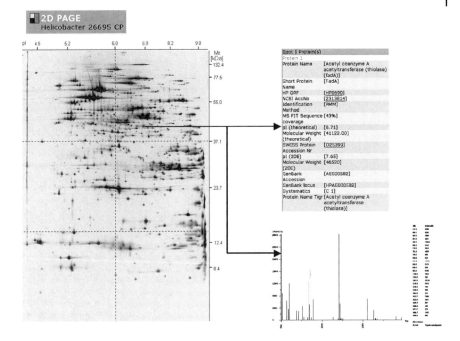

Fig. 3 2-D PAGE: *H. pylori* gel image. Protein spot information (annotation) on acetyltransferase and synthetically generated mass spectrum using peptide mass fingerprinting data.

For *H. pylori* 1563 protein coding genes are available with information on locus, gene length, protein length, GC content, pI/M_r values, cellular roles, GenBank/SWISS-PROT ID, and putative identification (protein name).

To retrieve data from these functional classification databases we have written a web form which enables the formulation of multicriteria queries by extending search criteria to refine queries interactively. Additionally, a list of genes given, for instance, in an EXCEL table, can be copied and pasted in a text area. All records matched to this list are displayed. Finally, records can be exported as tab-separated files. The opportunity to ask complex questions and the retrieval of information on genes, proteins and metabolism is permitted by FUNC_CLASS.

4.3.5
Data analysis and visualization

A major challenge in bioinformatics for proteome research is intelligent data analysis and visualization. We use the R language which comprises a large number of data analysis and graphic tools that are available as packages or functions. Thus far, there is no task for data analysis needed for proteome research that is not solvable in R. Currently, this language has also gained importance for microarray analysis. Additionally, R is characterized by relative simplicity in software function

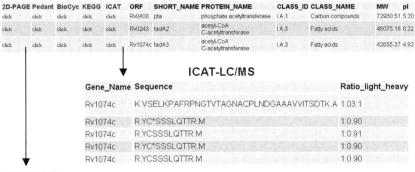

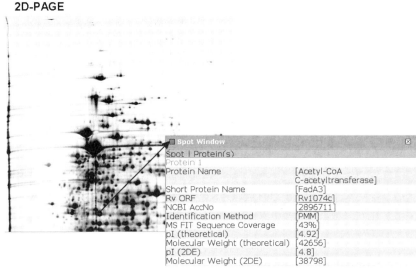

Fig. 4 FUNC_CLASS database: search for all acetyltransferase proteins with a $M_r > 40$ kDa for *M. tuberculosis* and output search results matching the search criteria (top); ICAT-LC/MS databases results (middle); 2D-PAGE database showing the annotation and the location of the protein in the 2-DE gel of *M. tuberculosis* (bottom).

calls for accomplishing complex tasks such as clustering, model fitting, statistical testing, plotting *etc*. Using the RMySQL package we can retrieve data from our MySQL databases and transfer these data into R data frames (matrix). RMySQL also enables the querying of databases using SQL statements. Thus, the retrieval of

4.3 Results and discussion | 71

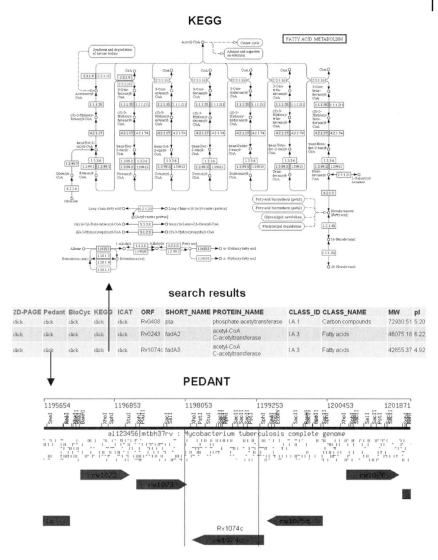

Fig. 5 The role of acetyltransferase (ORF: Rv1074c) within the fatty acid oxidation pathway of *M. tuberculosis* generated by KEGG (top). The position of the Rv1074c gene within the genomic sequence provided by the DNA viewer of the PEDANT system (bottom).

selected data can be provided. To generate a theoretical 2-DE pattern (virtual 2-DE plot) based on the theoretical pI/M_r values of proteins stored in our databases, an R-program is generated by means of a Perl-written CGI-script running on the web server. R dynamically creates the 2-DE plot as an image file (png format) that can be visualized directly *via* a browser. The virtual 2-DE patterns of *M. tuberculosis* or *H. pylori* are additionally overlaid with mouse-sensitive crosses showing those

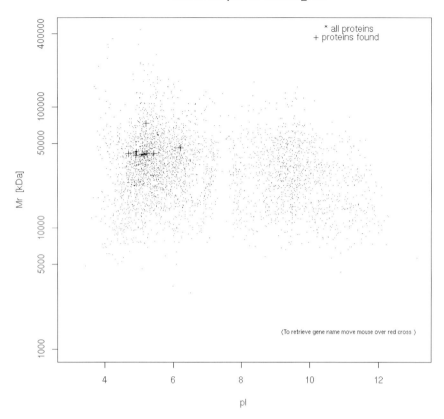

Fig. 6 Virtual 2-DE pattern of *M. tuberculosis*: all proteins with theoretical pI/M_r values are overlaid with mouse-sensitive markers (crosses) showing those protein entries found by the search criteria. Using the language R all graphics are dynamically created.

proteins that match the search criteria formulated by SQL statements. Thus, one can easily evaluate where specific protein spots are expected on a 2-DE gel and which proteins are outside of the analysis window of 2-DE.

The full capabilities of our interconnected proteome databases can only be demonstrated online. Here, we show some examples to gain information on microbial proteomes. One can search for information on proteins in 2-DE gel images of selected species by clicking on identified spots (Fig. 3). If peptide mass fingerprinting (PMF) data are available a synthetically generated MS spectrum is displayed for the stored PMF-list. Using functional classification database FUNC_CLASS, proteome relevant queries can be formulated such as: search for all acetyltransferase proteins with a molecular mass > 40 kDa for *M. tuberculosis* (Fig. 4, top). Thus, all entries in the database matching this query are displayed with their interconnections to the ICAT-LC/MS database (Fig. 4, middle) and to the

2D-PAGE database showing where the proteins are located in the 2-DE gel (Fig. 4, bottom). The role of the acetyltransferase within the fatty acid oxidation pathway of *M. tuberculosis*, automatically generated by the KEGG server, is depicted in Fig. 5, top. The position of the Rv1074c gene within the genomic sequence is shown by the PEDANT system (Fig. 5, bottom). The virtual 2-DE pattern (Fig. 6) created by all known proteins of *M. tuberculosis* is overlaid with markers (crosses) showing those protein entries found by the search criteria.

4.4
Concluding remarks

A web-accessible system of proteome databases has been developed that comprises 2-DE/MS, ICAT-LC/MS and a functional classification database for microbial proteins. These local databases are also linked to a variety of external public genomic and metabolic databases. Hence, researchers can gather information on genes, proteins and metabolic pathways. Different access methods, such as clicking on a spot in a 2-D gel or asking complex questions were developed to obtain customized information. A LIMS is used as a central repository for experimental and processed data. A part of these data are transferred into the web-accessible 2D-PAGE database. Applying the language R, data analysis and data visualization tasks may be performed dynamically. The power of R fulfills the requirements of a high-level dynamic data analysis and data visualization tool for proteome data. Furthermore, the proteome database system, especially the 2D-PAGE database, serves as a template for a prototype of a European Proteome Database of Pathogenic Bacterial.

The authors would like to express their thanks to Dr. Ursula Zimny-Arndt and Monika Schmid for 2-DE and MS analysis. This work was partially funded by the German Federal Ministry for Eduction and Research (contract number: 031U107C/031U207C) and by the European Union (EBPnetwork – contract number: QLRT-1999-31536).

4.5
References

[1] Patterson, S. D., Aebersold, R. H., *Nat. Genet.* 2003, *33*, 311–323.

[2] Hiscock, D., Upton, C., *Bioinformatics* 2002, *16*, 484–485.

[3] Oh, J. M. C., Hanash, S. M., Teichroew, D., *Electrophoresis* 1999, *20*, 766–774.

[4] Babnigg, G., Giometti, C. S., *Proteomics* 2003, *3*, 584–600.

[5] Cho, S. Y., Park, K.-S., Shim, J. E., Kwon, M.-S. et al., *Proteomics* 2002, *2*, 1104–1113.

[6] Appel, R. D., Bairoch, A., Sanchez, J. C., Vargas, J. R. et al., *Electrophoresis* 1999, *17*, 540–546.

[7] Taylor, C. F., Paton, N. W., Garwood, K. L., Kirby, P. D. et al., *Nat. Biotechnol.* 2003, *21*, 247–254.

[8] Nyman, T., Rosengren, A., Syyrakki, S., Pellinen, T. et al., *Electrophoresis* 2001, *22*, 4375–4382.

[9] Mollenkopf, H. J., Jungblut, P. R., Raupach, B., Mattow, J. *et al.*, *Electrophoresis* 1999, *20*, 2172–2180.

[10] Pleißner, K.-P., Eifert, T., Jungblut, P. R., *Comp. Funct. Genom.* 2002, *3*, 97–100.

[11] Mollenkopf, H. J., Mattow, J., Schaible, U. E., Grode, L. *et al.*, *Methods Enzymol.* 2002, *358*, 242–256.

[12] Jungblut, P. R., Bumann, D., *Methods Enzymol.* 2002, *358*, 307–316.

[13] Prehm, J., Jungblut, P. R., Klose, J., *Electrophoresis* 1987, *8*, 562–572.

[14] Ihaka, R., Gentleman, R., *J. Computational Graphical Statistics* 1996, *5*, 299–314.

[15] Gygi, S. P., Rist, B., Gerber, S. A., Turecek, F. *et al.*, *Nat. Biotechnol.* 1999, *10*, 994–999.

[16] Schmidt, F., Schmid, M., Jungblut, P. R., Mattow, J. *et al.*, *J. Am. Soc. Mass Spectrom.* 2003, *14*, 943–956.

[17] Frishman, D., Albermann, K., Hani, J., Heumann, K. *et al.*, *Bioinformatics* 2001, *17*, 44–57.

5
A targeted proteomics approach to the rapid identification of bacterial cell mixtures by matrix-assisted laser desorption/ionization mass spectrometry*

Bettina Warscheid and Catherine Fenselau

A proteomic approach to the rapid identification of bacteria is presented, which relies on the solubilization of a limited number of proteins from intact cells combined with on-probe tryptic digestion. Within 20 min, complete cleavage products of a limited set of bacterial proteins with molecular masses of about 4–125 kDa were obtained by on-probe digestion with immobilized trypsin. Bacterial peptides suitable for unimolecular decomposition analysis were generated within 5 min, and the sequence information obtained allowed identification of abundant proteins, and accordingly, their bacterial sources *via* searches in the NCBI database. Analysis of fragmentation products was also shown to allow for identification of bacterial peptides identical in mass but differing slightly in amino acid sequence by manual data analysis. In this work, *Bacillus subtilis* 168, *B. globigii*, *B. sphaericus* 14577, *B. cereus* T, and *B. anthracis* Sterne were examined, and various cold shock proteins were identified in all species. In addition, DNA-binding, 60 kDa-heat shock, surface-related and other stress-protective proteins were identified in the bacterial cell digests, and species-specific tryptic peptides could be generated from each of the *Bacillus* species studied. Bacterial peptides could be analyzed with greater sensitivity and mass accuracy than the parent proteins. The applicability of this targeted proteomics approach to the rapid identification of *Bacillus* species was further established by analyzing binary cell mixtures.

5.1
Introduction

The ability of mass spectrometry to provide rapid and reliable identification of microorganisms based on the desorption/ionization and detection of characteristic biological molecules was demonstrated more than 25 years ago [1], when untreated

---

* Originally published in Proteomics 2004, 10, 2877–2892

bacterial samples were placed directly into an electron impact instrument. The advent of MALDI-TOF instruments has provided improved sensitivity and speed [2, 3], and the potential has been realized for monitoring airborne microorganisms in public spaces and on the battle field by MS.

MALDI is not sufficiently reproducible in automated analysis of uncontrolled field samples to support library searching as the link between spectrum and species. There is general agreement that the ions detected above mass 4000 with MALDI represent translated proteins from microorganisms [2, 3]. It has been also proposed [4], that the protein molecular ions detected in such spectra can be matched against protein masses predicted from genome sequences to provide microorganism identifications that are independent of the conditions and reproducibility of the experiment. This proteomics approach has been further refined to accommodate post-translational modifications (PTM) [5] and to incorporate significance testing [6, 7]. Recently, this application of bioinformatics has been extended by devising methods for brief *in situ* production of peptides, and databases to provide identification of proteins and species based on peptide mass maps [8, 9]. However these strategies are not applied as readily to spectra of proteins from mixtures of microorganisms, or from samples contaminated by microorganisms in the biosphere.

We have proposed the combination of *in situ* proteolytic digestion with mass spectrometry-based microsequencing of product peptides to provide identification of parent proteins and thus of parent microorganisms with database searches based on these sequences [10, 11]. It could be shown that this approach is most productive when only a limited set of proteins is digested to provide a limited number of peptides [9–11]. In this situation individual peptide molecular ions can be observed with sufficient ion abundances to provide collision-induced dissociation or metastable fragmentation spectra of fragment ions that reveal sequence information. This approach has been initially demonstrated with *Bacillus* spores and their mixtures [10, 11], exploiting the selective solubility of the abundant family of spore proteins known as small acid-soluble proteins. In the present report, we refine this approach and extend it to vegetative Gram-positive bacteria, where the challenge is greater to solubilize only a limited set from among the many proteins accessible in these growing cells.

5.2
Materials and methods

5.2.1
Chemicals

Trifluoroacetic acid (TFA), methanol (MeOH), acetonitrile (ACN), ammonium bicarbonate (NH_4HCO_3), and α-cyano-4-hydroxy-cinnamic acid (CHCA) were purchased from Sigma (St.Louis, MO, USA). Trypsin immobilized on agarose beads was purchased from Pierce (Rockford, IL, USA) and washed three times with NH_4HCO_3 buffer (25 mM NH_4HCO_3, pH = 7.5) before use.

5.2.2
Bacillus strains

B. cereus strain T and B. globigii strain 9372 were originally obtained from H. O. Halvorson and the U.S. Army Laboratories at Dugway Proving Ground, Nevada, respectively. Additionally, the wild-type B. subtilis strain EMG 168, B. sphaericus strain ACTT 14577, B. cereus T, and B. anthracis Sterne, a non-pathogenic strain widely used as a vaccine for animals and livestock, were studied. All Bacillus strains were grown by standard techniques described elsewhere [12], and purified by mild salt detergent washes to obtain vegetative cell samples of $\geq$ 99% purity [13, 14]. Subsequently, some Bacillus cells were directly prepared for mass spectral analysis while the remaining material was lyophilized and stored at $-20°C$. To obtain cells with forespores of $\geq$ 95% purity, cell culturing was prolonged without adding fresh nutrients. Cells were analyzed by microscopy and harvested when forespores were developed.

5.2.3
Vegetative cell digestion

Bacillus strains were suspended in a 1:1 mixture of MeOH and 25 mM NH_4HCO_3 resulting in a final concentration of 2.5 mg cells/mL. Aliquots of 0.8 µL of bacterial cell suspensions were directly placed on the MALDI sample plate, and samples were allowed to air-dry for a minute. Subsequently, 1 µL of immobilized trypsin in 25 mM NH_4HCO_3 was deposited on each sample to allow for *in situ* proteolytic digestion of the protein subset solubilized from whole cells. To prevent solvent evaporation, the prepared samples were covered with a humidified chamber (100% relative humidity) at room temperature. Proteolytic cleavage reactions were stopped by adding 0.1% TFA after 5 to 20 min, and samples were allowed to air-dry. For peptide analysis by MALDI-TOF MS, 10 mM CHCA in a mixture of 70% ACN and 0.1% aqueous TFA was used as matrix. Subsequently, bacterial cell digests were directly analyzed with no need of any sample clean-up. In control experiments, multiple tryptic digests were generated from Bacillus strains not exposed to any temperature downshifts and from cells stored at $-20°C$ in parallel, and corresponding MALDI spectra showed no significant change in mass and relative abundance of protonated tryptic peptides.

5.2.4
MALDI-TOF MS and unimolecular decomposition product ion analysis

MALDI spectra were obtained with an AXIMA-CFR time-of-flight instrument (Kratos Analytical supplied by Shimadzu Biotech, Manchester, UK) with delayed extraction operated in the reflectron positive ion mode. Using the curved-field reflectron (CFR) of the instrument, unimolecular decomposition products from isolated protonated peptides can be detected in a single analysis step. The MALDI-TOF MS and unimolecular decomposition spectra presented are a sum of 100 to 150 laser shots rastered over the sample surface with a laser power of 40 arbitrary units (range of laser power 0–180, where 0 represents minimum and 180 maximum transmission). For unimolecular decomposition analysis, protonated peptide ions were isolated with an ion gate set to a 10–15 Da window. After correct isolation of protonated peptides was performed, enhanced fragmentation of parent ions was induced by increasing the laser power by 40–50%. A mixture of standard peptides was used for external calibration.

5.2.5
Database searches and identification of *Bacillus* species

Extended sequence information on various protein families such as ribosomal proteins, cold shock proteins (CSPs), and heat shock proteins (HSPs) is retrievable for *B. subtilis* strain 168 and *B. anthracis* strain A2012 in the NCBI databases, since their genomes are known. Additionally, a few CSPs and HSPs are archived for *B. cereus* T and *B. thuringiensis* subs. Kurstaki. Besides database entries for DNA-binding proteins in *B. globigii* and a 125 kDa surface layer protein in *B. sphaericus* 14577, little information on their protein content is available. Taxonomic information on spore-forming *Bacillus* species and the highly conserved protein family of small, acid-soluble proteins (SASPs), present only in their spores, has been considered in the evaluation of database search results accomplished in this study, and can be summarized as follows: *B. anthracis*, *B. mycoides*, *B. thuringiensis*, and *B. cereus*, are closely related species, and collectively compose the *Bacillus cereus* group [15]. They share α/β-type SASPs; SASP-2 and the γ-type SASP categorized for *B. cereus* T have been found to be also expressed in *B. thuringiensis* [10, 16]. In the literature, *B. globigii* is described as producing dark pigments when grown on a tyrosine-containing medium and is therefore also referred to as *B. subtilis* variety *niger* and *B. atrophaeus* [17, 18]. While a rather close relationship of the *Bacillus* species *globigii* and *subtilis* is suggested, we have recently shown that sequence homology of the SASP-1 in *B. globigii* and in *B. stearothermophilus* is extensive [10, 11, 19]. Moreover, sequences of the γ-type SASPs were confirmed to be identical in the two latter species [19], albeit differentiation of various strains of *B. subtilis* is feasible based on their unique γ-SASPs [20]. To the best of our knowledge, SASPs in *B. sphaericus* 14577 have not been studied so far.

Protonated tryptic peptide and product ion masses were used for searches in the NCBInr database taxonomically restricted to bacteria (eubacteria) with MASCOT sequence query [21], and parameters were usually set as follows: enzyme: trypsin, missed cleavages: 0, protein mass: unrestricted, product ion matches: b- and y-type ions, average peptide ion mass tolerance: $\pm$ 0.3 Da, average product ion mass tolerance: $\pm$ 1.5 Da. In general, unimolecular decomposition spectra showed intense and well-resolved ion signals, and product ions with signal intensities at least 5% above the noise level were generically used for database searches. In agreement with the cut-off score provided by MASCOT, the protein, and according to its source, the microorganism was considered as tentatively identified, if a score of 69 or higher was achieved. If partial sequence information obtained from bacterial peptides resulted in no database search hit, additional MASCOT search queries were conducted by allowing up to 1 and 2 missed cleavage sites and by stepwise increasing peptide and product ion mass tolerances to $\pm$ 1.5 Da and $\pm$ 2.0 Da, respectively.

5.3
Results and discussion

5.3.1
On-probe tryptic digestion of bacterial cells

5.3.1.1 *Bacillus subtilis* 168

The wild-type species *B. subtilis* 168 was considered as a genetically amenable, non-pathogenic model system on which to elaborate the potential of on-probe tryptic cell digestion for the rapid identification of bacteria by partial peptide sequencing combined with searches in public databases. To lyse cells and selectively solubilize bacterial proteins, *B. subtilis* 168 was suspended in different solvents such as deionized water, an aqueous solution of 5% TFA or a mixture of 25 mM NH_4HCO_3 and methanol (50:50, v/v). Among these, the latter solvent system was shown to be most suitable for protein solubilization from intact cells, and in particular, for solubilization and tryptic digestion of proteins with molecular masses above 10 000 Da as shown by MALDI-TOF MS analyses (data not shown). After 2.5 mg of vegetative cells were suspended in the adapted solvent system, a small aliquot of the crude cell suspension was directly placed on the MALDI sample plate, and proteolytic digestion was carried out on-probe with trypsin immobilized on agarose beads.

Generally, molecular masses of proteins in bacteria scatter from about 4 kDa to more than 100 kDa, and MALDI is commonly known to discriminate against the desorption/ionization of larger molecules in a mixture in favor of smaller components. With regard to that, tryptic digestion of selectively solubilized bacterial proteins from whole cells is advantageous as peptides with molecular masses below 4000 Da are usually generated, and can be analyzed with superior sensitivity and mass accuracy by MALDI-TOF MS analysis. Representative MALDI spectra of on-probe tryptic digests generated from *B. subtilis* 168 after cells were treated with a 1:1 mixture of NH_4HCO_3 and methanol are shown in Fig. 1. In general, no enzyme autolysis peaks were observed in MALDI spectra and protonated peptides with relative signal intensities suitable for PSD analysis could be generated by proteolysis with trypsin for 5 min. When the digestion time is extended to 20 min, superior S/N and additional well resolved peptide ions from bacterial proteins were observed in the *m/z* range of 1000–3100 Da. As the digestion time was extended from 20 to *e.g.* 45 min, no significant change in the extent of protein proteolysis from bacterial cells, and thus no additional protonated peptides, could be observed in the recorded MALDI spectra (data not shown).

5.3.1.2 *Bacillus globigii* and *sphaericus* 14577

In addition to *B. subtilis* 168, vegetative cells of the *Bacillus* species *globigii* and *sphaericus* 14577 were studied to demonstrate the general applicability of on-probe tryptic digestion to different microorganisms. The genomes of these species are both unsequenced, and hence they represent unknown samples to test our proposed strategy of partial sequencing of bacterial peptides for the rapid identification of *Bacilli*. Only a few entries in public protein databases can be found that list

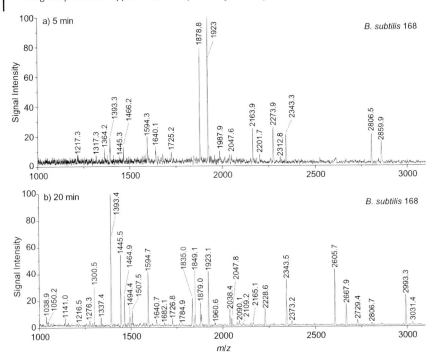

Fig. 1 MALDI-MS spectra of tryptic digest generated on-probe from B. subtilis 168 by solubilization of bacterial proteins with a 1:1 mixture of 25 mM ammonium bicarbonate buffer and methanol and enzymatic proteolysis for (a) 5 min, and (b) 20 min.

B. sphaericus 14577 and B. globigii as bacterial sources of proteins of known sequence. In our laboratory, complete sequences of three small, acid-soluble proteins have been recently determined in free endospores of the latter species by *de novo* sequencing using MS/MS [19]. Since B. globigii, also referred to as B. subtilis variety *niger* and B. atrophaeus [17, 18], was shown to share not only two α/β-type SASPs very similar in sequence but also the γ-type SASPs identical in amino acid sequence with B. stearothermophilus [19], a close taxonomic relationship of these two species can be anticipated.

Tryptic digests from vegetative cells of B. globigii and B. sphaericus 14577 were generated on the MALDI plate in 20 min or less using the same protocol as described for B. subtilis 168. A characteristic set of well resolved protonated peptides that differ significantly in masses was revealed for each species in the mass range of 1000–3300 Da (Fig. 2). Protonated peptides of 1141.3, 1393.5, 1507.5, 1726.5, 1923.0, and 2343.5 Da were observed in the tryptic digests from both species, allowing for a mass tolerance of ≤ 0.3 Da (Figs. 1b and 2a).

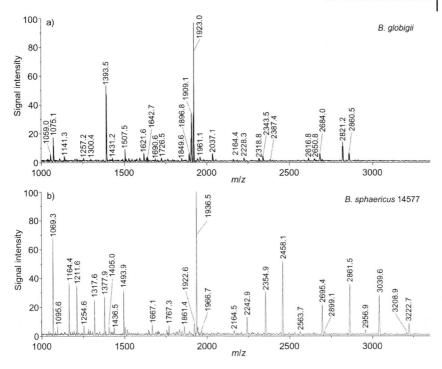

Fig. 2 MALDI-MS spectra of tryptic digests generated on-probe from (a) *B. globigii* and (b) *B. sphaericus* 14577 by solubilization of bacterial proteins using a 1:1 mixture of 25 mM ammonium bicarbonate buffer and methanol and enzymatic proteolysis for 20 min.

5.3.1.3 *Bacillus cereus* T and *anthracis* Sterne

The four closely related *Bacillus* species *mycoides, thuringiensis, cereus* and *anthracis* comprise the *B. cereus* group [15]. Differentiation of *B. thuringiensis* from *B. cereus* is only established based on the plasmids encoding for the insecticidal toxins [2], and the bacterial spores of these two species share α/β- and γ-type SASPs identical in mass and amino acid sequence [10, 16]. For the animal and human pathogen *B. anthracis* strain Ames, putative chromosomal virulence and surface proteins related to pathogenicity and important as potential targets for vaccines and drugs have been recently reported; however, their homologues have been revealed in *B. cereus*, highlighting the extent of inter-species similarity [23]. In contrast, the toxin genes *pXO1* and *pXO2* causing anthrax are found only in *B. anthracis* and not in any other member of the *B. cereus* group. Noteworthy, *B. anthracis* Sterne, used as surrogate for *B. antracis* Ames in this study, is a nonpathogenic strain, and misses the pXO2 plasmid [24]. In a previous study, we have shown that free endospores of *B. anthracis* Sterne can be clearly discriminated from *B. cereus* T, *B. thuringiensis* subs. Kurstaki, and *B. mycoides* (for the latter species data unpublished) *via* selective solubilization and *in situ* tryptic digestion of SASPs followed by partial sequencing

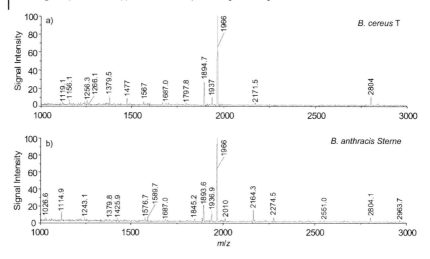

Fig. 3 MALDI-MS spectra of tryptic digests generated on-probe from (a) *B. cereus* T and (b) *B. anthracis* Sterne by solubilization of bacterial proteins using a 1:1 mixture of 25 mM ammonium bicarbonate buffer and methanol and enzymatic proteolysis for 20 min.

of unique bacterial peptides combined with searches against public databases [10], and the same strategy was shown to be applicable to the rapid identification of *Bacillus* spore species in mixtures [10, 11].

In this work, tryptic digests of *B. cereus* T and *B. anthracis* Sterne were prepared on-probe after solubilization of bacterial proteins with 25 mM NH_4HCO_3 and methanol (1:1, v/v) from intact vegetative cells, and a small set of distinctive protonated peptides could be observed in the MALDI-TOF MS spectra (Fig. 3). As expected for these closely related bacteria, some protonated tryptic peptides of, e.g., 1379.5, 1937.0, 1966.0 and 2804.0 Da were detected in tryptic digests of both species applying a mass tolerance of ≤ 0.3 Da. More importantly, MALDI spectra also revealed well resolved protonated peptides distinctive in mass, which could be specifically assigned to *B. cereus* T or to *B. anthracis* Sterne. Moreover, these specific bacterial peptides did not match any of the protonated peptides observed in trypic digests from the other *Bacillus* species studied here, except for the peptide detected at m/z 2164.3 in the tryptic digest of *B. anthracis* Sterne.

Since *B. anthracis* is an endospore-forming bacterium that can cause inhalational anthrax lethal for humans, we focused on additional experiments on the analysis of cells with developed forespores, and thus already in the process of turning to free endospores by cell lysis of the mother cells. To compare vegetative and forespore-containing cells of *B. anthracis* Sterne, tryptic digestion was performed under comparable experimental conditions. In these experiments, bacterial proteins were selectively solubilized from vegetative and forespore-containing cells using an aqueous solution of 5% TFA followed by on-probe tryptic digestion of the obtained cell suspensions for 20 min. While some redundancy is seen in MALDI spectra of tryptic digests from vegetative cells of *B. anthracis* Sterne following bacterial pro-

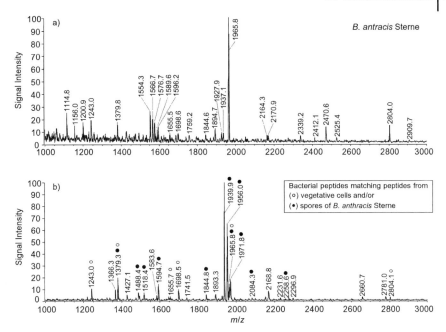

Fig. 4 MALDI-MS spectra of tryptic digests generated on-probe from (a) vegetative cells and (b) cells with developed forespores of B. anthracis Sterne by solubilization of bacterial proteins using an aqueous solution of 5% TFA and enzymatic proteolysis for 20 min.

tein solubilization with different solvent systems, additional protonated peptides of 1156.0, 1200.9, 1554.3, 1566.7, 1596.2 and 2470.6 Da were uniquely detected using 5% aqueous TFA. S/N were slightly lower (Figs. 4a and 3b). Interestingly, MALDI spectra of tryptic digests from forespore-containing cells of B. anthracis Sterne displayed peptides that match masses of peptides from SASPs, which have been previously reported for free endospores by our laboratory [10]. These proteolysis products are indicated in Fig. 4b. The identity of SASPs in cells with developed forespores could be confirmed by unimolecular decomposition of tryptic peptides of, e.g., 1518.4, 1594.7, 1939.9 and 1956.0 Da (data not shown). Based on this, conclusive evidence is provided that this highly preserved protein family is expressed early on and in high abundance within the process of cell transformation. In addition, protonated peptides of lower abundance could be observed, which match peptides observed in tryptic digests from vegetative cells of B. anthracis Sterne (Figs. 3b and 4a).

5.3.2
Partial sequencing for rapid identification of bacterial cells

5.3.2.1 **Unimolecular decomposition product ion analysis (UDPIA)**
To move beyond peptide mapping, and to make this approach applicable to the rapid identification of bacterial cell mixtures, UDPIA of distinct bacterial peptides was per-

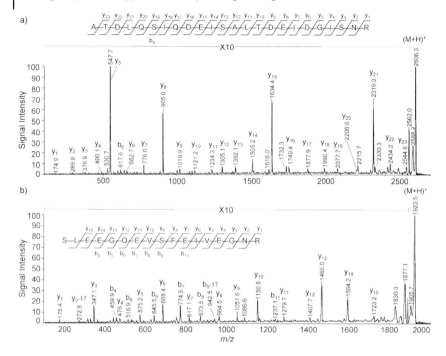

Fig. 5 Fragmentation spectra of protonated tryptic peptides of 2606.3 (a) and 1923.5 Da (b) generated by on-probe tryptic digestion of *B. subtilis* 168 cells following protein solubilization with a 1:1 mixture of 25 mM ammonium bicarbonate buffer and methanol.

formed. Protonated tryptic peptides were isolated with an ion gate set to a ± 10–15 Da window and followed by recording of unimolecular decomposition products. In most cases, extended metastable decay of protonated bacterial peptides was induced by higher laser power, resulting in MALDI spectra with well resolved fragment ions. To exemplify the high quality and extent of sequence information achievable by UDPIA, fragmentation spectra of protonated tryptic peptides of 2606.3 Da and 1923.5 Da from on-probe cell digests of *B. subtilis* 168 are presented in Fig. 5. Product ions observed match b- and y-type ions as indicated in the spectra, while the latter refer to Y"-type ions according to the nomenclature introduced by Roepstorff and Fohlmann [25]. In accordance with previous studies [10, 11, 26], formation of y-type ions was strongly favored, and particularly when the basic residue arginine occurs at the C-terminal site. Moreover, cleavage of amide bonds containing glutamic acid and aspartic acid residues resulted in y-type ions of high abundance [27, 28]. These residues also facilitated peptide fragmentation along the entire backbone, most strikingly perceived in this work with the protonated peptide of 2606.3 Da (Fig. 5a).

5.3.2.2 Identification of bacteria

For the rapid identification of bacteria in mixtures, the identity of parent proteins, and accordingly, their bacterial source was determined with high confidence using sequence-specific peptide information and generic algorithms for searches in public databases. Unimolecular decomposition data were usually not interpreted for database searches with MASCOT sequence query, making the process of bacterial identification automatable. Using this strategy and fragmentation data shown in Fig. 5, *B. subtilis* 168 could be determined as the bacterial source based on identification of its flagellin protein of 23.63 kDa and cold shock protein D (CSPD) of 7.31 kDa with MASCOT scores of 232 and 185, respectively. If fragmentation data of adequate quality resulted in protein identification, MASCOT scores of 150 were achieved on average. In several cases, however, high quality fragmentation data could be achieved by PSD analysis of bacterial peptides, such as the tryptic peptide of 1621.6 Da from *B. globigii* cells, and yet no positive matches were retrieved from database searches. This was particularly true for bacteria of unknown genomes, reflecting the incompleteness of current genome and proteome databases. In addition, proteins identical in amino acid sequence or exhibiting extended sequence homologies are likely to be expressed within and across bacteria, and predominantly in species of close taxonomic relationship. Consequently, the detection of reliable species-specific proteins and peptides unique in mass and amino acid sequence from whole bacteria is challenging, and relies on proper protein solubilization, complete protein digestion, and sensitive mass spectrometric analysis *via* MALDI. In the present work, we aim to identify bacterial proteins and protein families of high abundance and specificity, from which tryptic peptides can be quickly generated and partially sequenced for the identification of bacterial cells and their mixtures.

5.3.3
Identification of proteins and protein-families in bacterial cell digests

5.3.3.1 Flagellin and surface layer protein precursor

Post-source decay (PSD) analysis of seven protonated tryptic peptides from cells of *B. subtilis* 168, combined with database searching, resulted in the identification of the flagellin protein catalogued for this species in public databases with a total MASCOT score of 736 and sequence coverage of 39%. Bacterial peptides were measured with masses of 1883.9 and 1956.0 Da, which match distinctive tryptic peptides of the falgellin protein to within 0.1 Da (Tab. 1).

The rapid identification of the extracellularly located flagellin protein *via* on-probe cell digestion and partial peptide sequencing is appealing as the genes of this protein family have already been targeted as biomarkers in detection, population genetics, and epidemiological analysis of bacteria [29]. Furthermore, these proteins usually exist in as many as 20 000 copies *per* cell, composing the filament of the flagellum in motile bacteria [30], and thus represent naturally amplified biomarkers suitable to detect bacteria at potentially low concentrations. Eubacterial flagellin proteins feature a central domain that can vary considerably in both sequence and size across species [31, 32], while 'hydrophobic heptad repeat' sequences have been postulated, enabling

Tab. 1 Set of tryptic peptides generated on-probe from B. subtilis 168 (I) and B. sphaericus 14577 (II) that led to the identification of a unique flagellin protein (I) and a surface layer protein precursor (II), respectively (average masses are shown)

[M+H]$^+$		Span	Sequence	MALDI	Score
Obs.	Calc.				
I) gi|16080589 Flagellin protein (32,63 kDa) from B. subtilis 168					
1445.5	1445.6	220–232	VVDEAINQVSSQR	PSD	80
1464.8	1464.7	3–15	INHNIAALNTLNR	PSD	77
1883.9	1884.0	202–219	FADNAADTADIGFDAQLK	MS	–
1956.0	1956.1	183–201	EADGSIAALHSVNDLDVTK	MS	–
2228.6	2228.4	242–262	LEHTINNLSASGENLTAAESR	PSD	109
2605.6	2605.8	106–129	ATDLQSIQDEISALTDEIDGISNR	PSD	232
2667.9	2667.9	65–89	NSQDGISLIQTAEGALTETHAILQR	PSD	107
2993.9	2993.4	278–304	NNILSQASQAMLAQANQQPQNVLQLLR	PSD	131
Sequence Coverage: 39% Total MASCOT Score: 736					
II) gi|585994 Surface layer protein precursor (125,23 kDa) from B. sphaericus 14577					
1069.2	1069.2	196–204	QDFAVVFSR	PSD	103
1211.5	1211.4	185–195	TNLNPNAPITR	PSD	79
1377.9	1377.6	165–177	SYLEIAVANGVIK	MS	–
2354.9	2354.6	73–94	AEAATIFTNALELEAEGDVNFK	PSD	128
2458.0	2457.7	45–68	EAVQSLVDAGVIQGDANGNFNPLK	PSD	188
3039.6	3039.4	137–164	ILVDAFELEGEGDLSEFADASTVKPWAK	MS	–
3222.7	3222.5	98–127	ADAWYYDAIAATVENGIFEGVSATE-FAPNK	MS	–
Sequence Coverage: 10% Total MASCOT Score: 498					

the formation of α-helical coiled-coils in the highly conserved N- and C-terminal regions of eubacterial flagellins [30]. Two protonated tryptic peptides, INHNIAALNTLNR and NNILSQASQAMLAQANQQPQNVLQLLR, located in these conserved regions of the flagellin protein were observed in the present work, and comprise the amino acids 3–15 and 278–304, respectively. Since MALDI spectra of tryptic digests from B. subtilis 168 following protein solubilization with an aqueous solution of 5% TFA lacked these two peptides, while they are present in high abundance when NH_4HCO_3/MeOH (1:1, v/v) is used, the postulated hydrophobic character of these sequences could be supported by mass spectrometric analysis.

While the genome of B. subtilis 168 has been fully sequenced and a wealth of information on its proteome is available, only a few database entries can be retrieved for B. sphaericus 14577. For example, the gene encoding for a 125 kDa surface-layer protein from B. sphaericus 2362 has been sequenced [33]. Generally, surface-layer (S-layer) proteins can constitute up to 15% of the total protein content of a cell in the exponential growth phase [34]. These proteins are promising candidates for bacterial cell differentiation, because they exist in multiple copies, and differ significantly in molecular mass and amino acid sequence across species and strains [35]. In the case of *B. sphaericus*, the S-layer protein is already synthesized during vegetative growth [36] and upon completion of exponential growth there is a burst of expression [37].

In this work, various well resolved protonated peptides with high ion signal intensities that match peptides from theoretical tryptic digests of the surface layer protein precursor of 125.23 kDa could be observed in MALDI spectra of tryptic digests from *B. sphaericus* 14577 as summarized in Tab. 1 (Fig. 2b). In addition, four protonated tryptic peptides were partially sequenced with success, and allowed for unambiguous identification of the S-layer protein precursor and, according to its bacterial source, *B. sphaericus* 14577 with an overall score of 498 and 10% sequence coverage (Tab. 1). Since most of its tryptic peptides are lysine-terminated, with lower propensities of protonation than arginine-containing peptides in the MALDI process [38], only a small set of its potential tryptic peptides was observed. Noteworthy, protonated peptides from the S-layer protein precursor were entirely lacking in MALDI spectra of tryptic digests from *B. sphaericus* 14577 following protein solubilization with an aqueous 5% TFA solution (data not shown), indicating a very hydrophobic character of this protein located in the peptidoglycan network of Gram-positive bacteria. This observation is in agreement with previous comparative analysis of amino acids in S-layer proteins, which report a generally high content of hydrophobic amino acids [34].

In general, the presence of S-layer and flagellin proteins is common but not essential in bacteria. For example, *B. cereus* ATCC 14579 has been reported to be devoid of a S-layer [39], and *B. anthracis* is missing the sigma factor *sig*D, which is essential for the expression of the flagellum operon [40]. Moreover, a motile bacterium has also been characterized, which carries a flagellum operon similiar to *B. anthracis*, but lacks a *sig*D gene [42].

5.3.3.2
Cold shock and cold shock-like proteins

Various members of the cold shock protein (CSP) family could be identified in on-probe tryptic digests from the *Bacillus* species studied in this work, as summarized in Tab. 2. These CSPs consist of two main structural sequence elements that often differ only in few amino acids within and across *Bacillus* species. CSPs with molecular masses above 6 kDa feature additional sequences at the *N*- and *C*-termini, compared with CSPs smaller in size.

CSPs with average molecular masses of 4977, 4993, 7309 and 7365 Da could be assigned to *B. subtilis* 168, achieving MASCOT scores from 107 to 185 and sequence coverages of 25–94% (Tab. 2). Among these, the CSPD of 7309 Da has been previously identified in a cellular extract from *B. subtilis* 168 by MALDI-PSD analysis [26]. *B. globigii* could be distinguished from *B. subtilis* 168 based on the identification of its specific CSP of 4934 Da. In addition, a protonated tryptic peptide of 1909.1 Da was detected in cell digests from *B. globigii* and not *B. subtilis* 168, and the CSP of 6274 Da could be determined based on partial sequence information and database searches. Interestingly, *B. subtilis* 168 was categorized as a bacterial source for this protein, indicating -

Tab. 2 Compilation of tryptic peptides of cold shock and cold shock-like proteins from different *Bacillus* species[a, b]

Protein accession no.	Mass/Da	Peptide sequence	Mass/Da	Peptide sequence	Mass/Da	Seq. Cov.	Species listed in public databases
gi|1402739	4977	GFGFIEVEGQDDVFVHFSAIQGEGFK[c, d]	2861.1	TLEEGQAVSFEIVEGNR	1879.0	94%	B. subtilis 168
gi|1402741	4993	GFGFIEVEGQDDVFVHFSAIQGEGFK[c, d]	2861.1	TLEEGQSVSFEIVEGNR	1895.0	57%	B. subtilis 168
gi|729219	6294	GFGFIEVEGQDDVFVHFSAIQGEGFK[c, d]	2861.1	CLEEGQAVSFEIVEGNR	1880.1	45%	Sporosarcina globispora
gi|16079252	7309	GFGFIEVEGGDDVFVHFTAIEGDGYK	2807.0	SLEEGQEVSFEIVEGNR[c]	1923.0	25%	B. subtilis 168
gi|416834	7365	GFGFIEVEGQDDVFVHFSAIQGEGFK[c, d]	2861.1	TLEEGQAVSFEIVEGNR	1879.0	63%	B. subtilis 168
gi|456238	6274	GFGLIEVEGQDDVFVHFSAIQGEGFK	2827.1	TLEESQAVSFEIVEGNR[c, f, g]	1909.1	29%	B. subtilis 168
gi|1402735	4934	GSGFIER EAGDDVFVHFSAIQGEGFK	765.8 2054.2	SLEEGESVSFDVEQGQR	1897.0	37%	B. globigii
gi|1402737	5007	GFGFIEVEGGEDVFVHFSAIQGEGFK	2804.1	TLEEGQEVTFEVEQGNR	1966.1	94%	B. cereus T
gi|21401461	6684	GFGFIEVEGGEDVFVHFSAIQGEGFK	2804.1	TLEEGQEVTFEVEQGNR	1966.1	72%	B. anthracis A2012
gi|21399520	7196	GFGFIEMEGADDVFVHFSAIQGEGYK	2852.2	ALEEGQEVSFDITEGNR	1895.0	26%	B. anthracis A2012
gi|2493762	7200	GFGFIEMEGSEDVFVHFSAIQSDGYK	2898.2	ALEEGQEVSFDITEGNR	1895.0	26%	B. cereus T
gi|21397653	7550	GFGFIER EDGDDVFVHFSAIQQDGYK	825.9 2171.3	SLEEGQQVEFDIVDGAR	1893.0	25%	B. anthracis A2012
gi|2493763	7305	GFGFIER EDGDDVFVHFSAIQQDGYK	825.9 2171.3	SLEEGQQVEFDIVDGAR	1893.0	26%	B. cereus T
gi|2140290	7239	GFGFIEVEGGDDVFVHFSAIQGDGFK	2776.0	TLEEGQEVSFEIVEGNR[e]	1937.1	25%	B. anthracis A2012

a) Tryptic peptides in bold letters were identified *via* partial sequencing and database searching
b) Average molecular masses are shown. Bacterial peptides also detected in:
c) *B. globigii*
d) *B. sphaericus* 14577
e) *B. cereus* T
f) Peptide not observed in MALDI spectra of tryptic digests from *B. subtilis* 168 but in *B. globigii*
g) The peptide (TLEENATVSFEIVEGNR) identical in amino acid sequence was identified by manual interpretation of PSD data

extended sequence homologies or, less likely, the expression of a protein identical in sequence in *B. globigii*. Additionally, the tryptic peptide GFGFIE-VEGQDDVFVHFSAIQGEGFK of 2861.1 Da was identified in on-probe digests from *B. sphaericus* 14577 cells with a score of 167, and this sequence is redundant in CSPs of 4977, 4993, 7365 and 6294 Da listed for *B. subtilis* 168 and *Sporosarcina globispora*, respectively (Tab. 2). Nonetheless, CSPs identified by PSD analysis of bacterial peptides in tryptic digests from *B. cereus* T and *B. anthracis* Sterne differ sufficiently in their amino acid sequences in order to discriminate them against CSPs known for other bacterial species. CSPs with molecular masses of 5007, 7200, 7305 Da listed for *B. cereus* and CSPs of 6684, 7196 and 7550 Da catalogued for *B. anthracis* could not be distinguished based on the tryptic peptides observed in their MALDI spectra. Moreover, the tryptic peptide (TLEEGQEVSFEIVEGNR) of 1937.1 Da was observed in cell digests from both *B. cereus* T and *B. anthracis* Sterne, and led to the identification of the cold shock-like protein of 7239 Da, currently listed for the latter species in databases. Interestingly, a protonated tryptic peptide of 1936.5 Da was observed with very high ion signal intensities in the MALDI spectra of *B. sphaericus* 14577 digests, which differs by only 0.6 Da from the mass of the TLEEGQEVSFEIVEGNR peptide discussed above. The latter bacterial protein of 7239 Da from *B. cereus* and *B. anthracis* Sterne could be excluded in *B. spaericus* by partial peptide sequencing.

One of the most significant advantages of analysis of bacterial peptides by partial sequencing, rather than mass mapping of protein precursors, is the ability to distinguish proteins or peptides that are similar or even identical in nominal mass, but differ sufficiently in their amino acid sequences. For example, CSPA, CSPC and the ribosomal protein RL29 in *Escherichia coli* K-12 could not be distinguished by MALDI-TOF MS based on mass [42], and the latter protein was only identified by high-resolution and mass accuracy measurements using a MALDI-FT MS instrument [43] or by partially sequencing the extracted protein [44].

In general, members of this widespread and highly conserved protein family exist in multiple copies in bacteria [46]. For example, *B. subtilis* contains three *csp* genes, and CSPB, CSPC and CSPD have been reported to compose one of the most highly accumulating protein groups in this bacterium, in response to a temperature downshift [46, 47]. It is proposed that these proteins play important roles in the adaptation of cells to low temperature conditions [48] by, *e.g.*, keeping critical mRNAs accessible for the ribosomes under these conditions [50]. Nonetheless, CSPs are also present in bacterial cells at 37°C [50, 51]. Depletion of CSPs leads to compromised and deregulated protein synthesis [50, 52].

We have not attempted to induce overexpression of CSPs by cell growth at low temperature conditions in any way. In fact, *Bacilli* were carefully grown at 37°C without exposing them to any temperature shifts. After harvest, cells were quickly purified by mild salt washes and immediately prepared for mass spectrometric analysis. Vegetative cells that were not directly used in experiments were lyophilized and stored at −20°C. In control experiments, on-probe tryptic digests from the same *Bacillus* species resulted in MALDI spectra with no significant change in mass and signal intensity of protonated peptides using fresh cells and cells that have been stored at −20°C from one day to several years, respectively (data not shown). CSPs have been located in cytosolic spaces surrounding the central nucleoid in the bacterial cell [53], and the recognition of tryptic peptides of these abundant proteins in this work indicates that cytoplasmic proteins are solubilized and accessible for enzymatic digestion.

5.3.3.3 Ribosomal proteins

This protein family comprises high copy number proteins and can account for more than 20% of the total cellular proteome in bacteria [54]. In this work, only the 30S ribosomal protein S6 with an average molecular mass of 11.125 kDa could be tentatively identified based on partial sequence information on the protonated tryptic peptide of 2373.3 Da (QIVNVQSDAAAVQEFDR) from *B. subtilis* 168 with a rather low score of 71 and sequence coverage of 22%. Since most ribosomal proteins are very basic with p*I*-values around 11 [55], their propensity to become charged by proton transfer reactions during the MALDI process is high [56], and ribosomal proteins have been detected with high ion signal intensities by the analysis of *E. coli* cells with MALDI-TOF MS [42, 43]. Consequently, identification of such a limited number of ribosomal proteins was not anticipated. It is even more surprising, as various acidic CSPs with p*I*-values around 5 were successfully identified in this work. For some ribosomal proteins present in *E. coli*, PTMs involving, for example, methylation, acetylation or loss of the *N*-terminal methionine residue have been reported [42, 43, 57, 58], which may impede their identification.

In addition, theoretical tryptic digests of known ribosomal proteins in *B. subtilis* 168 and other species were performed allowing for no partial cleavages, and a high number of proteolysis products with molecular masses below 1000 Da was generated. Generally, peptides containing less than five amino acid residues are found to result in no reliable identification of protein precursor *via* sequence queries in public databases. Only a few tryptic peptides from ribosomal proteins are predicted to feature molecular masses of 1000–3000 Da, and about half of them are lysine-terminated species. Since protonation rates are enhanced for arginine-containing peptides [38], protonated peptides with lysine at the *C*-terminus were often not observed in the MALDI spectra of tryptic digests from bacterial extracts [26].

5.3.3.4 DNA-binding proteins

Another group of proteins identified in tryptic digests generated on-probe from *Bacilli* belongs to the family of protective DNA-binding proteins, as summarized in Tab. 3. In MALDI spectra of the tryptic digests from both *B. subtilis* 168 and *B. globigii*, two peptides at *m/z* values of 1393.5 (IQLIGFGNFEVR) and 1507.5 (AVDSVFDTILDALK) were identified in this work. From these, the nonspecific DNA-binding protein HBsu of 9884 Da in *B. subtilis* 168 and the DNA-binding protein HB of 9897 Da in *B. globigii* were determined as parent proteins with MASCOT scores of 150 and 106, respectively. The former protein has been observed before by proteomic analysis of a cell extract from *B. subtilis* spores [59]. A

Tab. 3 Compilation of DNA-binding proteins tentatively identified in cell digests of *Bacillus* species studied in this work (average masses are shown)

DNA-binding protein	Mass (Da)	Tryptic peptide observed	[M+H]$^+$ (Da)	Total score	Seq Cov.	Species listed in public databases	Detected in
gi|80156	9897	IQLIGFGNFEVR	1393.5	256	28%	B. globigii	B. globigii B. subtilis 168
gi|16079336	9884	AVDSVFDTILDALK	1507.5	256	28%	B. subtilis 168	B. globigii B. subtilis 168
gi|21399426	9642	VQLIGFGNFEVR	1379.8	94	13%	B. anthracis A2012	B. cereus T B. anthracis Sterne

moderate sequence coverage of 28% was obtained for each of the two DNA-binding proteins identified in this work; however, their amino acid sequences differ only in their *N*-terminal amino acids (Tab. 3), resulting in a single unique tryptic peptide from each, both with masses below 450 Da as shown by *in silico* digestion. Unimolecular decomposition analysis of protonated peptides of 1379.8 Da from tryptic digests of the *B. cereus* group members *anthracis* Sterne, and *cereus* T resulted in the identification of the arginine-terminated peptide (VQLIGFGNFEVR) of the DNA-binding protein of 9642 Da by database searches. Due to extended sequence homologies of DNA-binding proteins within and across prokaryotic species, proteins listed for 13 different bacterial species were additionally retrieved. Several DNA-binding proteins have been previously identified by MS analysis of *E. coli* cells [42] and their fractionated cell extracts [60].

5.3.3.5 Heat shock proteins

The family of 60 kDa heat shock proteins (HSP60) is composed of abundant essential proteins in most bacteria [61]. HSP60 proteins are well studied for their role as chaperone facilitators of protein folding and in rescuing the cell from stress conditions [62]. HSP60 are ubiquitous in eubacterial cells, and their gene sequences have been studied as universal targets for identification of various *Staphylococcus* species [63].

In our studies, PSD data on bacterial peptides of 2231.4, 2661 and 1845.0 Da from on-probe tryptic digests of *B. anthracis* Sterne and *B. cereus* T strongly indicated the presence of bacterial HSP60 proteins, and MASCOT scores in the range of 66 to 118 could be obtained in database searches. Interestingly, peptides from these ubiquitous proteins were observed in these two members of the *B. cereus* group, and not in others studied in this work. Extended sequence homologies within this highly conserved protein family are well known, and impeded the unambiguous identification of the HSP60 and the 60 kDa chaperonin Cpn60, and accordingly, their specific bacterial sources. Nonetheless, taking into account the sum of database search results the presence of HSP60 proteins with molecular masses in the range of 32.77 to 42.21 kDa and 19.67 to 19.71 kDa could be confirmed, and the closely related *Bacillus* species *anthracis, thuringiensis* and *cereus* were retrieved as potential bacterial sources.

5.3.3.6 Other stress related proteins and the prosthetic group of an acyl-carrier protein

Additional proteins, not classifiable in one of the former protein groups, could be tentatively identified based on partial sequence information on tryptic peptides from cell digests of *B. subtilis* 168 and *B. anthracis* Sterne. Since these proteins appear to be unique in their amino acid sequences, they provide helpful information for the species-specific identification of bacteria.

A small 20.63 kDa subunit of the alkyl hydroperoxide reductase was identified with a total score of 459 and sequence coverage of 27% in on-probe tryptic digests from *B. subtilis* 168 (Tab. 4). Enzymes such as alkyl hydroperoxide reductase, as well as DNA-binding proteins, provide mechanisms that can destroy oxidizing agents, *e.g.* alkyl hydroperoxides and hydrogen peroxide. Consequently, these gene products cause significant resistance to oxidative stress, and they are active in growing bacterial cells [64, 65]. In contrast to growing cells, inactivation of genes coding for alkyl hydroperoxide reductase had no effect on hydrogen peroxide resistance of *B. subtilis* spores, and thus the proteins have been postulated to play no role in spore protection [66]. In addition, the prosthetic group of an acyl carrier protein of 8.51 kDa, involved in the bacterial fatty acid metabolism, and the helix-turn-helix mercury resistance protein of 28.29 kDa were identified as putative species-specific parent proteins in tryptic cell digests of *B. anthracis* Sterne (Tab. 4). The latter protein belongs to the large family of helix-turn-helix proteins, which represent regulatory proteins in both prokaryotes and eukaryotes. Neither of these two proteins has been previously reported in mass spectrometric studies of whole bacterial cells.

5.3.4
Analysis of a 1:1 mixture of *B. globigii* and *B. sphaericus* 14577

On-probe digestion of bacterial cells, following adequate protein solubilization, combined with partial sequence analysis and database searching, was developed to provide rapid species-specific identification of bacteria in mixtures. To provide evidence for the applicability of this approach to vegetative cell mixtures, we analyzed a 1:1 mixture of *B. sphaericus* 14577 and *B. globigii*. Tryptic peptides unique to each *Bacillus* species were observed in MALDI spectra of the cell mixture digest (Fig. 6a). For example, six tryptic peptides of 1069.2, 1211.5, 1377.7, 2242.8, 2457.9 and 3039.5 Da from the S-layer protein precursor specific for *B. sphaericus* 14577 were observed by MALDI-TOF MS. On the other hand, tryptic peptides of 1643.1, 1896.7, 1909.1, 2616.9 and 2821.0 Da match those bacterial peptides in mass that were only observed in cell digests from *B. globigii*. Since bacterial peptides were adequate in ion abundance and sufficiently resolved with a mass resolution of about 4000 at m/z 1900, defined as full width at half maximum, unimolecular decomposition analysis could be readily performed. Interestingly, automated database searching, based on PSD

Tab. 4 Set of tryptic peptides generated on-probe from *B. subtilis* 168 (I) and *B. anthracis* Sterne (II & III), which led to the identification of unique proteins (average masses are shown)

[M+H]$^+$		Span	Sequence	MALDI	Score
Obs.	Calc.				
I) gi|16081061 Alkyl hydroperoxide reductase (small subunit) of 20.63 kDa from *B. subtilis* 168					
1494.4	1494.6	18–30	NGEFIDVTNEDLK	MS	–
1594.5	1594.7	107–120	NFDVLDEETGLADR	PSD	138
2047.8	2048.3	63–80	GTFIIDPDGVIQTVEINAGGIGR	PSD	173
2342.5	2342.6	121–143	ELEDLQEQYAALKEL-GVEVYSVSTDTHFVHK	PSD	148
II) gi|21401834 pp-binding domain of an acyl-carrier protein of 8.51 kDa from *B. anthracis* A2012					
1576.7	1576.8	16–30	LGVEETEVVPAASFK	MS	–
1589.7	1589.8	63–77	IATVGDAVTYIESHL	PSD	153
III) gi|21397448 HTH_MERR, helix-turn-helix, mercury resistance protein of 28.29 kDa from *B. anthracis* A2012					
1698.6	1069.2	122–135	GFDFSHNPYEEEAR	PSD	119

analysis of the tryptic peptide of 1909.1 Da (Fig. 6b), led to the identifcation of the peptide TLEE**SQA**VSFEIVEGNR from the major cold shock protein of 6274 Da listed for *B. subtilis* 168. *B. subtilis* was not a component in the mixture. In contrast, manual interpretation of the obtained fragmentation spectrum presented in Fig. 6b revealed the presence of the tryptic peptide TLEE**NAT**VSFEIVEGNR identical in mass but differing in three amino acids in its sequence. This slight change in amino acid sequence appeared to be most evident in masses of the predicted y_6- and y_7-ions. Those associated with the former identification, 1349.5 and 1221.4 Da, were not observed in the PSD spectrum, but ions supporting the latter sequence, 1322.6 and 1251.9 Da, are (Fig. 6b). This result demonstrates that *B. globigii* can be distinguished from *B. subtilis* 168 based on a single peptide most likely to be derived from an abundant CSP specific to the former species.

Tab. 5 Overview of protein families and individual proteins identified in on-probe tryptic digests from the *Bacillus* species studied in this work

Protein family/protein(s)	B. subtilis 168	B. globigii	B. sphaericus 14577	B. cereus T	B. anthracis Sterne
Surface-related	X		X		
Cold shock	X	X	X	X	X
Ribosomal	X				
DNA-binding	X	X		X	X
60 kDa heat shock				X	X
Stress related, acyl-carrier	X				X

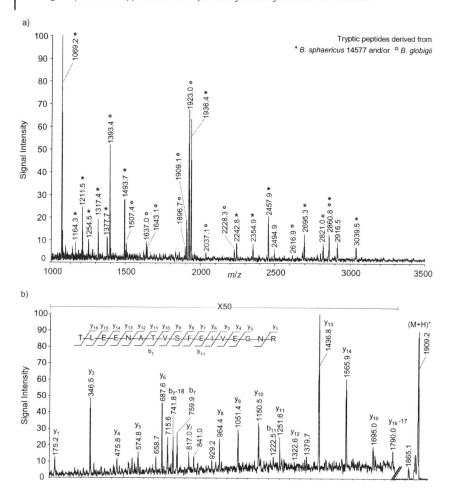

Fig. 6 (a) MALDI-MS spectrum of the tryptic digest generated on-probe from a 1:1 vegetative cell mixture of B. sphaericus 14577 and B. globigii following protein solubilization with a mixture of 25 mM ammonium bicarbonate buffer and methanol (50:50, v/v). Tryptic peptides that match bacterial peptides observed in MALDI spectra of cell digests from single species are marked as shown in the spectrum, and a mass tolerance of ±1 Da was allowed. (b) Fragmentation spectrum of the tryptic peptide of 1909.2 Da from B. globigii.

5.4
Concluding remarks

Additional experiments demonstrated that the strategy presented is also applicable to other binary mixtures of *Bacilli* (data not shown). An overview of the protein families and additional individual proteins identified in this work from five different Bacillus species is given in Tab. 5, and may provide useful information for further studies with an even wider range of bacterial species. In our ongoing research we intend to focus on even more complex mixtures and extreme ratios of Gram-positive bacteria, as well as the analysis of Gram-negative bacteria, and on optimization of the sensitivity and throughput.

In support of this and other proteomics-based approaches for the rapid identification of biohazards, we encourage the continued elucidation and the centralized archiving of bacterial genomes.

We thank Dr. Yetrib Hathout, Dr. Patrick Pribil, and Dr. Jeffrey R. Whiteaker for helpful discussions regarding this work, and Prof. Robert J. Cotter for the use of the AXIMA-CFR mass spectrometer. Financial support was provided by the German Research Society (Deutsche Forschungsgemeinschaft, DFG) and the U.S. Defense Advanced Research Projects Agency (DARPA, Office of Special Programs).

5.5
References

[1] Anhalt, J. P., Fenselau, C., *Anal. Chem.* 1975, *47*, 219–225.

[2] Fenselau, C., Demirev, P. A., *Mass Spectrom. Rev.* 2001, *20*, 157–171.

[3] Lay, O. J. Jr., *Mass Spectrom. Rev.* 2001, *20*, 172–194.

[4] Demirev, P. A., Ho, Y.-P., Ryzhov, V., Fenselau, C., *Anal. Chem.* 1999, *71*, 2732–2738.

[5] Demirev, P. A., Lin, J. S., Pineda, F. J., Fenselau, C., *Anal. Chem.* 2001, *73*, 4566–4573.

[6] Pineda, F. J., Lin, J. S., Fenselau, C., Demirev, P. A., *Anal. Chem.* 2000, *72*, 3739–3744.

[7] Pineda, F. J., Antoine, M. D., Demirev, P. A., Feldman, A. B. et al., *Anal. Chem.* 2003, *75*, 3817–3822.

[8] Yao, Z.-P., Demirev, P. A., Fenselau, C., *Anal. Chem.* 2002, *74*, 2529–2534.

[9] English, R. D., Warscheid, B., Fenselau, C., Cotter, R. J., *Anal. Chem.* 2003, *75*, 6886–6893.

[10] Warscheid, B, Fenselau, C., *Anal. Chem.* 2003, *75*, 5618–5627.

[11] Warscheid, B, Jackson, K., Sutton, C., Fenselau, C., *Anal. Chem.* 2003, *75*, 5608–5617.

[12] Hathout, Y., Demirev, P. A., Ho, Y-P., Bundy, J. et al., *Appl. Environ. Microbiol.* 1999, *65*, 4313–4319.

[13] Nicholson, W. L., Setlow, P., in: Harwood, C. R., Cutting, S. M. (Eds.), *Molecular biological methods for Bacillus*, John Wiley, Chichester, England, 1990, p. 391–450.

[14] Jenkinson, H. F., Sawyer, W. D., Mandelstam, J., *J. Gen. Microbiol.* 1981, *123*, 1–16.

[15] Helagson, E., Oksatd, O., Caugant, D., Johansen, H., *Appl. Environ. Microbiol.* 2000, *66*, 2672–2630.

[16] Hathout, Y., Setlow, B., Cabrera-Martinez, R. M., Fenselau, C., Setlow, P., *Appl. Environ. Microbiol.* 2003, *69*, 1100–1107.

[17] Nakamura, L. K., *Int. J. Syst. Bacteriol.* 1989, *39*, 295–300.

[18] Priest, F. G., in: Sonenshein, A. L., Hoch, J. A., Losick, R. (Eds.), *Bacillus subtilis and other Gram-positive Bacteria*, American Society for Microbiology, Washington, DC 1993, pp. 3–16.

[19] Whiteacker, J. R., Warscheid, B., Pribil, P., Hathout, Y., Fenselau, C., *J. Mass Spectrom.* 2004, in press.

[20] El-Helow, E. R., *FEMS Microbiol. Lett.* 2001, *196*, 119–122.

[21] http://www.matrixscience.com/cgi/index.pl?-page=./home.html.

[22] Turnbull P. C. B., Hutson, R. A., Ward, M. J., Jones, M. N., Quinn, C. P., *J. Appl. Bact.* 1992, *72*, 21–28.

[23] Read, T. D., Peterson, S. N., Tourasse, N., Baillie, L. W., et al., *Nature* 2003, *423*, 81–86.

[24] Mock, M., Fouet, A., *Annu. Rev. Microbiol.* 2001, *55*, 647–671.

[25] Roepstorff, P., Fohlman, J., *J. Biomed. Mass. Spectrom.* 1984, *11*, 601.

[26] Harris, W. A., Reilly, J. P., *Anal. Chem* 2002, *74*, 4410–4416.

[27] Wattenber, A., Organ, A. J., Schneider, K., Tyldesley, R. et al., *J. Am. Soc. Mass Spectrom.* 2002, *13*, 772–783.

[28] Schilling, B., Wang, W., McMurray, J. S., Medzihradszky, K., *Rapid Commun. Mass Spectrom.* 1999, *13*, 2174–2179.

[29] Winstanley, C., Morgan, A. W., *Microbiology* 1997, *143*, 3071–3084.

[30] Morgan, D. G., Khan, S., *Bacterial flagella* in Encyclopedia of Life Science, Nature Publishing Group 2001; www.els.net.

[31] Joys, T. M., *Can. J. Microbiol* 1988, *34*, 452–458.

[32] Wilson, D. R., Beveridge, T. J., *Can. J. Microbiol.* 1993, *39*, 451–472.

[33] Bowditch, R. D., Baumann, P., Yousten, A. A., *J. Bacteriol.* 1998, *174*, 758–764.

[34] Sleytr, U. B., Messner, P., Pum, D., Sara, M., *Mol. Microbiol.* 1993, *10*, 911–916.

[35] Sidhu, M. S., Olsen, I., *Microbiology* 1997, *143*, 1039–1052.

[36] Lewis, L. O., Yousten, A. A., Murray, R. G. E., *J. Bacteriol.* 1987, *169*, 72–79.

[37] Broadwell, A. H., Baumann, P., *Appl. Environ. Microbiol.* 1986, *52*, 758–764.

[38] Krause, E., Wenschu, H., Jungblut, P. R., *Anal. Chem.* 1999, *71*, 4160–4165.

[39] Kotiranta, A., Haapasalo, M., Kari, K., Kerosuo, E. et al., *Infect. Immun.* 1998, *66*, 4895–4902.

[40] West, J. T., Estacio, W., Marquez-Magana, L., *J. Bacteriol.* 2000, *182*, 4841–4848.

[41] Glaser, P., Frangeoul, L., Buchrieser, C., Rusniok, C. et al., *Science* 2001, *294*, 849–852.

[42] Ryzhov, V., Fenselau, C., *Anal. Chem.* 2001, *73*, 746–750.

[43] Jones, J. J., Stump, M. J., Fleming, R. C., Lay, J. O., Wilkins, C. L., *Anal. Chem.* 2003, *75*, 1340–1347.

[44] Reid, G. E., Shang, H., Hogan, J. M., Lee, G. U., McLuckey, S. A., *J. Am. Chem. Soc.* 2002, *124*, 7353–7362.

[45] Goldstein, J., Pollitt, N. S., Inouye, M., *Proc. Natl. Acad. Sci. USA* 1990, *87*, 283–287.

[46] Kaan, T., Homuth, G., Mader, U., Bandow, J., Schweder, T., *Microbiology* 2002, *148*, 3441–3455.

[47] Graumann, P. L., Marahiel, M. A., *Arch. Microbiol.* 1996, *166*, 293–300.

[48] Graumann, P. L., Marahiel, M. A., *J. Mol. Microbiol. Biotechnol.* 1999, *1*, 203–209.

[49] Schindler, T., Graumann, P. L., Perl, D., Ma, S. et al., *J. Biol. Chem.* 1999, *274*, 3407–3413.

[50] Perrin, C., Guimont, C., Bracquart, F. M., Gaillard, J. L., *Curr. Microbiol.* 1999, *39*, 342–347.

[51] Hebraud, M., Guzzo, J., *FEMS Microbiol. Lett.* 2000, *190*, 29–34.

[52] Graumann, P., Wendrich, T. M., Weber, H. M., Schroder, K., Marahiel, M. A., *Mol. Microbiol.* 1997, *25*, 741–756.

[53] Weber, M. H. W., Volkov, A. V., Fricke, I., Marahiel, M. A., Graumann, P. L., *J. Bacteriol.* 2001, *21*, 6435–6443.

[54] Bremer, H., Dennis, P. P., in: Neidhardt, F. D. (Ed.), *Escherichia coli and Salmonella: Cellular and Molecular Biology*, ASM Press, Washington, DC 1996, pp. 167–182.

[55] Noller, H. F., Nomura, M., in: Neidhardt, F. D. (Ed.), *Escherichia coli and Salmonella*, ASM Press, Washington, DC 1996, pp. 167–182.

[56] Cohen, S., Chait, B., *Anal. Chem.* 1996, *68*, 31–37.

[57] Arnold, R. J., Reilly, J. P., *Anal. Biochem.* 1999, *269*, 105–112.

[58] Wittmann, H. G., *Annu. Rev. Biochem.* 1982, *51*, 155–183.

[59] Dickinson, D. N., La Duc, M. T., Haskins, W. E., Gornushkin, I. et al., *Appl. Environ. Microbiol.* 2004, *70*, 475–482.

[60] Dai, Y., Li, L., Roser, D. C., Long, S. R., *Rapid Commun. Mass Spectrom.* 1999, *13*, 73–78.

[61] Karlin, S., Mrazek, J., *J. Bacteriol.* 2000, *182*, 5238–5250.

[62] Sigler, P., Zhaohui, X., Rye, H. S., Burston, S. G. et al., *Annu. Rev. Biochem.* 1998, *67*, 581–608.

[63] Goh, S. H., Potter, S, Wood, J. O., Hemmingsen, S. M. et al., *J. Clin. Microbiol.* 1996, *4*, 818–823.

[64] Antelmann, H., Engelmann, R., Schmid, R., Hecker, M., *J. Bacteriol.* 1996, *178*, 6571–6578.

[65] Imlay, J. A., Linn, S., *J. Bacteriol.* 1987, *169*, 2967–2976.

[66] Casillas-Martinez, L., Setlow, P., *J. Bacteriol.* 1997, *179*, 7420–7425.

6
Protein identification and tracking in two-dimensional electrophoretic gels by minimal protein identifiers*

Jens Mattow*, Frank Schmidt*, Wolfgang Höhenwarter, Frank Siejak, Ulrich E. Schaible and Stefan H. E. Kaufmann

Protein identification by matrix-assisted laser desorption/ionization mass-spectrometry peptide mass fingerprinting (MALDI-MS PMF) represents a cornerstone of proteomics. However, it often fails to identify low-molecular-mass proteins, protein fragments, and protein mixtures reliably. To overcome these limitations, PMF can be complemented by tandem mass spectrometry and other search strategies for unambiguous protein identification. The present study explores the advantages of using a MALDI-MS-based approach, designated minimal protein identifier (MPI) approach, for protein identification. This is illustrated for culture supernatant (CSN) proteins of *Mycobacterium tuberculosis* H37Rv after separation by two-dimensional gel electrophoresis (2-DE). The MPI approach takes into consideration that proteins yield characteristic peptides upon proteolytic cleavage. In this study, peptide mixtures derived from tryptic protein cleavage were analyzed by MALDI-MS and the resulting spectra were compared with template spectra of previously identified counterparts. The MPI approach allowed protein identification by few protein-specific signature peptide masses and revealed truncated variants of mycobacterial elongation factor EF-Tu, previously not identified by PMF. Furthermore, the MPI approach can be employed to track proteins in 2-DE gels, as demonstrated for the 14 kDa antigen, the 10 kDa chaperone, and the conserved hypothetical protein Rv0569 of *M. tuberculosis* H37Rv. Furthermore, it is shown that the power of the MPI approach strongly depends on distinct factors, most notably on the complexity of the proteome analyzed and accuracy of the mass spectrometer used for peptide mass determination.

---

* Originally published in Proteomics 2004, 10, 2927–2941

Proteomics of Microbial Pathogens. Edited by Peter R. Jungblut and Michael Hecker
Copyright © 2007 WILEY-VCH Verlag GmbH & Co. KGaA, Weinheim
ISBN: 978–3–527–31759–2

6.1
Introduction

Proteomics allows parallel separation and characterization of proteins from highly complex biological samples. Classically, it combines protein separation by 2-DE and protein characterization by MS, most notably by MALDI-MS peptide mass fingerprinting (PMF) [1–5] and ESI-MS/MS [6, 7]. MALDI-MS PMF is widely used for identification of proteins from organisms with completely sequenced genomes. It is based on mass determination of peptides derived by enzymatic or chemical protein cleavage and comparison of the experimentally derived peptide masses against theoretical peptide mass data stored in protein databases. Although MALDI-MS PMF is undoubtedly an effective tool for protein identification, it has repeatedly been questioned as the method of choice for high-throughput applications, with MS/MS being reserved for the analysis of * These authors contributed equally to this work.post-translational modifications [8]. This notion is based on the fact that PMF often fails to identify low-molecular-mass proteins and protein mixtures reliably, especially at low protein amounts. Accordingly, MALDI-MS PMF may be complemented by sequence generating methods, such as PSD MALDI-MS [9], ESI-MS/MS [6, 7], and MALDI-MS/MS [10], for unequivocal protein identification.

Mycobacterium tuberculosis H37Rv is an ideal model organism for proteomics, as its genome encoding about 4000 proteins is completely sequenced [11, 12]. We have performed several proteome analyses on virulent and attenuated mycobacterial strains, which were primarily aimed at identifying proteins that play a role in mycobacterial pathogenicity and persistence in the host [13–19]. Recently, we established the subproteome of culture supernatant (CSN) proteins of *M. tuberculosis* H37Rv by combination of high-resolution 2-DE, MALDI-MS PMF, ESI-MS/MS, and *N*-terminal sequencing by Edman degradation [18]. In the course of this study, 430 distinct spots from 2-DE patterns of *M. tuberculosis* H37Rv CSN were analyzed by MALDI-MS PMF and 446 MALDI spectra were generated. The study led to the identification of 137 different proteins and their assignment to 381 distinct spots. Seventy-nine (58%) of the identified proteins were detected in more than one protein spot, most likely due to differential post-translational modifications and/or protein truncation. Certain proteins, most notably heat shock proteins, were identified in multiple spots widely spread across the 2-DE gel. In addition, many spots were unambiguously shown to represent protein mixtures. Minor spot components were often represented by only few peptide masses insufficient for identification by MALDI-MS PMF.

Based on these observations we tested whether it is possible to identify proteins reliably by a single or few protein-specific signature peptide mass(es) upon database independent comparisons of MALDI spectra. It has recently been suggested by Schmidt *et al.* [20] that a small set of protein-specific signature peptide masses generated by MALDI-MS, designated as minimal protein identifier (MPI), should allow protein identification. The MPI approach takes into consideration that proteins yield characteristic abundant peptides upon proteolytic cleavage. Additionally to the MPI approach, there are other strategies aiming at the identification of proteins by database independent comparisons of mass spectra [21–25]. These approaches are based on the evaluation of the similarity of mass spectra. In contrast, using the MPI approach mass spectra are systematically screened for selected protein-specific peptide masses not taking into account their overall similarity. The present study determines the technical feasibility and limitations of the MPI approach, as exemplified by the identification of M. tuberculosis H37Rv CSN proteins from high-resolution 2-DE gels. A total of 446 MALDI spectra generated to establish the M. tuberculosis H37Rv CSN subproteome were compared with spectra of previously identified counterparts. The MPI approach allowed us to propose and establish protein identity between two samples. We demonstrate the power of the MPI approach, in particular for the identification of low-molecular-mass proteins and protein fragments. As an example we show that the MPI approach facilitated identification of low-molecular-mass fragments of mycobacterial elongation factor EF-Tu (Tuf; Rv0685) previously not identified by MALDI-MS PMF. In addition, the MPI approach can be applied to efficiently track proteins in 2-DE gels, as exemplified for the 14 kDa antigen (HspX; Rv2031c), the 10 kDa chaperone (GroES; Rv3418c), and the conserved hypothetical protein Rv0569 of M. tuberculosis H37Rv.

To assess the feasibility of the MPI approach, the uniqueness of peptide masses in the M. tuberculosis H37Rv proteome was investigated by calculating the incidences of representative peptide masses in a theoretical dataset comprising the entirety of tryptic peptides (with m/z 500–4000) of all predicted proteins of M. tuberculosis H37Rv. These calculations emphasized that the efficacy of the MPI approach strongly depends on distinct factors, most notably on the complexity of the proteome investigated, the separation and dynamic range of the 2-DE technique used for protein separation and the accuracy of the mass spectrometer used for peptide mass determination. Provided that the mass spectrometer used allows peptide mass determination with high accuracy, the MPI approach is suitable for reliable protein identification based on few protein-specific signature peptide masses.

6.2
Materials and methods

6.2.1
Materials

M. tuberculosis H37Rv was purchased from the American Type Culture Collection (Manassas, USA). The 2-DE equipment was purchased from WITA (Teltow, Germany). ZipTipC$_{18}$ pipette tips were obtained from Millipore (Bedford, MA, USA). Trypsin (porcine sequencing-grade modified trypsin) used for in-gel digestion of proteins was obtained from Promega (Madison, WI, USA). α-Cyano-4-hydroxycinnamic acid (CHCA) and 2,5-dihydroxybenzoic acid (DHB) were purchased from Sigma-Aldrich (Taufkirchen, Germany). Other chemicals were delivered from Bio-Rad (Munich, Germany) or Merck (Darmstadt, Germany). Peptide solutions were concentrated using a Speed Vac concentrator (Savanta, Hicksville, USA).

MALDI-MS was carried out with a Voyager Elite MALDI-TOF mass spectrometer (Applied Biosystems, Framingham, MA, USA) with delayed extraction. MALDI-MS/MS was carried out with a 4700 Proteomics Analyzer (Applied Biosystems).

6.2.2
Two-dimensional electrophoresis

2-DE samples of *M. tuberculosis* H37Rv CSN were prepared as described [18]. Protein separation by 2-DE was performed as a combination of carrier ampholyte IEF and SDS-PAGE [26]. The size of the applied 2-DE gels was 23 × 30 cm. IEF was performed in rod gels containing 9 M urea, 3.5% acrylamide, 0.3% piperazine diacrylamide, and a total of 4% carrier ampholytes pH 2–11. Protein samples were applied at the anodic side of the IEF gels and focused under nonequilibrium pH gradient electrophoresis conditions (8870 Vh). For analytical and preparative investigations 0.75 or 1.5 mm thick gels were used, respectively. For analytical investigations, 100 μg of protein sample were applied. For preparative experiments, we applied 500 μg of protein sample. SDS-PAGE was performed in gels containing 15% acrylamide using the IEF gels as stacking gels. For analytical and preparative investigations, proteins were visualized following electrophoresis by either silver staining [27] or Coomassie Blue G-250 (CBB G-250) staining [28].

6.2.3
Mass spectrometry

MALDI-MS analysis of gel-separated proteins was performed as described [13, 17]. Spots were excised from preparative CBB G-250-stained 2-DE gels and proteins were in-gel digested using trypsin. Resulting peptides were desalted and concentrated using either self-developed peptide-collecting devices [29] or ZipTipC$_{18}$ pipette tips [17], and subsequently analyzed by MALDI-MS. The mass spectrometer was operated in the reflector mode using presettings previously described [15]. CHCA and DHB were used as matrices. A mass accuracy of 30–50 ppm was obtained by internal calibration of the spectra using synthetic peptides with known molecular mass and/or peptides occurring due to autolysis of trypsin as internal markers [24]. In some cases, sequence information was generated for unambiguous protein identification by MALDI-MS/MS performed as described [25].

6.2.4
Compilation of a theoretical dataset comprising all tryptic peptides of *M. tuberculosis* H37Rv

The genome sequence of *M. tuberculosis* H37Rv contains about 4000 protein-coding genes [12], of which 3924 were described initially [11]. For generation of a dataset comprising all theoretical tryptic peptides of *M. tuberculosis* H37Rv with an *m/z* ratio between 500–4000 the amino acid (AA) sequences of the latter were downloaded from the FTP site of the Sanger Institute (ftp://ftp. sanger.ac.uk/pub/tb/sequences/TB.pep) and cleaved *in silico* with trypsin by MS-Digest [http://prospector. ucsf.edu/ucsfhtml4.0/msdigest.htm]. Acetylation of protein N-termini (AcetN), pyroglutamate formation at N-terminal glutamine of peptides (pyroGlu), oxidation of methionine (Met-ox), and modification of cysteine by acrylamide (Cys-am) were considered as possible modifications. One missed cleavage site was allowed. The generated information (AA sequence, MH+, considered peptide modifications, *etc*.) was extracted from the html site using a Perl-script and implemented in a MySQLFront database [http://mysqlfront.venturemedia.de/]. The program MS-Digest was also used to calculate the ChemScores of all theoretical tryptic peptides of *M. tuberculosis* H37Rv. For this purpose the default presettings and a Met-ox factor of 1 were applied. The ChemScores of experimentally observed tryptic peptides were deduced from the generated theoretical peptide dataset.

6.2.5
Comparison of MALDI spectra by the program MS-Screener

The program MS-Screener developed by Schmidt and Rack (MPI for Infection Biology, Berlin, Germany) can be used to compare mass spectra (MALDI-MS(/MS) as well as ESI-MS/MS spectra) on the basis of their peak lists (.dta, .pkm, .pkt, or .txt files), to recalibrate mass spectra, to determine and eliminate exogenous contaminant peaks, and to create matrices for cluster analyses. This program has recently been used to perform hypothesis-free in-depth cluster analyses of MALDI spectra from 2-DE-separated proteins of *Helicobacter pylori* [24, 25] and is available to academic users free of charge (http://www.mpiib-berlin.mpg.de/2D-PAGE/download.html). Using MS-Screener, a template spectrum can be compared simultaneously with a large dataset of spectra resulting in a ranked list of co-occurring masses. Searches can be performed with a user-defined mass tolerance specified in Da or ppm.

6.2.6
Determination of exogenous contaminant masses

For determination of common exogenous contaminant masses related to matrix components, keratins, trypsin or the dye used for staining of preparative 2-DE gels (CBB G-250), we first generated a list comprising all possible masses in the m/z range from 500 to 4000 in 0.1 Da intervals. This list was subsequently compared with the peak lists (.pkm files) of all 446 MALDI dataset spectra and identical mass values were counted using MS-Screener. Keratins of human skin and dandruff, most notably keratin 1 (hair α keratin; NCBInr.10.2.2003; acc. # 17318569), type II cytoskeletal 6D keratin (acc. # 1346347), keratin 9 (NCBInr.10.2.2003; acc. # 4557705), and keratin 10 (acc. # 21961605) were frequently encountered exogenous contaminants. Mass peaks observed in $\geq$ 20 (4.5%) of the MALDI dataset spectra that matched tryptic cleavage products of these keratins, autolytic cleavage products of porcine trypsin (NCBInr.10.2.2003; acc. # 136429), or CBB G-250 were removed from the spectra and corresponding peak lists prior to further analysis. To investigate whether observed mass peaks were associated with the matrix rather than with analyzed peptide mixtures, matrix reference spectra were generated.

6.2.7
Generation of template spectra

For generation of template spectra, protein spots selected for high intensity were excised from a preparative 2-DE gel of *M. tuberculosis* H37Rv CSN, digested with trypsin and subsequently analyzed by MALDI-MS PMF. Following identification, unassigned mass peaks and peaks matching exogenous contaminants were removed from the spectra and corresponding peak lists (.pkm files). The resulting adjusted spectra were used as templates to be compared with the dataset.

6.3
Results and discussion

6.3.1
Proteome analysis of *M. tuberculosis* H37Rv CSN

We recently performed a comprehensive proteome analysis of *M. tuberculosis* H37Rv CSN [18]. Analytical 2-DE gels with a size of 23 × 30 cm resolved about 1250 CSN protein species, 430 of which were analyzed by MALDI-MS PMF and in part also by ESI-MS/MS and/or N-terminal sequencing by Edman degradation. The analysis revealed 137 distinct mycobacterial proteins, 42 of which had previously been described as secreted proteins. The dataset analyzed in this study comprised all 446 MALDI spectra recorded to establish the *M. tuberculosis* H37Rv CSN subproteome.

6.3.2
Determination of exogenous contaminant masses by MS-Screener

Mass peaks related to matrix components, human keratins, trypsin, and the dye used for staining of preparative 2-DE gels (CBB G-250) are frequently observed in MALDI spectra [24, 25, 30–33]. In the present study, 123 mass peaks were observed in $\geq$ 20 spectra (Tab. 1), 92 of which could be related to exogenous contaminants. In total, the 446 dataset spectra contained 16 657 mass peaks including 5528 matching exogenous contaminants. The elimination of the latter prior to further analysis resulted in a dataset containing 11 129 experimental mass peaks.

As summarized in Tab. 1, 37 frequently observed masses correlate with peptide masses assignable to keratins of human skin and dandruff, namely keratin 1, type II cytoskeletal 6D keratin, keratin 9, and keratin 10. The most commonly observed keratin-related peptide masses matched keratin 1 (m/z observed: 1475.8 $\pm$ 0.05; 1179.6 $\pm$ 0.05; 1277.7 $\pm$ 0.05), keratin 9 (m/z observed 1307.7 $\pm$ 0.05), and keratin 10 (m/z observed: 1707.7 $\pm$ 0.05), respectively. Although the porcine trypsin used for enzymatic cleavage of proteins is derivatized to prevent autolysis, we often observed peaks matching autolytic cleavage products of trypsin, most notably peptides with an m/z of 842.5 $\pm$ 0.05 and 2211.1 $\pm$ 0.05. We also frequently observed peaks matching CBB G-250, such as m/z 634.3 $\pm$ 0.05, 662.3 $\pm$ 0.05, and 832.3 $\pm$ 0.05. Furthermore, 13

Tab. 1. Mass peaks frequently observed in the MALDI spectra of the dataset

a) CHCA-specific

m/z observed (± 0.05)	Frequency absolute[a]	Frequency %[a]	MH+ theor.[b]
524.1	12	4.9	524.12
542.2	28	11.5	[c]
542.3	25	10.2	[c]
545.1	30	12.3	[c]
545.2	104	42.6	[c]
550.1	18	7.4	550.10
568.1	199	81.6	568.13
573.0	28	11.5	[c]
573.1	31	12.7	[c]
712.2	77	31.6	[c]
726.2	21	8.6	[c]
876.3	19	7.8	[c]
985.3	26	10.7	[c]

b) DHB-specific

m/z observed (± 0.05)	Frequency absolute[d]	Frequency %[d]	MH+ theor.[b]
530.5	74	36.6	530.50
558.5	75	37.1	558.53
574.9	63	31.2	[c]
582.3	23	11.4	[c]
586.5	46	22.8	586.55
594.5	9	4.5	594.47
599.3	39	19.3	[c]
614.6	12	5.9	614.57
615.4	22	10.9	[c]
622.5	21	10.4	622.50
646.3	41	20.3	[c]
650.5	11	5.4	650.57
697.3	27	13.4	[c]
699.2	48	23.8	[c]
699.3	20	9.9	[c]
748.9	10	5.0	748.91
752.3	37	18.3	[c]
752.4	29	14.4	[c]
798.3	20	9.9	[c]
816.3	74	36.6	[c]
874.5	21	10.4	[c]
881.3	37	18.3	[c]
973.5	18	8.9	[c]
996.5	26	12.9	[c]
996.6	65	32.2	[c]

c) CBB G-250-specific

m/z observed (± 0.05)	Frequency absolute[e]	Frequency %[e]	MH+ theor.[f]
634.2	37	8.3	634.29
634.3	171	38.3	634.29
662.3	285	63.9	662.31
804.2	26	5.8	804.30
804.3	181	40.6	804.30
832.3	298	66.8	832.31

d) Trypsin-specific

m/z observed (± 0.05)	Frequency absolute[e]	Frequency %[e]	MH+ theor.
842.5	281	63.0	842.51
870.5	26	5.8	870.54
1045.6	32	7.2	1045.56
1838.7	26	5.8	1838.88
1940.9	27	6.1	1940.94
2211.0	21	4.7	2211.11
2211.1	258	57.8	2211.11
2211.2	27	6.1	2211.11
2225.1	117	26.2	2225.12
2225.2	30	6.7	2225.12
2239.1	30	6.7	2239.14
2283.2	21	4.7	2283.18
2284.1	20	4.5	2283.18
2284.2	32	7.2	2283.18
2298.2	23	5.2	2297.20

e) Keratin-specific

m/z observed (± 0.05)	Frequency absolute[e]	Frequency %[e]	Putative source	MH+ theor.
599.3	41	9.2	Keratin 9	599.28
607.2	27	6.1	Keratin 1	607.28
697.3	27	6.1	Keratin 9	697.33
712.3	66	14.8	Keratin, type II	712.34
874.5	23	5.2	Keratin 1	874.50
973.5	21	4.7	Keratin 1	973.53
995.5	23	5.2	Keratin 10	995.52
1109.5	29	6.5	Keratin 10	1109.49
1165.6	26	5.8	Keratin, type II/ Keratin 10	1165.59/ 1165.59
1179.6	118	26.5	Keratin 1	1179.60
1234.7	53	11.9	Keratin 10	1234.68
1277.7	98	22.0	Keratin 1	1277.63/ 1277.71

e) Keratin-specific (Continued)

m/z observed (± 0.05)	Frequency absolute[e]	Frequency %[e]	Putative source	MH+ theor.
1302.7	41	9.2	Keratin 1	1302.70/ 1302.72
1307.6	25	5.6	Keratin 9	1307.68
1307.7	81	18.2		
1357.7	48	10.8	Keratin 1/ Keratin, type II/ Keratin 10	1357.70/ 1357.73/ 1357.72
1383.7	69	15.5	Keratin 1	1383.69
1393.7	29	6.5	Keratin 1	1393.73
1434.8	36	8.1	Keratin 10	1434.77
1475.7	70	15.7	Keratin 1	1475.75/ 1475.79
1475.8	131	29.4	Keratin 1	1475.75/ 1475.79
1493.7	49	11.0	Keratin 10	1493.74
1493.8	21	4.7	Keratin 10	1493.74
1523.8	33	7.4	Keratin 1/ Keratin, type II	1523.79/ 1523.81
1638.8	36	8.1	Keratin 1	1638.86
1638.9	42	9.4	Keratin 1	1638.86
1707.7	78	17.5	Keratin 10	1707.77
1707.8	54	12.1	Keratin 10	1707.77
1716.8	31	7.0	Keratin 1	1716.85
1791.6	22	4.9	Keratin 9	1791.73
1837.7	41	9.2	Keratin 9	1837.97
1851.7	58	13.0	Keratin 9	1851.93
1994.0	58	13.0	Keratin 1	1993.98
2383.9	42	9.4	Keratin 1	2383.95
2384.0	35	7.8	Keratin 1	2383.95
2705.1	32	7.2	Keratin 9	2705.16
2705.2	26	5.8	Keratin 9	2705.16

f) Matching *M. tuberculosis* proteins

m/z observed (± 0.05)	Frequency absolute[e]	Frequency %[e]	Putative source	MH+ theor.
1162.5	69	15.5	HspX	1162.55
1162.6	57	12.8	HspX	1162.55
1176.6	33	7.4	HspX	1176.57
1884.7	27	6.1	HspX	1884.94
1427.7	51	11.4	GroES	1427.72
1619.9	21	4.7	GroES	1619.85
1775.9	34	7.6	GroES	1775.95
2342.2	24	5.4	GroES	2342.25

g) Of unknown origin

m/z observed (± 0.05)	Frequency absolute[e]	Frequency %[e]
529.2	29	6.5
684.3	40	9.0
777.2	56	12.6
854.3	37	8.3
867.5	39	8.7
870.3	29	6.5
1092.6	20	4.5
1142.6	21	4.7
1193.6	22	4.9
1256.7	24	5.4
1320.6	59	13.2
1410.7	33	7.4
1458.7	20	4.5
1503.9	20	4.5
1505.8	21	4.7
1593.8	22	4.9
1601.8	21	4.7
1714.9	29	6.5
2004.0	21	4.7
2083.0	24	5.4
2106.1	25	5.6
2177.1	27	6.1

a) CHCA subdataset: 244 = 100%
b) Deduced from matrix reference spectrum
c) Not detected in matrix reference spectrum
d) DHB subdataset: 202 = 100%
e) Entire dataset: 446 = 100%
f) Deduced from CBB G250 reference spectrum

CHCA- and 25 DHB-specific mass peaks were frequently observed. For example, > 40% of all spectra generated using CHCA comprised peaks with an m/z of 545.2 ± 0.05, 568.1 ± 0.05, and 712.2 ± 0.05. In the case of DHB, the three most common matrix-specific peaks were m/z 530.5 ± 0.05, 574.9 ± 0.05, and 816.3 ± 0.05.

6.3.3
The MPI approach revealed HspX-specific peptide masses in multiple spectra

Forty-nine protein spots of differential electrophoretic mobility located in the acidic, low-molecular-mass range of the *M. tuberculosis* H37Rv CSN 2-DE standard pattern were previously shown to contain the mycobacterial 14 kDa antigen (HspX; Rv2031c) [18]. To further analyze the 2-DE distribution of HspX, we compared an adjusted MALDI spectrum of spot 5_85, identified as HspX based on 13 HspX-specific peptide masses (Fig. 1), with the 446 MALDI spectra previously recorded to establish the *M. tuberculosis* H37Rv CSN subproteome. The comparisons were

6.3 Results and discussion

performed using the program MS-Screener. The tolerated mass error was 0.1 Da. As depicted in the lower part of Fig. 1, 142 dataset spectra derived from the analysis of 131 distinct spots contained at least one of the 13 searched for HspX-specific peptide masses. As shown in Fig. 2A, the majority ($n = 96$) of the spots putatively containing HspX according to the MS-Screener analysis are located in the acidic, low-molecular-mass range (section 5) of the *M. tuberculosis* H37Rv CSN 2-DE pattern. The remaining spots putatively comprising HspX were located in section 1 (14 spots), section 2 (2 spots), section 3 (15 spots), and section 6 (4 spots). The number of observed HspX-specific peptide masses per spectrum was between 1 and 6. Except for the template spectrum, none of the dataset spectra contained ≥ 10, and only 2 spectra contained ≥ 5 HspX-specific peptide masses. The most commonly detected HspX-specific peptide mass was m/z 1162.55 ± 0.1. This peptide mass, matching the AAs 91–100 of HspX (theoretical MH+: 1162.5533 Da), was observed in 125 dataset spectra derived from the analysis of 114 distinct spots. Other frequently observed HspX-related peptide masses were m/z 1884.92 ± 0.1 Da ($n = 35$), 666.394 ($n = 17$), 1868.92 ± 0.1 ($n = 16$), and 1122.52 ± 0.1 ($n = 12$). Note, that if a high proportion of matches observed were random, one would expect them to affect all HspX-specific masses equally. In addition, one would expect that the spots affected by the matches were evenly distributed over the entire 2-DE gel. In contrast, we predominantly observed certain HspX-specific peptide masses in the dataset spectra, most notably 1162.55 ± 0.1, and the vast majority of the affected spots were localized in the acidic, low-molecular-mass range of the gel. This suggests that a significant number of matches detected by MS-Screener were valid.

Forty-nine of the 131 spots putatively containing HspX according to the MS-Screener analysis were previously identified as HspX by MALDI-MS PMF, ESI-MS/MS and/or N-terminal sequencing (Fig. 2A') [18]. These underlying data are available via our mycobacterial proteome 2D-PAGE database (http://www.mpiib-berlin.mpg.de/2D-PAGE/index-2DPAGE.html). In addition, 19 of the 82 remaining spots putatively comprising HspX were subjected to MALDI-MS/MS. For 8 of these spots the proposed HspX identity was confirmed (Fig. 2A). The MS/MS results were included in our mycobacterial proteome database. A representative MS/MS spectrum, derived from the analysis of peptide m/z 1162.55 of spot 5_8 is depicted in Fig. 3. In the case of the other spots analyzed by MS/MS, the mass spectra were of poor quality and thus could not be related to any protein. The identity of the remaining spots putatively comprising HspX could not be reviewed due to the low intensity of the observed HspX-specific peptide masses and/or due to the presence of neighboring peptide masses within the precursor mass window. Note, that in all cases where a peptide mass of m/z 1162.55 ± 0.1 was analyzed by MS/MS, the proposed HspX identity was confirmed, provided that the generated spectra were of sufficient quality. Although we did not observe any false-positive assignments as yet, this does not necessarily mean that all putative HspX-specific peptide masses are indeed HspX-derived. In particular, we assume that, contrary to MS-Screener predictions, those 17 spots putatively comprising HspX whose MALDI spectra did not show the peptide mass m/z 1162.55 ± 0.1 do not contain HspX. As depicted in Fig. 2A, all these putative false-positive spots were localized

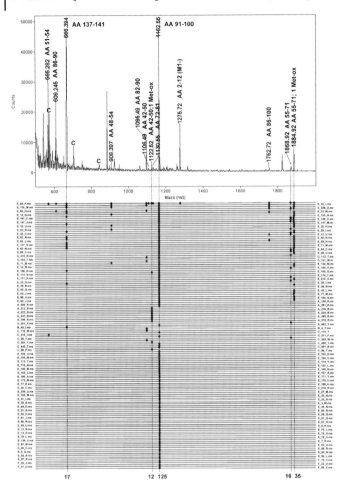

Fig. 1 Results of the MS-Screener analysis concerning HspX. The upper part of the figure depicts the MALDI spectrum of spot 5_85 identified as HspX, with 13 matching peptide masses labeled. Numbers given in brackets specify the AA residues (start and end) of the theoretical peptides matched. Mass peaks marked by C matched exogenous contaminants. The template spectrum was compared with the dataset consisting of 446 MALDI spectra derived from the analysis of 430 distinct protein spots from 2-DE patterns of *M. tuberculosis* H37Rv CSN by the program MS-Screener. The analysis revealed 142 dataset spectra derived from the analysis of 131 distinct spots comprising at least one of the 13 searched for HspX-specific peptide masses. These are depicted as horizontal lines in the lower part of the figure. For reasons of clarity, the y-axes of the dataset spectra are not displayed. Only matches observed at a tolerated mass error of ± 0.1 Da are marked by rhombi. Dashed vertical lines indicate the 5 most commonly observed HspX-specific peptide masses. The most commonly ($n = 125$) observed HspX-specific peptide mass was m/z 1162.55 ± 0.1 Da. Note, that this mass was observed in the spectra of nearly all spots putatively containing HspX.

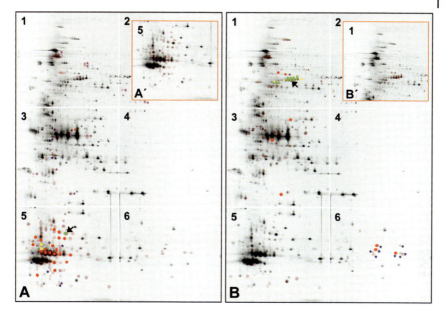

Fig. 2 2-DE distribution of all spots putatively comprising (A) HspX and (B) Tuf according to the MS-Screener analyses, and all spots previously identified as (A′) HspX and (B′) Tuf by MALDI-MS, ESI-MS/MS, and/or N-terminal sequencing. The respective spots are bordered. The template spots are marked by arrows. Red dots indicate spots whose MALDI spectra contained 3–4 of the searched protein-specific masses. Yellow dots symbolize between 5 and 9, and green dots $\geq$ 10 protein-specific masses. Spots designated by blue asterisks were previously unidentified and identified as HspX and Tuf, respectively, upon MS-Screener analysis by MS/MS. In the case of HspX (A), the 17 spots "putatively containing HspX" whose MALDI spectra did not contain the characteristic HspX-specific peptide mass m/z 1162.55 $\pm$ 0.1 are marked by purple asterisks. We assume that these spots localized in the mean or higher M_r range of the 2-DE pattern of *M. tuberculosis* H37Rv CSN are false-positives. MS/MS analyses of the respective peptide masses assignable to HspX were not possible due to low peak intensities. In the case of spot 3_211 marked by a red asterisk it is also very likely that the assignment of m/z 1162.55 $\pm$ 0.1 to HspX is false-positive. This spot was identified by MALDI-MS PMF as putative dehydrogenase (Rv3389c) and the peptide mass m/z 1162.55 $\pm$ 0.1 is also assignable to this protein. Unfortunately, the peptide could not be analyzed by MS/MS. Please note that for reasons of clarity asterisks are not superimposed on the spots. Figures 2A′ and B′, depicting all spots previously identified as HspX and Tuf, respectively are overlaid upon uninteresting parts of the 2-DE patterns containing no spots putatively representing these proteins. Please also note, that the 2-DE standard pattern of *M. tuberculosis* H37Rv CSN is depicted in the form of six sectors with horizontal and vertical overlaps. An undivided 2-DE standard pattern of *M. tuberculosis* H37Rv CSN proteins with M_r and pI labels has recently been published [18].

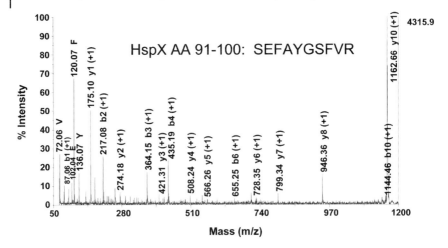

Fig. 3 The figure depicts the MALDI-MS/MS spectrum of the peptide m/z 1162.55 of spot 5_8 representing the AAs 91–100 of HspX (theoretical MH+: 1162.5533 Da). Please note, that in the MALDI PMF of spot 5_8 this mass was the only one assignable to HspX. Nonetheless, the proposed identity was unambiguously confirmed. For reasons of clarity, only the 19 mass peaks assignable to immonium ions, b-ions and y-ions, respectively, of the AAs 91–100 of HspX are marked in the MS/MS spectrum.

in the mean or higher M_r range (sections 1, 2, and 3) of the *M. tuberculosis* H37Rv CSN 2-DE standard pattern. Unfortunately, MS/MS analyses of the putative false-positive HspX-specific peptide masses were not possible due to low peak intensities.

As described in Section 3.5, 81 of the 255 640 tryptic peptides (of m/z 500–4000) that can be deduced from the 3924 predicted proteins of *M. tuberculosis* H37Rv have a theoretical monoisotopic protonated mass (MH+) of 1162.5533 ± 0.1 Da. In addition to HspX, so far three of these 81 proteins have been identified in the CSN fraction of *M. tuberculosis* H37Rv, namely N-acetyl-γ-glutamyl-phosphate reductase (ArgC; Rv1652; identified in spot 2_34), a heat shock protein (ClpB; Rv0384c; identified in spots 1_403-1_406), and a putative dehydrogenase (Rv3389c; identified in spot 3_211). To evaluate whether the high frequency of the peptide mass m/z 1162.55 ± 0.1 Da may be due to the presence of these proteins, we checked the MALDI spectra of the respective spots for the presence of a peptide mass of m/z 1162.55 ± 0.1 Da. An appropriate peptide mass of very low intensity was solely detected in spot 3_211 identified as putative dehydrogenase (Rv3389c) by MALDI-MS PMF (Fig. 2A), strongly suggesting that in this case the assignment of m/z 1162.55 ± 0.1 to HspX is false-positive. Regrettably, due to its low peak intensity, this peptide could not be analyzed by MS/MS.

The results of the MS-Screener analysis suggest that a high number of the spots represent protein mixtures rather than single proteins. In the previous proteome analysis of *M. tuberculosis* H37Rv CSN, 62 (16%) of all 381 identified 2-DE spots were unambiguously shown to comprise more than one protein component [18].

In contrast, 95 (72.5%) of all 131 spots putatively comprising HspX contained additional protein components. In-depth analyses revealed that even well separated and focused 2-DE spots can represent protein mixtures. Thus, caution is required in evaluating experimentally observed variations in spot intensity.

6.3.4
The MS-Screener analysis revealed truncated variants of Tuf previously not identified by PMF

The proteome analysis of M. *tuberculosis* H37Rv CSN revealed 14 distinct spots representing Tuf [18] which displayed an electrophoretic mobility (M_r 41–48 kDa; pI ~5) consistent with the theoretical M_r and pI of Tuf (44.4 kDa; 5.2) (Fig. 2B′). The MPI approach identified several ($n = 78$) additional spots putatively representing Tuf, 8 of which were verified upon MS-Screener analysis by ESI-MS/MS (Fig. 2B). The identity of the remaining spots putatively comprising Tuf could not be evaluated by MS/MS due to low peak intensities of the Tuf-specific peptide masses. The 8 newly identified Tuf variants showed a very low M_r (10–13 kDa) and a pI ranging between 7 and 9, due to N-terminal protein truncation. The identification of these Tuf variants by MALDI-MS PMF failed because of their small size (only ~1/4 of the full-length protein) resulting in a small number of matching peptides ($n \leq 4$) and a very low sequence coverage ($\leq 17\%$). All Tuf-related masses observed in the MALDI spectra of these spots matched C-terminal Tuf peptides located in the sequence range between AAs 292–376 of the full-length protein.

The results of the MS-Screener analysis concerning Tuf are depicted in Fig. 4. The template spectrum derived from the analysis of spot 1_362 comprised 19 Tuf-specific peptide masses. The MS-Screener analysis revealed 99 dataset spectra from 92 distinct spots comprising at least one of the 19 searched for Tuf-specific peptide masses. Eleven spectra contained ≥ 10, and 3 spectra contained 5–9 Tuf-specific peptide masses. The most commonly observed ($n = 56$) Tuf-specific peptide mass was m/z 1357.72 ± 0.1. This mass matches the AAs 372–383 of Tuf. Note, that in the template spectrum the m/z 1357.72 had a relatively low intensity (4036 counts). The template spectrum comprised 14 mass peaks of higher relative intensity, 5 of which were Tuf-specific and 7 of which could be assigned to exogenous contaminants. The most intense Tuf-specific mass peak (m/z 1404.64) had a relative intensity of 39 235 counts but was observed in only 20 spectra of the dataset. Thus, high intensity protein-specific peptide masses are not necessarily the most attractive signature peptide masses. The utility of a mass peak as a representative signature peptide mass is influenced by the location of the corresponding peptide within the protein because terminal protein sequences are candidates for truncation.

To demonstrate the general applicability of the MPI approach, further proteins were analyzed, namely the 10 kDa chaperone (GroES; Rv3418c) and the conserved hypothetical protein Rv0569 of M. *tuberculosis* H37Rv. The results of these analyses are available via our mycobacterial proteome 2D-PAGE database (http://www.mpiib-berlin.mpg.de/2D-PAGE/supp-mat/mattow_MPI_ supplementary_material.pdf).

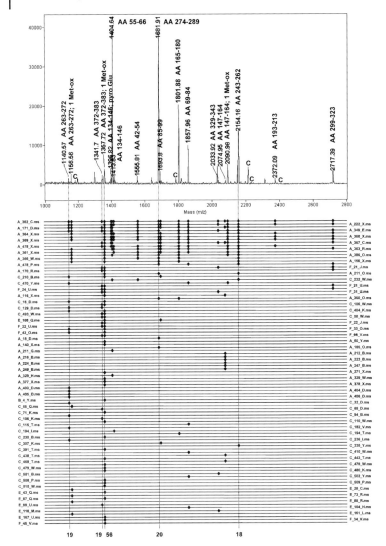

Fig. 4 Results of the MS-Screener analysis concerning Tuf. The upper part of the figure depicts the adjusted template MALDI spectrum of spot 1_362 identified as Tuf, with 18 Tuf-specific peptide masses labeled. The comparison of this spectrum with the dataset revealed 99 MALDI spectra (depicted as horizontal lines in the lower part of the figure) derived from the analysis of 92 different spots containing at least one Tuf-specific peptide mass. These included 24 spectra from 22 distinct spots containing ≥ 3 Tuf-specific peptide masses. Eight spots with spectra comprising ≥ 3 Tuf-specific peptide masses had previously not been identified by conventional MALDI-MS PMF and displayed an electrophoretic mobility (M_r 10–13 kDa; pI 7–9) markedly different from the expected one of full-length Tuf (M_r 44.4 kDa; pI 5.2). The targeted ESI-MS/MS analysis of these 8 spots confirmed the proposed Tuf identity. Symbols used correspond to legend of Fig. 1.

6.3.5
Frequency of tryptic peptides with similar m/z ratios in M. tuberculosis H37Rv

Certain peptide masses, such as m/z 1162.55 ± 0.1 assignable to HspX, were detected in many dataset spectra. Thus, it was important to analyze whether experimental mass peaks showing a similar m/z ratio necessarily represent identical peptides. To this end, we determined the frequency of tryptic peptides with similar m/z ratios in M. tuberculosis H37Rv. Note, that the frequency of tryptic peptides with a similar m/z ratio strongly depends on the proteome complexity and the tolerated mass error. The crucial role of accurate peptide mass determination in protein identification strategies employing PMF has already been pointed out [34, 35].

Tab. 2 illustrates the frequency of 25 peptide masses for all tryptic peptides of M. tuberculosis H37Rv considering a possible mass error (Δm) of 0.1, 0.05, 0.01, and 0.001 Da, respectively. The masses investigated represent the theoretical MH+ values of selected tryptic peptides of mycobacterial 14 kDa antigen (HspX; Rv2031c), (Rv0685), peptidyl-prolyl cis-trans isomerase (PpiA; Rv0009), 10 kDa chaperone (GroES; Rv3418c), and the conserved hypothetical protein Rv0569. As shown in Tab. 2, the number of matching peptides strongly correlates with the tolerated mass error. For example, 81 of all 255 640 theoretical tryptic peptides of M. tuberculosis H37Rv have a theoretical MH+ of 1162.5533 ± 0.1 Da, but only 48 peptides show this MH+ ± 0.05 Da. The number of matching theoretical tryptic peptides of M. tuberculosis H37Rv decreases significantly when a more restrictive mass error of ± 0.01 Da (11 matches) or ± 0.001 Da (6 matches) is applied. On average, 76.7 ($\Delta m = \pm$ 0.1 Da), 45.2 ($\Delta m = \pm$ 0.05 Da), 11.1 ($\Delta m = \pm$ 0.01 Da), and 3.2 ($\Delta m = \pm$ 0.001 Da) matching theoretical tryptic peptides were observed for the 25 peptide masses investigated.

We also investigated whether the detection of two protein-specific peptide masses may represent a valid criterion for unequivocal protein identification. As depicted in Tab. 2, 666.4051 Da, 1122.5101 Da, 1162.5533 Da, 1868.9428 Da, and 1884.9377 Da are the theoretical MH+ values of tryptic peptides of HspX. All ten possible pairs of these peptide masses were formed, and analyzed for the number of those proteins of M. tuberculosis H37Rv which possess theoretical peptides of both masses. As shown in Tab. 3, on average 5.4 ($\Delta m = \pm$ 0.1 Da), 2.4 ($\Delta m = \pm$ 0.05 Da), 1.2 ($\Delta m = \pm$ 0.01 Da), and 1 ($\Delta m = \pm$ 0.001 Da) of the 3924 predicted proteins of M. tuberculosis H37Rv showed theoretical tryptic peptides matching both masses.

Our data suggest that in the case of M. tuberculosis H37Rv and microorganisms with proteomes of comparable size (comprising ~4000 proteins), 1–2 experimentally determined matching peptide masses are insufficient for unambiguous protein identification unless they were recorded with an exceptionally high mass accuracy of about 0.001 Da/1 ppm. At present this can only be achieved by Fourier transform ion cyclotron resonance (FTICR) instruments, whereas the currently available state-of-the-art MALDI-TOF(/TOF) instruments with delayed extraction allow peptide mass determination with a mass error of about 3–10 ppm [36–38]. The MALDI mass spectrometer used in this study is equipped to determine peptide masses with an accuracy of about 30–50 ppm.

Tab. 2 Searches at different mass tolerances against all theoretical tryptic peptides of M. tuberculosis H37Rv using single theoretical MH+ values corresponding to peptides of mycobacterial proteins indicated

Source	Peptide mass	Mass tolerance (Da)			
		±0.10	±0.05	±0.01	±0.001
HspX	666.4051	82	46	7	7
	1122.5101	66	33	3	1
	1162.5533	81	48	11	6
	1868.9428	64	33	10	6
	1884.9377	89	52	25	1
GroES	1427.7171	83	51	9	1
	1523.8182	77	42	6	2
	1619.8532	96	60	16	5
	1775.9543	59	35	10	3
	2342.2455	52	31	5	1
PpiA	1067.6002	69	34	6	2
	1305.7167	85	49	13	4
	1601.7924	71	39	12	3
	1749.8594	54	36	10	4
	1757.8935	61	42	5	3
Tuf	1140.5723	76	38	11	1
	1341.7220	84	47	14	2
	1357.7150	83	53	12	1
	1681.9237	57	30	5	3
	2154.1770	46	22	4	1
Rv0569	799.5042	93	46	8	8
	929.5460	130	93	38	2
	1109.5340	92	65	15	6
	1163.5697	77	55	7	3
	1269.7068	90	51	16	3
∅		76.7	45.2	11.1	3.2

The number of theoretical peptides matching is indicated.

6.3.6
In mass spectrometry it is important to consider detection probabilities of proteins and peptides

All experimental peptides with an m/z ratio of 1162.5533 ± 0.1 Da further analyzed by MS/MS ($n = 19$) were identified as AAs 91–100 of HspX. Thus, contrary to theoretical predictions, our experimental data suggest that a single protein-specific peptide mass can indeed facilitate protein identification. Regarding the specificity of the peptide with an m/z of 1162.5533 ± 0.1 in unequivocally identifying HspX in a number of spots where its mass was the sole HspX constituent in PMF, it becomes

Tab. 3 Searches at different mass tolerances against all theoretical tryptic peptides of M. tuberculosis H37Rv using pairs of theoretical MH+ values corresponding to commonly detected peptides of HspX

Peptide Masses	Mass tolerance (Da)			
	±0.10	±0.05	±0.01	±0.001
666.4051 / 1122.5101	4	1	1	1
666.4051 / 1162.5533	6	4	1	1
666.4051 / 1868.9428	2	1	1	1
666.4051 / 1884.9377	2	1	1	1
1122.5101 / 1162.5533	5	2	1	1
1122.5101 / 1868.9428	4	3	1	1
1122.5101 / 1884.9377	2	2	1	1
1162.5533 / 1868.9428	2	2	1	1
1162.5533 / 1884.9377	4	1	1	1
1868.9428 / 1884.9377	23	7	3	1
∅	5.4	2.4	1.2	1

The number of M. tuberculosis H37Rv proteins with peptides matching both masses is indicated.

evident that certain peptide masses are valid as signature peptide masses beyond theoretical considerations. How can this apparent inconsistency be explained? What reasons are responsible for AAs 91–100 of HspX being the only tryptic peptide in the experimental dataset with an m/z of 1162.5533 ± 0.1 even though the M. tuberculosis H37Rv proteome comprises 81 theoretical peptides of this mass?

One explanation is the fact that we investigated a mycobacterial subproteome of reduced complexity. According to proteome analyses only about 150 proteins of M. tuberculosis H37Rv are secreted or released without N-terminal signal sequence [18, 39, 40]. However, according to bioinformatic predictions the number of secreted proteins of M. tuberculosis H37Rv may add up to 860 [41, 42]. In addition, it has to be noted that protein identification and hence peptide detection is restricted by the analytical window and dynamic range of the 2-DE or chromatographic technique applied for protein separation prior to MS(/MS) analysis. The 2-DE technique used allows separation of proteins with a pI ranging from 4 to 11.5 and an M_r between 6 and 140 kDa [43]. However, for proteins with an $M_r > 100$ kDa and a p$I > 10$, the resolution efficiency of this technique is limited. In the previous subproteome analysis of M. tuberculosis H37Rv CSN, none of the 137 identified proteins had a theoretical $M_r > 100$ kDa, and only 3 (2%) had a theoretical pI value > 10 [18]. Thus, about 85% ($n = 3341$) of the 3924 predicted proteins of M. tuberculosis H37Rv fall into the effective separation range of the applied 2-DE technique (pI 4–10; M_r 6–100 kDa). In addition, not all predicted genes are necessarily translated constituting a further potential reduction. We assume that 10–20% of all predicted proteins of *M. tuberculosis* H37Rv are putatively detectable in our model. Accordingly, the significance of a single protein-specific peptide mass must be much higher than initially expected based on theoretical frequencies of peptide masses.

In addition, it is important to note that only a fraction of the theoretical tryptic peptides of a given protein is detectable by MALDI-MS. This fact is central to the MPI approach as well as to the ChemScore developed by Parker [44] which describes the theoretical detection probability of peptides by MALDI-MS. The ChemScores of peptides range between 0 and 100. The higher the ChemScore of a peptide, the more likely it is detected by MALDI-MS. The ChemScore is calculated on the basis of a combination of chemical properties, in particular the AA composition of a peptide and the AA sequences spanning the cleavage sites. Furthermore, a series of user-adjustable parameters are considered to reflect the experimental conditions, such as the fact that certain peptides are more efficiently generated by trypsin digestion or more efficiently detected by MALDI-MS PMF. Peptides containing AA residues with high pK values are most likely protonated. In contrast, certain peptides are likely to be affected by post-translational modifications or cannot be detected by MALDI-MS because their m/z ratio is outside the detectable m/z range.

We determined the ChemScores of all theoretical tryptic peptides of *M. tuberculosis* H37Rv as well as of an experimental subset consisting of 981 different peptides from 92 distinct mycobacterial CSN proteins identified without ambiguity. Figures 5A and B show the normalized frequencies of the theoretical and experimentally observed tryptic peptides in defined ChemScore windows. As depicted, tryptic peptides with low ChemScores ($<$ 10) were significantly underrepresented in the experimental dataset. Nearly half (45.2%) of all theoretical tryptic peptides of *M. tuberculosis* H37Rv showed a ChemScore $<$ 5. In contrast, only 9.4% of the experimentally observed peptides exhibited an equally low ChemScore. About 16% of the theoretical peptides have a ChemScore of 100 whereas nearly half (49.8%) of the experimentally observed ones have this score. Thus, the ChemScore indeed adequately reflects the detection probability of a peptide by MALDI-MS. This strongly suggests that a substantial number of the theoretical tryptic peptides of *M. tuberculosis* H37Rv are undetectable or unlikely to be detected by MALDI-MS. This factor also affects the significance of experimentally observed signature peptide masses, which is higher than deduced from theoretical calculations. The significance of individual peptide masses as signature masses rises proportionally to the reduction of putatively detectable theoretical peptides generated from an investigated proteome.

We consider the detection probability of peptides an important feature for future evaluation of MALDI-MS PMF data. For example, it is conceivable to take into account only matches affecting theoretical peptides with a ChemScore $\geq$ 5. Our data suggest that such an approach would markedly decrease the potential for false-positive matches. Provided that our data are representative it would nearly cut in half the number of theoretical tryptic peptides, at the same time having little impact on the number of correctly assigned peptide masses as only a few of the experimental peptides (9.4%) exhibit a ChemScore $<$ 5.

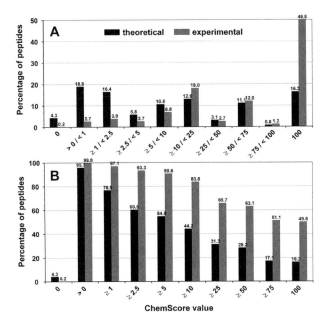

Fig. 5 Normalized frequencies of theoretical and experimentally observed tryptic peptides of *M. tuberculosis* H37Rv peptides in defined ChemScore windows. As a prerequisite a dataset comprising all theoretical tryptic peptides with an m/z between 500 and 4000 ($n = 255\,640$) of all predicted proteins ($n = 3924$) *M. tuberculosis* H37Rv was generated. The ChemScores of the theoretical tryptic peptides of *M. tuberculosis* H37Rv as well as of an experimental subset comprising 981 different peptides from 92 distinct mycobacterial CSN proteins identified without ambiguity were subsequently determined using the program MS-Digest. In (A) the frequencies of the theoretical and experimentally observed tryptic peptides of *M. tuberculosis* H37Rv peptides within defined narrow ChemScore windows are compared. In contrast, (B) depicts the frequencies in a cumulative manner.

Beside the ChemScore, additional factors which may further reduce the number of theoretical tryptic peptides putatively detectable by MALDI-MS have not been investigated. Note, that the experimental and theoretical datasets investigated in this study were not parameter-free. Accordingly, insights gained in this study are not necessarily fully applicable to other organisms and/or experimental protocols (2-DE, MALDI-MS).

6.4
Concluding remarks

We have employed the MPI approach introduced by Schmidt *et al.* [20] for targeted identification of 2-DE-separated CSN proteins of *M. tuberculosis* H37Rv. This approach is based on comparing experimentally derived proteolytic peptide mass maps of proteins recorded by MALDI-MS to those of previously identified counterparts. We demonstrate that the MPI approach may be used to complement MALDI-MS PMF, particularly for the identification of low-molecular-mass proteins and protein fragments. In analogy to the ChemScore developed by Parker [44], which is calculated on the basis of chemical properties, the MPI approach considers the detection probability of peptides by MALDI-MS. In contrast, MALDI-MS PMF is based on correlating experimental and theoretical mass data stored in protein sequence databases without taking into account which peptides are efficiently generated by proteolytic cleavage and likely to be detected by MALDI-MS. The application of the MPI approach facilitated identification of a series of N-terminally truncated low-molecular-mass fragments of mycobacterial Tuf previously not identified by MALDI-PMF. As demonstrated for the 14 kDa antigen, the 10 kDa chaperone, and the conserved hypothetical protein Rv0569 of *M. tuberculosis* H37Rv the MPI approach can be adopted to efficiently track the 2-DE distribution of a given protein.

We performed a series of calculations pertaining to the theoretical frequency of similar peptide masses. These suggest that in the case of *M. tuberculosis* H37Rv with a proteome comprising about 4000 proteins and 255 000 theoretical tryptic peptides already few experimentally determined peptide masses allow protein identification, provided that they are recorded with an exceptionally high mass accuracy of about 1 ppm. The MALDI-TOF mass spectrometer used allowed peptide mass determination with an accuracy of 30–50 ppm, insufficient for unambiguous protein identification by few peptides. Thus, we complemented the MPI approach by MS/MS, to verify or discard a proposed protein identity. Contrary to theoretical considerations, the vast majority of protein identities proposed by the MPI approach were confirmed, provided that the generated MS/MS data were of sufficient quality.

Our data suggest that the actual significance of experimentally observed peptide masses markedly exceeds theoretical predictions, in particular when a protein is highly abundant. In our experimental model this was partly due to the fact that we investigated the culture supernatant proteins of *M. tuberculosis* H37Rv, a mycobacterial subproteome of limited complexity. However, we also demonstrated that a very high proportion of the theoretical tryptic peptides of *M. tuberculosis* H37Rv are undetectable or unlikely to be detected by MALDI-MS.

Screening of MALDI spectra for characteristic peptide masses of a given protein only allows directed protein identification. However, it is conceivable to modify the MPI approach to allow hypothesis-free protein identification. As suggested by Schmidt *et al.* [20] a comprehensive database can be generated comprising experimentally determined peptide mass fingerprints of gel-separated and/or recombinant proteins. Such a database may be used to evaluate MALDI spectra of proteins to be identified, instead of correlating experimental data with entire protein sequence databases. Here, we show that such an approach can markedly decrease the potential for false-positive assignments and thus increase the significance of protein identification results. The ChemScore adequately reflects the detection probability of peptides by MALDI-MS. Accordingly, theoretically assessed ChemScores can also be employed for generation of virtual peptide mass maps, comprising only those peptide masses of proteins that are putatively detected with greatest sensitivity. It is evident that the MPI approach and the ChemScore are complementary in their final goal to increase the confidence in protein identification.

The authors wish to thank the BMBF (Competence Network "Neue Methoden zur Erfassung des Gesamtproteoms von Bakterien") and the WHO (Global Program for Vaccines and Immunization – Vaccine Research and Development) for their financial support, Alexander Rack (MPI for Infection Biology, Berlin, Germany), Till Eifert, and Sven Büttner (Algorithmus, Berlin, Germany) for their bioinformatic support, and Dr. Bernd Thiede and Dr. Guido Hegasy (MPI for Infection Biology, Berlin, Germany) for critical reading of the manuscript.

6.5
References

[1] Henzel, W. J., Billeci, T. M., Stults, J. T., Wong, S. C. et al., *Proc. Natl. Acad. Sci. USA* 1993, *90*, 5011–5015.

[2] James, P., Quadroni, M., Carafoli, E., Gonnet, G., *Biochem. Biophys. Res. Commun.* 1993, *195*, 58–64.

[3] Mann, M., Hojrup, P., Roepstorff, P., *Biol. Mass Spectrom.* 1993, *22*, 338–345.

[4] Yates III, J. R., Speicher, S., Griffin, P. R., Hunkapiller, T., *Anal. Biochem.* 1993, *214*, 397–408.

[5] Pappin, D. J. C., Hojrup, P., Bleasby, A. J., *Curr. Biol.* 1993, *3*, 327–332.

[6] Fenn, J. B., Mann, M., Meng, C. K., Wong, S. F., Whitehouse, C. M., *Science* 1989, *246*, 64–71.

[7] Mann, M., Wilm, M., *Anal. Chem.* 1994, *66*, 4390–4399.

[8] Banks, R. E., Dunn, M. J., Hochstrasser, D. F., Sanchez, J. C. et al., *Lancet* 2000, *356*, 1749–1756.

[9] Spengler, B., Kirsch, D., Kaufmann, R., Jaeger, E., *Rapid Commun. Mass Spectrom.* 1992, *6*, 105–108.

[10] Medzihradszky, K. F., Campbell, J. M., Baldwin, M. A., Falick, A. M. et al., *Anal. Chem.* 2000, *72*, 552–558.

[11] Cole, S. T., Brosch, R., Parkhill, J., Garnier, T. et al., *Nature* 1998, *393*, 537–544.

[12] Camus, J. C., Pryor, M. J., Medigue, C., Cole, S. T., *Microbiology* 2002, *148*, 2967–2973.

[13] Jungblut, P. R., Schaible, U. E., Mollenkopf, H. J., Zimny-Arndt, U. et al., *Mol. Microbiol.* 1999, *33*, 1103–1117.

[14] Jungblut, P. R., Müller, E. C., Mattow, J., Kaufmann, S. H., *Infect. Immun.* 2001, *69*, 5905–5907.

[15] Mattow, J., Jungblut, P. R., Schaible, U. E., Mollenkopf, H. J. et al., *Electrophoresis* 2001, *22*, 2936–2946.

[16] Mattow, J., Jungblut, P. R., Müller, E. C., Kaufmann, S. H., *Proteomics* 2001, *1*, 494–507.

[17] Mollenkopf, H. J., Mattow, J., Schaible, U. E., Grode, L. et al., *Methods Enzymol.* 2002, *358*, 242–256.

[18] Mattow, J., Schaible, U. E., Schmidt, F., Hagens, K. et al., *Electrophoresis* 2003, *24*, 3405–3420.

[19] Schmidt, F., Donahoe, S., Hagens, K., Mattow, J. et al., *Mol. Cell Proteomics* 2004, *3*, 24–42.

[20] Schmidt, F., Lueking, A., Nordhoff, E., Gobom, J. et al., *Electrophoresis* 2002, *23*, 621–625.

[21] Wan, K. X., Vidavsky, I., Gross, M. L., *J. Am. Soc. Mass Spectrom.* 2002, *13*, 85–88.

[22] Alfassi, Z. B., *J. Am. Soc. Mass Spectrom.* 2003, *14*, 262–264.

[23] Alfassi, Z. B., *J. Am. Soc. Mass Spectrom.* 2004, *15*, 385–387.

[24] Schmidt, F., Schmid, M., Jungblut, P. R., Mattow, J. et al., *J. Am. Soc. Mass Spectrom.* 2003, *14*, 943–956.

[25] Krah, A., Schmidt, F., Becher, D., Schmid, M. et al., *Mol. Cell. Proteomics* 2003, *2*, 1271–1283.

[26] Klose, J., Kobalz, U., *Electrophoresis* 1995, *16*, 1034–1059.

[27] Jungblut, P. R., Seifert, R., *J. Biochem. Biophys. Methods* 1990, *21*, 47–58.

[28] Doherty, N. S., Littman, B. H., Reilly, K., Swindell, A. C. et al., *Electrophoresis* 1998, *19*, 355–363.

[29] Otto, A., Thiede, B., Müller, E. C., Scheler, C. et al., *Electrophoresis* 1996, *17*, 1643–1650.

[30] Harris, W. A., Janecki, D. J., Reilly, J. P., *Rapid Commun. Mass Spectrom.* 2002, *16*, 1714–1722.

[31] Karty, J. A., Ireland, M. M., Brun, Y. V., Reilly, J. P., *J. Chromatogr. B* 2002, *782*, 363–383.

[32] Parker, K. C., Garrels, J. I., Hines, W., Butler, E. M. et al., *Electrophoresis* 1998, *19*, 1920–1932.

[33] Vestling, M. M., Murphy, C. M., Fenselau, C., *Anal. Chem.* 1990, *62*, 2391–2394.

[34] Clauser, K. R., Baker, P., Burlingame, A. L., *Anal. Chem.* 1999, *71*, 2871–2882.

[35] Eriksson, J., Chait, B. T., Fenyo, D., *Anal. Chem.* 2000, *72*, 999–1005.

[36] Belov, M. E., Anderson, G. A., Wingerd, M. A., Udseth, H. R. et al., *J. Am. Soc. Mass Spectrom.* 2004, *15*, 212–232.

[37] Loboda, A. V., Ackloo, S., Chernushevich, I. V., *Rapid Commun. Mass Spectrom.* 2003, *17*, 2508–2516.

[38] Smith, R. D., *Trends Biotechnol.* 2002, *20*, S3–S7.

[39] Braunstein, M., Belisle, J. T., in: Hatfull, G. F., Jacobs, W. R. Jr. (Eds.), *Molecular Genetics of Mycobacteria*, ASM Press, Washington, DC 2000, pp. 203–220.

[40] Rosenkrands, I., King, A., Weldingh, K., Moniatte, M. et al., *Electrophoresis* 2000, *21*, 3740–3756.

[41] Saleh, M. T., Fillon, M., Brennan, P. J., Belisle, J. T., *Gene* 2001, *269*, 195–204.

[42] Gomez, M., Johnson, S., Gennaro, M. L., *Infect. Immun.* 2000, *68*, 2323–2327.

[43] Jungblut, P. R., Bumann, D., Haas, G., Zimny-Arndt, U. et al., *Mol. Microbiol.* 2000, *36*, 710–725.

[44] Parker, K. C., *J. Am. Soc. Mass Spectrom.* 2002, *13*, 22–39.

7
Continued proteomic analysis of *Mycobacterium leprae* subcellular fractions*

Maria Angela M. Marques, Benjamin J. Espinosa, Erika K. Xavier da Silveira,, Maria Cristina V. Pessolani, Alex Chapeaurouge, Jonas Perales,, Karen M. Dobos, John T. Belisle, John S. Spencer and Patrick J. Brennan

Recently the sequence of the *Mycobacterium leprae* chromosome, the only known obligate intracellular mycobacterium, was completed. It has a dramatic reduction in functional genes, with a coding capacity of only 49.5%, the lowest one so far observed among bacterial genomes. The leprosy bacillus seems to preserve a minimal set of genes that allows its survival in the host. The identification of genes that are actually expressed by the bacterium is of high significance in the context of mycobacterial pathogenesis. In this current study, a proteomic approach was undertaken to identify the proteins present in the soluble/cytosol and membrane subcellular fractions obtained from armadillo derived *M. leprae*. Proteins from each fraction were separated by two-dimensional gel electrophoresis (2-DE) and identified by mass spectrometry. A total of 147 protein spots were identified from 2-DE patterns and shown to comprise products of 44 different genes, twenty eight of them corresponding to new proteins. Additionally, two highly basic proteins (with $pI > 10.0$) were isolated by heparin affinity chromatography and identified by N-terminal sequencing. This study constitutes the first application of proteomics to a host-derived *Mycobacterium*.

7.1
Introduction

Mycobacterium leprae, the causative agent of leprosy, is a slow growing, obligate, intracellular pathogen. The nine-band armadillo is the only source of appreciable quantities of the bacteria for purposes of scientific studies; the mouse foot pad and nude mouse models provide small amounts of highly viable bacilli. Leprosy is a

---

* Originally published in Proteomics 2004, 10, 2942–2953

Proteomics of Microbial Pathogens. Edited by Peter R. Jungblut and Michael Hecker
Copyright © 2007 WILEY-VCH Verlag GmbH & Co. KGaA, Weinheim
ISBN: 978-3-527-31759-2

chronic infectious disease that mainly affects the skin and peripheral nerves. Over the last few years, the number of registered leprosy patients worldwide has drastically decreased due to the implementation of World Health Organization recommended multidrug therapy. However, leprosy still represents a major public health problem in countries such as India and Brazil [1]. In addition, a continuous stream of new leprosy patients is constantly being recognized in leprosy endemic areas, raising questions as to the source of these new infections.

Without some form of preclinical tests for leprosy, perplexing questions such as the sources of infection, transmission of the disease, the incubation period, and the true extent of leprosy infection, cannot be resolved. Accordingly, there is considerable interest in *M. leprae*-specific proteins and peptides as skin test reagents or for use in *in vitro* blood tests for evidence of infection. To date, there are many other unresolved questions about leprosy, such as the predilection of *M. leprae* for nerve cells, its slow growth, and obligate intracellular habitat. Yet another consideration is that *M. leprae* is the only available *in vivo* derived *Mycobacterium*, and knowledge of its proteome could provide clues to the minimum mycobacterial gene set and minimal array of proteins necessary for intracellular survival but which are incompatible with extracellular survival. Precise analysis of the *M. leprae* proteome is required in order to comprehend these specific attributes of *M. leprae*.

The proteome of a microbe represents the functional complement of the genetic information. Initially, the definition of *M. leprae* proteins had essentially an immunological aspect driven by the search for key antigens as diagnostic reagents or vaccine components. Indeed, many of the major protein antigens were defined using sera from leprosy patients or monoclonal antibody reagents produced against the major proteins in subcellular fractions [2–5]. Young *et al.*, [6] in pioneering work, constructed a *M. leprae* recombinant DNA expression library in λgt11 phage and monoclonal antibodies were used to identify genes encoding proteins with M_rs of 65 kDa, 36 kDa, 28 kDa, 18 kDa, and 12 kDa [6]. Subsequent studies have demonstrated that two of these proteins belong to the heat shock family; 65 kDa, with homology *Escherichia coli* GroEL [7], and 18 kDa with homology to low M_r heat shock protein [8]. In addition, a 28 kDa protein was characterized as a superoxide dismutase enzyme [9]. Recombinant DNA technology became an important tool in the identification of many other *M. leprae* proteins (Tab. 1). Studies complementary to the initial genetic approach, using classical biochemical methodology, resulted in the identification of the major native proteins of *M. leprae* purified from armadillo tissues [10, 11]. This work confirmed the marked expression of the heat shock proteins previously described by Young *et al.* [6], in addition to providing a chemical definition of GroES, the most abundant protein of *M. leprae*. Also, these studies culminated with the characterization of the major membrane proteins (MMP) of *M. leprae*, namely, MMP-I (35 kDa) and MMP-II (22 kDa), the latter identified as bacterioferritin [12]. Minor proteins have since been identified by *N*-terminal sequencing of protein spots obtained by 2-DE [13].

Our early attempts to map proteins from different subcellular compartments of *M. leprae* by 2-DE resulted in the resolution of about 400 protein spots, but only a few of these proteins were identified by *N*-terminal sequencing and Western blot

Tab. 1 Proteins of *M. leprae* identified prior to proteomics techniques

Functional category	Sanger ID[a]	Previous protein denomination	Predicted M_r/pI	References
Virulence, detoxification, adaptation	ML0072	SodA, superoxide dismutase	23158/5.75	[6, 9]
	ML0317	GroEL2, 60 kDa chaperonin 2	56893/4.90	[2, 4, 7, 38]
	ML0380	GroES, 10 kDa Chaperonin	10800/4.67	[11, 39]
	ML0381	GroEL1, 60 kDa chaperonin 1	55817/4.75	[40]
	ML1683	Hlp, ML-LBP21 Probable histone-like protein	20970/12.38	[15, 41]
	ML1795	18 kDa antigen, heat shock protein	16707/4.88	[6, 8, 10]
	ML2042	AhpC, alkyl hydroperoxide reductase	21497/4.59	[42]
	ML2496	DnaK, 70 kDa heat shock protein (molecular chaperone)	66679/4.62	[43, 44]
Lipid metabolism	ML0097	Ag85A, mycolyltransferase	35411/5.14	[45, 46]
	ML0098	Ag85C, mycolyltransferase	31191/6.39	[46]
	ML2028	Ag85B, mycolyltransferase	34807/5.1	[45, 47]
	ML2402	Putative enoyl-CoA hydratase/isomerase	27517/5.49	[15]
Information pathways	ML0501	Antigen T5, aspartyl-tRNA synthetase	64632/4.99	[48]
	ML1842	30S ribosomal protein S5	22614/11.21	[15, 16]
	ML1877	EF-Tu, elongation factor EF-Tu	43668/5.28	[14, 49]
	ML1895	50S ribosomal protein L7/L12	13593/4.35	[42]
	ML1958	30S ribosomal protein S4	23568/10.78	[15, 16]
Cell wall and cell processes	ML0049	ESAT-6, possible secreted protein	10465/5.04	[18]
	ML0050	CFP-10, possible secreted protein	10964/4.11	[19]
	ML0158	34 kDa antigen	31374/5.54	[20]
	ML0234	LSR2 protein (15 kDa antigen)	12165/11.13	[50]
	ML0603	LpK, putative lipoprotein	39051/7.85	[51]
	ML0841	MMP-I, major membrane protein I	33652/4.89	[10, 52]
	ML0922	Ag84, immunogenic protein	28873/4.39	[53]
	ML2395	PRA, proline rich antigen	26296/7.67	[54, 55]
	ML2454	Possible hemagglutinin	20421/9.94	(to be published)

Tab. 1 Continued

Functional category	Sanger ID[a]	Previous protein denomination	Predicted M_r/pI	References
PE/PPE	ML0411	43L, serine-rich antigen	42467/4.2	[5, 56]
Intermediary metabolism and respiration	ML1479	ATP-dependent Clp protease proteolytic subunit	23622/4.99	[15]
	ML1684	3-isopropylmalate dehydratase small subunit	22137/5.63	[15]
	ML2038	Bfr, bacterioferritin	18263/4.37	[10–12]
	ML2059	Short-chain dehydrogenase/reductase family	23392/6.23	[15]
	ML2069	Malate synthase	80142/4.97	[15, 57]
	ML2198	PIII, CysA, thiosulfate sulfurtransferase	31095/5.00	[21]
	ML2703	Bifunctional thioredoxin reductase/thioredoxin	14428/7.62	[58, 59]
Regulatory proteins	ML0773	MtrA, putative two-component response regulator	25287/5.46	[14]
Conserved hypotheticals	ML1041	Conserved hypothetical protein	21186/4.73	[15]
	ML1289	Conserved hypothetical protein	18554/5.43	[14]

a) The Sanger ID refers to (http://www.sanger.ac.uk/Projects/M_leprae/, [17]) ting [14]. Recent attempts to define the nature of the interaction between Schwann cells and *M. leprae* led to the identification of a class of cationic proteins, *i.e.* the histone-like protein, the ribosomal proteins S4 and S5 [15, 16], and heparin-binding hemagglutinin adhesin protein (to be published). As summarized in Tab. 1, most of these proteins were identified prior to the initiation of the genome sequencing project on the Tamil Nadu (TN) strain of *M. leprae*. Recently the sequence of the *M. leprae* chromosome was completed, providing the information that the 3 268 203 bp genome contained only 1604 protein encoding genes representing only 49.5% coding capacity [17]. Most of the active genes are common to *M. tuberculosis* and other bacteria; however, 136 genes are considered unique to *M. leprae*. Recently, by using the information arising from definition of the *M. leprae* genome, our laboratory has produced the recombinant forms of several proteins and produced specific polyclonal or monoclonal antibodies [18–20]. In this way, we were able to detect and immunologically characterize the native ESAT-6 (ML0049) and CFP-10 (ML0050) proteins of *M. leprae*. In order to increase the possibility of detecting low copy number proteins, we have examined individually the three subcellular fractions of *M. leprae*, *i.e.*, the cytosolic/soluble, membrane, and cell wall fractions. In this paper, we describe the analysis of *M. leprae* membrane and soluble/cytosolic proteins by 2-DE and the identification of both new prominent and minor proteins by ESI-MS/MS, MALDI-MS PMF, and *N*-terminal sequencing.

7.2
Materials and methods

7.2.1
Preparation and separation of *M. leprae* proteins

M. leprae purified from armadillo spleens and livers was fractionated as described previously [10]. Briefly, *M. leprae* resuspended in 10 mM NH_4HCO_3 containing 1 mM PMSF was disrupted by intermittent probe sonication (MSE Soniprep 150, MSE-Sonyo; Integrated Services, Palisades Park, NJ, USA) for 30 cycles (60 s bursts/60 s of cooling). The whole sonicate was digested with 10 µg/mL of DNase and RNase for 1 h at 4°C. Cell walls were obtained by centrifugation at 27 000 × g for 30 min and the supernatant from this step was recentrifuged at 100 000 × g for 2 h to provide a sediment enriched in cell membrane. The supernatant, corresponding to the pool of proteins solubilized after extensive sonication of the bacterial cells, was designated the soluble/cytosolic fraction. Initially, different solubilization buffers were tested, such as: 7 M urea, 2 M thiourea, 4% CHAPS, 5 mM tributylphosphine (TBP), 40 mM Tris pH 6.8, 0.5% ampholyte 3–10; and 7 M urea, 2 M thiourea, 1% ASB-14, 5 mM TBP, 0.5% carrier ampholyte pH 4–7. However, a buffer containing 8 M urea, 2% CHAPS, 20 mM DTT and 0.5% ampholyte pH 4–7 was the only one that gave a well resolved protein profile by 2-DE. For production of preparative 2-DE gels, 500 µg of soluble/cytosolic and membrane fraction were dried (Savant Instruments, Marietta, OH, USA), solubilized in solubilization buffer, and held at 4°C overnight. Samples were in-gel rehydrated and run on linear 13 cm, pH 4–7 IPG strips (Amersham Biosciences, Piscataway, NJ, USA). IEF of the proteins on the IPG strips was performed with an IPGphor System (Amersham Biosciences) using the following program: 50 V for 13.5 h; increased to 500 V over 1 h; increased to 1000 V and held for 1 h; increased to 4000 V over 1.5 h and held for 1 h; increased to 8000 V over 2 h and held for 6 h. The IPG strips were subsequently equilibrated in 5 mL of 50 mM Tris-HCl (pH 8.8), 6 M urea, 30% glycerol, 2% SDS, 65 mM DTT and placed on 15% SDS-PAGE gels (19.5 cm × 19.5 cm × 1.5 mm). Electrophoresis in the second dimension was performed at 15 mA *per* gel for approximately 8 h.

7.2.2
MS

Protein spots stained by Coomassie blue were excised from 2-DE gels and subjected to in-gel digestion [21] with modified trypsin (Roche Diagnostics, Indianapolis, IN, USA); the resulting peptides were extracted with 60% ACN in 0.1% TFA. Peptides were applied to a 0.2 × 50 mm C_{18} capillary reverse phase column (Michrom Bio-Resources, Auburn, CA, USA) and eluted with an increasing ACN gradient using a MicroPro capillary HPLC system (Eldex Laboratories, Napa, CA, USA). The reverse phase effluent was introduced directly into a Finnigan LCQ (Thermoquest, San Jose, CA, USA) electrospray mass spectrometer, and the peptides were analyzed by MS or MS/MS. The electrospray needle was operated at 4 kV with a sheath gas flow of N_2 at 40 psi and a capillary temperature of 200°C. MS/MS was automatically performed on the most dominant ion of the previous MS scan and the collision energy was set at 40%. MS/MS data of the peptides were matched to *M. leprae* proteins using the Xcalibur, Bioworks 3.1 turboSEQUEST software (ThermoFinnigan, San Jose, CA, USA). The software was set to consider the oxidation of methionine (+16.0) and the acrylamide modification of cysteine (+71.0). To identify proteins by MALDI-MS PMF, spots were excised from silver-stained 2-DE gels and proteins were digested in-gel with trypsin. The resulting peptides were concentrated and desalted using C_{18} ZipTips (Millipore, Bedford, MA, USA) and eluted onto a MALDI-TOF sample plate with 50% ACN. After partial air drying, 0.5 µL of MALDI matrix solution containing CHCA (Sigma, St. Louis, MO, USA) was added to each sample spot. Mass spectra were recorded in positive reflection mode using an Applied Biosystems (Foster City, CA, USA) MALDI-TOF Voyager-DE PRO mass spectrometer equipped with delayed ion extraction technology. The TOF was measured using the following parameters: 20 kV accelerating voltage, 150 ns delay, low mass gate of 700 Da and an acquisition mass range of 800–4000 Da. Accurate mass values were obtained by internally calibrating the recorded MALDI spectra using trypsin autolysis products as internal standards. The search program MS-FIT (http://prospector.ucsf.edu) was used to compare the generated MALDI-MS peptide mass fingerprint data with theoretical peptide mass data of all mycobacterial proteins stored in the NCBI database. The search parameters were protein M_r ranging from 1000 to 100 000 Da, trypsin digests (two missed cleavage sites), peptide mass tolerance of ±50 ppm, monoisotopic mass, cysteines modified by carbamidomethylation, methionine in oxidized form, and pyroglutamate formation at *N*-terminal glutamine of peptides.

7.2.3
Isolation of basic proteins of M. leprae

In order to enrich for basic proteins, the cytosolic fraction was loaded onto a Hi Trap Heparin-Sephadex column (Amersham Biosciences) pre-equilibrated with 50 mM sodium phosphate, pH 6.8. After washing the column with 100 mL of the same buffer, bound proteins were eluted with a 40 mL linear gradient of 0.05–1 M NaCl in the same buffer, at a flow rate of 0.5 mL/min. Basic protein containing fractions were analyzed by SDS/PAGE and stained with silver. For N-terminal sequencing, after SDS-PAGE, proteins were electrotransferred onto Immobilon-^{SQ} PVDF membranes (Millipore), visualized with Coomassie blue, excised and subjected to automated Edman degradation [22].

7.3
Results and discussion

Due to the limited amount of bacteria available from the infected armadillo model, the definition of the *M. leprae* proteome and consequently of the whole spectrum of proteins expressed *in vivo* by this obligate intracellular pathogen, constitutes a challenging research area. To increase the possibility of detecting low copy number proteins, *M. leprae* was subfractionated into its major subcellular compartments. Partial results are presented here corresponding to the cytosolic/soluble and membrane fractions. We have adopted a classical standardized protocol to prepare the *M. leprae* membrane fraction [14]. Proteins present in this fraction were separated by 2-DE (Fig. 1) and 71 spots were analyzed by ESI-MS/MS. Of these, 66 protein spots were identified corresponding to proteins derived from 29 different genes. Sixteen of them were new proteins in the context of the *M. leprae* proteome, since they had not been reported previously (Tab. 2). The proteins identified in the membrane fraction were analyzed with respect to their grand average of hydropathy scores using ProtParam (http://ca.expasy.org/tools/protparam.html) in which a score of > -0.4 indicates probability for membrane association.

Complementary analysis was performed by detecting probable transmembrane regions, signal sequence and lipoprotein signature using two relevant prediction servers PSORT (http://psort.nibb.jp) and TMPRED (http://www.embnet/sofware/TMPRED_form.html). Proteins with a lipoprotein signature were not identified. Typical membrane proteins related to lipid metabolism (ML2161, ML2162, ML2401, ML2461) and respiration (ML0315, ML1711) were identified (Tab. 2). ML0841 and ML2038 (major membrane proteins I and II, respectively) previously described to be located in the membrane fraction [10] were also detected. A few others were classified as probable membrane proteins (ML0131, ML2332). Additionally, some proteins expected to be only occasionally associated with the cell membrane such as those involved in protein synthesis (ML1845, ML1849, ML1895) or those that had been previously shown to be secreted/located in the cell wall, such as SodA (ML0072), the heat shock proteins 65 (chaperonin 2, ML0317), 18 kDa (ML1795), GroES (ML0380) and DnaK (ML2496), and elongation factor EF-Tu (ML1877) [14, 23], were also observed.

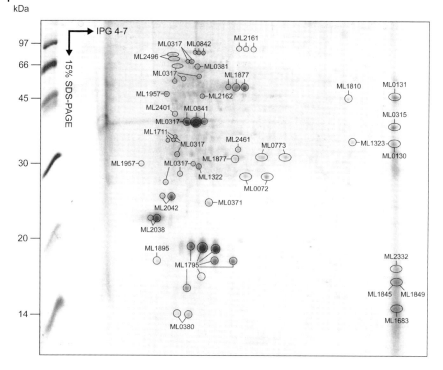

Fig. 1 2-DE analysis of *M. leprae* membrane proteins run on a 13 cm pH 4–7 IPG strip in the first dimension followed by the second dimension on a 15% SDS-PAGE gel. Membrane proteins (500 µg) were incorporated into IPG Drystrips by rehydration, separated by 2-DE and stained by Coomassie blue. Protein spots were excised, in-gel digested with trypsin, and analyzed and identified by MS/MS.

One hundred and seventy two spots were picked for analysis by ESI-MS/MS from 2-DE gels of the soluble/cytosolic fraction, resulting in the identification of 81 of them, which were shown to derive from 27 different genes (Fig. 2). Five additional spots were identified by MALDI-MS PMF as superoxide dismutase (ML0072) (5 peptide matches and 35% sequence coverage), bifunctional thioredoxin reductase/thioredoxin (ML 2703) (6 peptide matches and 15% sequence coverage) and GroEL2 (three fragments of the protein) (Fig. 2). We also performed heparin affinity chromatography on the soluble/cytosolic fraction to isolate highly basic proteins (with p*I* > 10.0), which are known to be difficult to separate by 2-DE. Two proteins were isolated (Fig. 3) and submitted to N-terminal sequencing, allowing their identification as the LSR2 protein (also known as 15 kDa antigen; ML0234), and the integration host factor (IHF; ML0540), a new protein in the context of the *M. leprae* proteome.

The complete list of the proteins identified in the soluble/cytosolic fraction is summarized in Tab. 3. Seventeen of them (shown in bold) were considered new to the *M. leprae* proteome database. Some of the proteins identified in this fraction were typical soluble proteins involved in intermediary metabolism, virulence,

Tab. 2 Proteins identified by ESI-MS/MS in the *M. leprae* membrane fraction

Sanger ID[a]	Gene	Protein	M,theor. (Da)	M, 2-DE (kDa)	pI theor.	PM	Seq. Cov. (%)	No. of spots in 2-DE	PREDICTED TM/SS[e]	GRAVY score[f]
ML0072[b]	sodA	Superoxide dismutase	23127	28	5.75	2–3	9.79–15.26	2	TM=113–130[c]	−0.309
ML0130		Conserved hypothetical protein	30641	33	7.67	5	18.31	1		−0.367
ML0131		Putative oxidoreductase	41226	45	7.7	9	40.7	1	TM=6–40[c]	0.071
ML0315[b]		Putative oxidoreductase	32738	40	7.38	7	40.34	1	TM=217–233[d], 119–152[c], 214–235[d]	0.005
ML0317[b]	groEL2	60 kDa chaperonin 2	56892	25–65	4.72	2–12	8.58–40.96	13	TM=403–424[c], 499–519[c]	−0.093
ML0380[b]	groES	10 kDa Chaperonin	10769	10	4.67	2	29.54–35.99	2		−0367
ML0381[b]	groEL1	60 kDa chaperonin 1	55817	60	4.75	10	31.72	1	TM=105–121[c]	0.101
ML0773	mtrA	Putative two-component response regulator	25287	30	5.46	4–5	31.09–43.24	2		−0.031
ML0841		Major membrane protein 1	33652	35	4.89	11–15	54.3–72.02	3	TM=271–287[d], 197–216[c], 265–288[c]	−0.025
ML0842		Conserved hypothetical protein	44538	78	5.61	7–9	25.86–31.98	3		0.008
ML1322	prcB	Proteasome beta-type subunit 2	30607	29	4.96	8	38.51	1	SS[c], TM= 101–117[d], 96–116[c], 139–163[c]	0.016
ML1323	prcA	Proteasome alpha-type subunit 1	28678	33	7.47	4–5	28.36–33.08	2	23–44[c]	−0.414

Tab. 2 Continued

Sanger ID[a]	Gene	Protein	M_r theor. (Da)	M_r 2-DE (kDa)	pI theor.	PM	Seq. Cov. (%)	No. of spots in 2-DE	PREDICTED TM/SS[e]	GRAVY score[f]
ML1683		Probable histone-like protein	20971	14	12.38	6	32.95	1		−0.330
ML1711[b]	eftA	Electron transfer flavoprotein alpha-subunit	31838	33	4.54	4–8	30.71–52.93	3	TM=77–93[d], 69–93[c]	0.325
ML1795[b]	hsp18	18 kDa Antigen	16708	16–18	4.88	4–8	32.37–55.41	7		0.120
ML1810	moxR	MoxR homolog	40811	45	6.52	10	44.82	1		0.041
ML1845	rpsH	30S ribosomal protein S8	14526	16	11.04	4	38.46	1	TM=92–113[c]	−0.156
ML1849	rplN	50S ribosomal protein L14	12249	16	10.56	4	39.04	1		−0.172
ML1877[b]	tuf	Elongation factor EF-Tu	43636	30–50	5.21	3–16	16.24–57.68	4	TM=94–111[c]	−0.267
ML1895[b]	rplL	50S ribosomal protein L7/L12	13593	18	4.35	4	32.85	1	TM=34–54[c]	0.120
ML1957	rpoA	Alpha-subunit of RNA polymerase	37703	29–46	4.46	4–12	30.71–45.64	2		−0.196
ML2038[b]	bfrA	Bacterioferritin	18264	23	4.37	4–6	40.71–51.83	2		−0.389
ML2042[b]	ahpC	Alkyl hydroperoxide reductase	21498	25	4.59	7	49.2	2		−0.176
ML2161[b]	fadB	Putative fatty acid oxidation complex alpha-subunit	75962	80	5.38	5–6	15.02–16.11	3	TM=105–124[c], 137–157[c], 323–340[c], 659–678[c]	−0.049

Tab. 2 Continued

Sanger ID[a]	Gene	Protein	M_t theor. (Da)	M_r 2-DE (kDa)	pI theor.	PM	Seq. Cov. (%)	No. of spots in 2-DE	PREDICTED TM/SS[e]	GRAVY score[f]
ML2162	fadA	Putative beta-ketoadipyl CoA thiolase	42314	45	4.87	5	27.32	1	TM=384–400[d], 69–88[c], 254–275[c], 355–376[c], 384–401[c]	0.058
ML2332		Conserved hypothetical protein	15497	18	10.42	7	62.2	1	TM=7–23[d], 4–25[c]	−0.068
ML2401[b]		Putative enoyl-CoA hydratase/isomerase	36595	40	4.6	6	26.15	1	TM=105–126[c], 135–155[c], 157–175[c]	0.102
ML2461	fadB2	3-hydroxyacyl-CoA dehydrogenase	30534	31	5.25	10	62.2	1	TM=193–209[d], 9–35[c], 191–211[c]	0.247
ML2496[b]	dnaK	70 kDa heat shock protein	66679	66–67	4.62	6–9	17.1–23.48	2		−0.402

a) The Sanger ID refers to (http://www.sanger.ac.uk/Projects/M_leprae/[17])
b) Protein also found in the soluble/cytosolic fraction
c) TMPRED software (http://www.embnet/software/TMPRED_form.html) used for identification
d) PSORT software (http://psort.nibb.jp) used for identification
e) TM, transmembrane/hydrophobic regions
f) According to Kyte and Doolite [60] and prot param software (http://ca.expasy.org/tools/protparam.html)

Novel identifications are shown in **bold**; M_t, theoretical M_r; M_r 2-DE, observed M_r in the 2-DE gel; PM, number of peptides matched; Seq. Cov. (%), percentage coverage in the protein; GRAVY, grand average of hydropathy; SS signal sequence; scores < −0.4 suggest greater probability for cytosolic origin

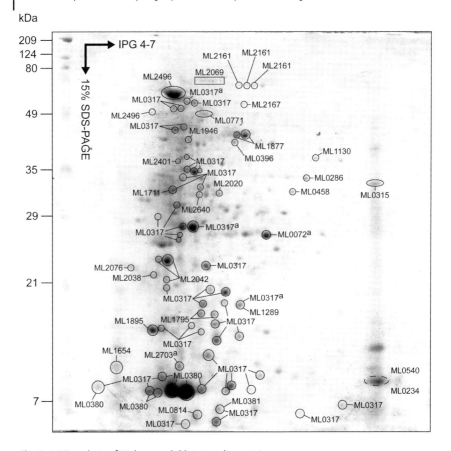

Fig. 2 2-DE analysis of *M. leprae* soluble/cytosolic proteins run on a 13 cm pH 4–7 IPG strip in the first dimension followed by the second dimension on a 15% SDS-PAGE gel. Soluble/cytosolic proteins (500 µg) were incorporated into IPG Drystrips by rehydration, separated by 2-DE and stained by Coomassie blue. Protein spots were excised, in-gel digested with trypsin, and analyzed and identified by MS/MS. Proteins identified by MALDI-TOF MS (a), and by N-terminal sequencing (ML0234, ML0540) are also indicated.

adaptation, detoxification and information pathways. This includes ML0234, ML0286, ML0380, ML0540, and ML2496. However, others clearly originate from the membrane and cell wall compartments of the bacteria. The contamination of this fraction with proteins originally located in the cell envelope was an expected finding since *M. leprae* subcellular fractions were prepared from lysates obtained by extensive sonication of bacterial cells. Although sonication is a well recognized drastic method for cell disruption, this method has been largely adopted by those in the mycobacterial field due to the highly resistant cell wall of these organisms. Actually, of the total proteins identified in the soluble/cytosolic fraction, 14 were also present in the membrane fraction as indicated in Tables 2 and 3.

Tab. 3 List of identified proteins in *M. leprae* soluble/cytosolic fraction

Sanger ID[a]	Gene	Protein	Ident. method	M_r theor. (Da)	M_r 2-DE (Da)	pI theor.	PM	Seq. Cov. (%)	No. of spots in 2-DE	Predicted TM/SS[e]	GRAVY score[f]
ML0072[b]	sodA	Superoxide dismutase	MALDI-MS	23127	28	5.75	5	35	1	TM=113–130[c]	−0.309
ML0234	lsr2	LSR2 protein (15 kDa antigen)	NTS	12164	10	11.13	–	–	–		−0.657
ML0286	fba	Putative fructose biphosphate aldolase	MS/MS	36741	35	5.68	2	9.66	1		−0.145
ML0315[b]		Putative oxidoreductase	MS/MS	32738	32	7.38	3	17.82	1	TM=217–233[d], 119–152[c], 214–235[c]	0.005
ML0317[b]	groEL2	60 kDa chaperonin 2	MS/MS	56892	5–57	4.72	2–18	4.07–40.65	43	TM=403–424[c], 499–519[c]	−0.093
ML0380[b]	groES	10 kDa chaperonin	MS/MS	10769	8–9	4.67	1–7	23–80	5	TM=105–121[c]	−0.367
ML0381[b]	groEL1	60 kDa chaperonin 1	MS/MS	55817	7	4.75	4	10.41	1	TM=160–176[d], 18–34[c], 156–181[c]	0.101
ML0396		Conserved hypothetical protein	MS/MS	40278	42	5.27	5	23.75	1		−0.150
ML0458		Putative oxidoreductase	MS/MS	32875	34	5.9	4	18.39	1	TM=212–232[c]	−0.013
ML0540	mihF	Putative integration host factor	NTS	11474	10	10.83	–	–	–		−0.471
ML0771	sahH	Putative s-adenosyl-homocysteine hydrolase	MS/MS	53987	55	5.05	4	14.73	1	TM=65–83[c]	−0.238
ML0814		Conserved hypothetical	MS/MS	8654	6	4.89	3	58.54	1		0.070

Tab. 3 Continued

Sanger ID[a]	Gene	Protein	Ident. method	M_r theor. (Da)	M_r 2-DE (Da)	pI theor.	PM	Seq. Cov. (%)	No. of spots in 2-DE	Predicted TM/SS[e]	GRAVY score[f]
ML1130	thrC	Threonine synthase	MS/MS	37353	40	6.1	4	23.99	1	TM=100–122[c]	0.158
ML1289		Conserved hypothetical	MS/MS	18555	19	5.43	2	13.64	1		0.013
ML1654	acpM	Acyl carrier protein (meromycolate extension)	MS/MS	12484	12	3.84	3	33.04	1		−0.217
ML1711[b]	eftA	Electron transfer flavoprotein (alpha) subunit	MS/MS	31838	32	4.54	6	35.34	1	TM=77–93[d], 69–93[c]	0.325
ML1795[b]	hsp18	18 kDa antigen	MS/MS	16708	18	4.88	6–7	37.84–44.59	2		0.120
ML1877[b]	tuf	Elongation factor EF-Tu	MS/MS	43636	45	5.21	3–5	11.5–21.21	2	TM=94–111[c]	−0.267
ML1895[b]	rplL	50S ribosomal protein L7/L12	MS/MS	13593	15	4.35	7	40	1	TM=34–54[c]	0.120
ML1946		Conserved hypothetical protein	MS/MS	34593	40	4.77	2	11.08	1	TM=199–225[c]	−0.035
ML2020		Conserved hypothetical protein	MS/MS	33141	33	5.17	3	11.42	1	TM=20–38[c]	−0.139
ML2038[b]	bfrA	Bacterioferritin	MS/MS	18264	22	4.37	3	20.75	1		−0.389
ML2042[b]	ahpC	Alkyl hydroperoxide reductase	MS/MS	21498	21–23	4.59	5–7	35.38–51.28	3		−0.176
ML2069	glcB	Malate synthase	MS/MS	80142	78	5.09	8	14.71	1	TM=364–381[c]	−0.250
ML2076		Conserved hypothetical protein	MS/MS	17323	23	4.32	1	7.41	1	TM=44–60[c, d]	−0.296
ML2161[b]	fadB	Putative fatty acid oxidation complex alpha subunit	MS/MS	75962	75	5.38	4	8.62	2	TM=105–124[c], 137–157[c], 323–340[c], 659–678[c]	−0.049

Tab. 3 Continued

Sanger ID[a]	Gene	Protein	Ident. method	M_r theor. (Da)	M_r 2-DE (Da)	p/ theor.	PM	Seq. Cov. (%)	No. of spots in 2-DE	Predicted TM/SS[e]	GRAVY score[f]
ML2167	pdc	Pyruvate (or indolepyrufate) decarboxylase	MS/MS	61114	61	5.41	2	6.4	1	TM=283–299[d], 73–91[c], 276–298[c], 405–429[c],	0.074
ML2401[b]		Putative enoyl-CoA hydratase/isomerase	MS/MS	36595	38	4.6	3	11.6	1	TM=105–126[c], 135–155[c], 157–175[c]	0.102
ML2496[b]	dnaK	70 kDa heat shock protein (molecular chaperone)	MS/MS	66679	70	4.62	27	59.25	2		−0.402
ML2640		Conserved hypothetical protein	MS/MS	34454	34	4.94	6	29.96	1	SS[d]	−0.281
ML2703	trxA/B	Bifunctional thioredoxin reductase/thioredoxin	MALDI-MS	14428	12	7.62	6	15	1	TM=13–30[c]	−0.060

a) The Sanger ID refers to (http://www.sanger.ac.uk/Projects/M_leprae/[17])
b) Protein also found in the membrane fraction
c) TMPRED software (http://www.embnet/software/TMPRED_form.html) used for identification
d) PSORT software (http://psort.nibb.jp) used for identification
e) TM, transmembrane/hydrophobic regions
f) According to Kyte and Doolite [60] and prot param software (http://ca.expasy.org/tools/protparam.html)

Novel identifications are shown in bold; NTS, N-terminal sequencing; M_n theoretical M_r; M_r 2-DE, observed M_r in the 2-DE gel; PM, number of peptides matched; Seq. Cov. (%), percentage coverage in the protein; ProtParam (hht://usexpasy.org/tols/prot param.html); scores < −0.4 suggest greater probability for cytosolic origin; GRAVY, grand average of hydropathy.

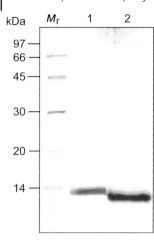

Fig. 3 SDS-PAGE of basic proteins purified by heparin affinity chromatography. Basic proteins eluted with a linear gradient 0.05–1 M NaCl were separated by 15% SDS-PAGE and visualized by silver-staining. Pooled fractions 64 to 72 correspond to IHF (lane 1), and pooled fractions 74 to 80 correspond to LSR2 (lane 2).

Fragments of the GroEL2 protein (ML0317) were identified in 46 spots spread throughout the gel, either alone or in combination with other proteins (Fig. 2, Tab. 3). This was not a surprise finding based on GroEL2 abundance as shown in previous reports [14] and its high instability, as evidenced by Western bloting [24], probably related to its recent characterization as a trypsin-like protease [25]. Besides GroEL2, many other proteins were identified in multiple spots (Fig. 2, Tab. 3). Moreover, the spots identified as ML0381 and ML2703 probably also correspond to protein fragments, suggesting a general status of protein degradation in the fractions under study. This probably can be explained by the already well-known low viability of bacteria isolated from armadillo tissues. Additionally, the standardized protocol for bacteria purification from infected tissues lasts approximately 2 d allowing further degradation to occur [10].

An advantage of proteome definition is that it allows confirmation of the existence of gene products predicted from DNA sequence, which is a major contribution to genomic science [26]. Accordingly, our study has revealed the expression of nine novel conserved hypothetical proteins corresponding to 31% of the new identified proteins (Tab. 4). Based on the information in the *M. leprae* genome database (http://www.sanger.ac.uk/Projects/M_leprae), the remaining novel proteins identified in the present study can be classified into various functional categories as shown in Tab. 4. Most of them are involved in intermediary metabolism. In terms of catabolic pathways, it is important to emphasize the identification of enzymes involved in fatty acid β-oxidation (encoded by the genes ML2401, *fadB2* and *fadA*). This finding reinforces the idea that intracellular mycobacterium derives most of its energy from the degradation of host derived lipids [27]. Additionally, the enzyme pyruvate decarboxylase (encoded by the *pdc* gene) was identified, indicating that *M. leprae* relies on alcohol fermentation as an alternative pathway to regenerate NAD+, given the loss by the bacterium of the genes that encode NADH dehydrogenase [17]. In terms of anabolic pathways, a surprising finding from the genome sequence is that *M. leprae* has the capacity to make most of its own components [17]. In this respect, we identified two enzymes involved in cysteine and threonine synthesis (encoded, respectively, by the genes *thrC* and *sahH*), indicating

Tab. 4 Functional distribution of new *M. leprae* proteins

Functional categories	*M. leprae* Sanger ID	*M. tuberculosis* Sanger ID	Expression by *M. tuberculosis*
Lipid metabolism	ML1654	Rv2244	P/T [29, 30]
	ML2161	Rv0860	P [29]
	ML2162	Rv0859	–
	ML2401	Rv1071c	–
	ML2461	Rv0468	P/T [29, 30]
Information pathways	ML0540	Rv1388	T [31]
	ML1845	Rv0718	T [31]
	ML1849	Rv0714	P/T [29, 30]
	ML1957	Rv3457c	P/T [29, 30]
Intermediary metabolism and respiration	ML0131	Rv2951c	P [29]
	ML0286	Rv0363c	P [32]
	ML0315	Rv0068	P [29]
	ML0458	–	–
	ML0771	Rv3248c	P [29]
	ML1130	Rv1295	P [29]
	ML1322	Rv2110c	P [29]
	ML1323	Rv2109c	P [29]
	ML1711	Rv3028c	P [29]
	ML2167	Rv0853c	–
Regulatory proteins	ML1810	Rv1479	P [29]
Conserved hypothetical proteins	ML0130	Rv2952	T [30]
	ML0396	Rv0046c	P [29]
	ML0814	Rv3208A	P [29]
	ML0842	Rv1464	P [29]
	ML1946	Rv1099c	P/T [29, 30]
	ML2020	Rv1896c	–
	ML2076	Rv1827	P [29]
	ML2332	Rv3718c	P[a]
	ML2640	Rv0146	–

a) http://www.mpiib-berlin.mpg.de/2D-PAGE/EBP-PAGE/index.html
Functional classification and ID refer to (http://www.sanger.ac.uk/Projects/M_leprae/, http://www.sanger.ac.uk/Projects/M_tuberculosis/, [17, 35]); P, gene expression demonstrated by proteomic analysis; T, gene expression demonstrated by transcription analysis.

that these pathways are active in the intracellular environment. Another interesting observation was the identification of the alpha and beta subunits composing the proteasome. The proteasome of *M. tuberculosis* was recently shown to serve as a defense against oxidative or nitrosative stress [28] and its expression in host-derived *M. leprae* reinforces its potential role in intracellular survival. Tab. 4 also shows that all novel identified proteins except ML0458 have hortologues in *M. tuberculosis* and that the majority of them have already been shown to be expressed by the tubercle bacillus in variable culture conditions [29–33].

The identification of 46 different proteins is a considerable achievement in the context of an organism so difficult to derive. Nevertheless, the considerable number of protein fragments identified here indicates that the source of *M. leprae*, its storage and the protocols used for its fractionation should be revised in future proteomic studies. Additionally, our results indicate that further purification steps, such as chromatographic processes or 2-DE with narrow-range IPG will be required in order to increase the chances to identify greater numbers of *M. leprae* proteins. Many chromatographic processes and electrophoretic prefractionation techniques in proteome analysis have been recently discussed by Righetii *et al.* [34], and some of them are now being applied to *M. leprae*.

7.4
Concluding remarks

The *M. tuberculosis* H37Rv genome is composed of 3924 individual genes [35], and approximately 400 proteins have been identified and mapped by 2-DE of cellular proteins and culture filtrates, corresponding to 10% of the predicted gene products (http://www.mpiib-berlin.mpg.de/2D-PAGE/EBP-PAGE/index.html, [28, 36, 37]). *M. leprae* TN consists of 1604 individual genes. Up to the onset of this work, *i.e.* the application of modern genomics, approximately 37 proteins had been identified. This work allowed the identification of 46 individual proteins (some previously defined) corresponding to 2.87% of all predicted gene products; 29 of the 46 are new proteins. Of interest, the expression of nine new conserved hypothetical proteins was demonstrated. Although we are still far from the identification of the predicted 1604 proteins of *M. leprae*, this work represents an appreciable contribution to *M. leprae* proteome definition and represents the first application of proteomics to a host-derived *Mycobacterium*.

We gratefully acknowledge Preston Hill, Ana G. C. Neves-Ferreira and Richard H. Valente for technical support. This work was supported by NIH, NIAID Contract NO1 AI 25469 and NIH, NIAID grant AI 47197 and PDTIS/FIOCRUZ. E. K. Xavier da Silveira was a recipient of a postdoctoral fellowship from CNPq/Brazil.

7.5
References

[1] World Health Organization, *Wkly. Epidemiol. Rec.* **2002**, *77*, 1–8.

[2] Gillis, T. P., Buchanan, T. M., *Infect. Immun.* 1982, *37*, 172–178.

[3] Britton, W. J., Hellqvist, L., Basten, A., Raison, R. L., *J. Immunol.* 1985, *135*, 4171–4177.

[4] Britton, W. J., Hellqvist, L., Garsia, R. J., Basten, A., *Clin. Exp. Immunol.* 1987, *67*, 31–42.

[5] Sathish, M., Esser, R. E., Thole, J. E., Clark-Curtiss, J. E., *Infect. Immun.* 1990, *58*, 1327–1336.

[6] Young, R. A., Mehra, V., Sweetser, D., Buchanan, T. *et al.*, *Nature* 1985, *316*, 450–452.

[7] Young, D., Lathigra, R., Hendrix, R., Sweetser, D., Young, R. A., *Proc. Natl. Acad. Sci. USA* 1988, *85*, 4267–4270.

[8] Booth, R. J., Harris, D. P., Love, J. M., Watson, J. D., *J. Immunol.* 1988, *140*, 597–601.

[9] Thangaraj, H. S., Lamb, F. I., Davis, E. O., Jenner, P. J. et al., *Infect. Immun.* 1990, *58*, 1937–1942.

[10] Hunter, S. W., Rivoire, B., Mehra, V., Bloom, B. R., Brennan, P. J., *J. Biol. Chem.* 1990, *265*, 14065–14068.

[11] Rivoire, B., Pessolani, M. C. V., Bozic, C. M., Hunter, S. W. et al., *Infect. Immun.* 1994, *62*, 2417–2425.

[12] Pessolani, M. C., Smith, D. R., Rivoire, B., McCormick, J. et al., *J. Exp. Med.* 1994, *180*, 319–327.

[13] Pessolani, M. C. V., Brennan, P. J., *Infect. Immun.* 1996, *64*, 5425–5427.

[14] Marques, M. A. M., Chitale, S., Brennan, P. J., Pessolani, M. C. V., *Infect. Immun.* 1998, *66*, 2625–2631.

[15] Marques, M. A. M., Mahapatra, S., Nandan, D., Dick, T. et al., *Microbes Infect.* 2000, *2*, 1407–1417.

[16] Marques, M. A., Mahapatra, S., Sarno, E. N., Santos, S., et al., *Braz. J. Med. Biol. Res.* 2001, *34*, 463–470.

[17] Cole, S. T., Eiglmeier, K. R., Parkhill, J., James, K. D. et al., *Nature* 2001, *4093*, 1007–1011.

[18] Spencer, J. S., Marques, M. A. M., Lima, M. C. B. S., Junqueira-Kipnis, A. P. et al., *Infect. Immun.* 2002, *70*, 1010–1013.

[19] Spencer, J. S., Kim, H., Marques, M. A. M., Gonzalez-Juarerro, M. et al., *Infect. Immun.* 2004, *72*, 3161–3170.

[20] Silbag, F. S., Cho, S.-N., Cole, S. T., Brennan, P. J., *Infect. Immun.* 1998, *66*, 5576–5579.

[21] Hellman, U., Wernstedt, C., Gonez, J., Heldin, C. H., *Anal. Biochem.* 1995, *224*, 451–455.

[22] Hawke, D. H., Harris, D. C., Shively, J. E., *Anal. Biochem.* 1985, *147*, 315–330.

[23] Sonnenberg, M. G., Belisle, J. T., *Infect. Immun.* 1997, *65*, 4515–4524.

[24] Gillis, T. P., Miller, R. A., Young, D. B., Khanolkar, S. R., Buchanan, T. M., *Infect. Immun.* 1985, *49*, 371–377.

[25] Portaro, F. C. V., Hayashi, M. A. F., Arauz, L. J., Palma, M. S. et al., *Biochemistry* 2002, *41*, 7400–7406.

[26] Humphery-Smith, I., Cordwell, S. J., Blackstock, W. P., *Electrophoresis* 1997, *18*, 1217–1242.

[27] Wheeler, P. R., Ratledge, C., in: Bloom, B. R. (Ed.), *Tuberculosis: Pathogenesis, Protection and Control*, American Society for Microbiology Press, Washington, DC, USA 1994, pp. 353–385.

[28] Darwin, K. H., Ehrt, S., Gutierrez-Ramos, J. C., Weich, N., Nathan, C. F., *Science* 2003, *302*, 1900–1902.

[29] Rosenkrands, I., King, A., Weldingh K., Moniatte, M. et al., *Electrophoresis* 2000, *21*, 3740–3756.

[30] Fisher, M. A., Plikaytis, B. B., Shinnick, T. M., *J. Bacteriol.* 2002, *184*, 4025–4032.

[31] Betts, J. C., Lukey, P. T., Robb, L. C., McAdam, R. A., Duncan, K., *Mol. Microbiol.* 2002, *43*, 717–731.

[32] Rosenkrands, I., Slayden, R. A., Crawford, J., Aagaard, C. et al. *J. Bacteriol.* 2002, *184*, 3485–3491.

[33] Jungblut, P. R., Muller, E. C., Mattow, J., Kaufmann, S. H., *Infect. Immun.* 2001, *69*, 5905–5907.

[34] Righetti, P. G., Castagna, A., Herbert, B., Reymond, F., Rossier, J. S., *Proteomics.* 2003, *3*, 1397–1407.

[35] Cole, S. T., Brosch, R., Parkhill, J., Garnier, T. et al., *Nature* 1998, *393*, 537–544.

[36] Jungblut, P. R., Schaible, U. E., Mollenkopf, H. J., Zimny-Arndt, U. et al., *Mol. Microbiol.* 1999, *33*, 1103–1117.

[37] Mattow, J., Schaible, U. E., Schmidt, F., Hagens, K. et al., *Electrophoresis* 2003, *24*, 3405–3420.

[38] Hunter, S. W., McNeil, M., Modlin, R. L., Mehra, V. et al., *J. Immunol.* 1989, *142*, 2864–2872.

[39] Mande, S. C., Mehra, V., Bloom, B. R., Hol, W. G., *Science* 1996, *271*, 203–207.

[40] Rinke de Wit, T. F., Bekelie, S., Osland, A., Miko, T. L. et al., *Mol. Microbiol.* 1992, *6*, 1995–2007.

[41] Shimoji, Y., Ng, V., Matsumura, K., Fischetti, V. A., Rambukkana, A., *Proc. Natl. Acad. Sci. USA* 1999, *96*, 9857–9862.

[42] Pessolani, M. C. V., Brennan, P. J., *Infect. Immun.* 1996, *64*, 5425–5427.

[43] Britton, W. J., Hellqvist, L., Baster, A., Raison, R. L., *J. Immunol.* 1985, *135*, 4171–4177.

[44] Mckenzie, K. R., Adams, E., Britton, W. J., Garsia, R. J., Basten, A., *J. Immunol.* 1991, *147*, 312–319.

[45] Pessolani, M. C. V., Brennan, P. J., *Infect. Immun.* 1992, *60*, 4452–4459.

[46] Rinke de Wit, T. F., Bekelie, S., Osland, A., Wieles, B. et al., *Infect. Immun.* 1993, *61*, 3642–3647.

[47] Thole, J. E., Schoningh, R., Janson, A. A., Garbe, T. et al., *Mol. Microbiol.* 1992, *6*, 153–163.

[48] Wieles, B., Spierings, E., van Noort, J., Naafs, B. et al., *Infect. Immun.* 1995, *63*, 4682–4685.

[49] Silbaq, F. S., Cho, S.-N., Cole, S. T., Brennan, P. J., *Infect. Immun.* 1998, *66*, 5576–5579.

[50] Sela, S., Thole, J. E., Ottenhoff, T. H., Clark-Curtiss, J. E., *Infect. Immun.* 1991, *59*, 4117–4124.

[51] Maeda, Y., Makino, M., Crick, D. C., Mahapatra, S. et al., *Infect. Immun.* 2002, *70*, 4106–4111.

[52] Winter, N., Triccas, J. A., Rivoire, B., Pessolani, M. C. et al., *Mol. Microbiol.* 1995, *16*, 865–876.

[53] Hermans, P. W., Abebe, F., Kuteyi, V. I., Kolk, A. H. et al., *Infect. Immun.* 1995, *63*, 954–960.

[54] Klatser, P. R., De Wit, M. Y., Kolk, A. H., *Clin. Exp. Immunol.* 1985, *62*, 468–473.

[55] Rinke de Wit, M. Y., Klatser, P. R., *J. Gen. Microbiol.* 1988, *134*, 1541–1548.

[56] Wieles, B., van Agterveld, M., Janson, A., Clark-Curtiss, J. et al., *J. Infect. Immun.* 1994, *62*, 252–258.

[57] Wheeler, P. R., Ratledge, C., *J. Gen. Microbiol.* 1988, *134*, 2111–2121.

[58] Wieles, B., van Soolingen, D., Holmgren, A., Offringa, R. et al., *Mol. Microbiol.* 1995, *16*, 921–929.

[59] Wieles, B., van Noort, J., Drijfhout, J. W., Offringa, R. et al., *J. Biol. Chem.* 1995, *270*, 24604–24606.

[60] Kyte, J., Doolite, R. F., *J. Mol. Biol.* 1982, *157*, 105–132.

8
CFP10 discriminates between nonacetylated and acetylated ESAT-6 of *Mycobacterium tuberculosis* by differential interaction*

Limei Meng Okkels, Eva-Christina Müller, Monika Schmid, Ida Rosenkrands, Stefan H. E. Kaufmann, Peter Andersen and Peter R. Jungblut

ESAT-6 (the 6 kDa early secreted antigenic target) protein species in short-term culture filtrate of *Mycobacterium tuberculosis* were separated in a 4–5 narrow range p*I* gradient two-dimensional gel electrophoresis (2-DE). Eight ESAT-6 protein species were analyzed in detail by peptide mass fingerprinting matrix-assisted laser desorption/ionization-mass spectrometry as well as by electrospray ionization-tandem mass spectrometry. An *N*-terminal Thr acetylation was identified in four species and a *C*-terminal truncation was identified in two species. In 2-DE blot overlay assays, the recombinant 10 kDa culture filtrate protein (CFP10) discriminated *N*-terminal acetylated and nonacetylated ESAT-6 by differential interaction, whereas removal of the *C*-terminal 11 residues of ESAT-6 had no effects thereon. This example shows that the access to the protein species level can be a prerequisite to understand regulation of protein-protein interaction.

8.1
Introduction

Since its identification in 1995, much effort has been made to elucidate the immunological and the biological properties of the early secreted antigenic target (ESAT)-6 of *Mycobacterium tuberculosis*. Initial studies demonstrated that this protein is a dominant T-cell determinant in *M. tuberculosis* infection of human and animals [1, 2]. Subsequent experimental vaccination studies using recombinant ESAT-6 or a single epitope from this antigen have shown that ESAT-6 induces partial protection against aerosol *M. tuberculosis* challenge in mice [3, 4]. In addition, recombinant *M. bovis* bacillus Calmette-Guerin (BCG) secreting ESAT-6 significantly enhanced protection against tuberculosis compared to wild-type BCG [5].

---

* Originally published in Proteomics 2004, 10, 2954–2960

Proteomics of Microbial Pathogens. Edited by Peter R. Jungblut and Michael Hecker
Copyright © 2007 WILEY-VCH Verlag GmbH & Co. KGaA, Weinheim
ISBN: 978-3-527-31759-2

The development of recombinant DNA technologies in *M. tuberculosis* complex has promoted functional studies of ESAT-6, and gene knock-out/knock-in experiments have revealed that ESAT-6 is essential for virulence of *M. tuberculosis* [6–8], and may be an effector protein of cytolysis [8].

Information about biochemical characteristics of ESAT-6 is still limited. Being a secreted protein, no signal peptide has been identified on ESAT-6, and it is still a puzzle how the ESAT-6 protein is translocated into the surrounding milieu. Bioinformatic analysis has suggested that the genes flanking *esat6* encode a novel secretion pathway of *M. tuberculosis* [9–11], and the indispensability of this functional unit in the secretion of ESAT-6 has recently been demonstrated by several research groups [5, 12, 13]. Yet, the molecular mechanisms underlying this type of transport still remain to be elucidated.

Another major question is whether ESAT-6 is modified following translation. In culture supernatants of *M. tuberculosis* H37Rv, so far five spots on 2-DE gels prepared by carrier ampholyte have been identified as ESAT-6 by mass spectrometry [14]. In short-term culture filtrate (ST-CF) of *M. tuberculosis* the ESAT-6 protein focused at two pIs within the range of 4–4.5. At each pI, ESAT-6 appeared as triplet spots [15]. Attempts by standard glycoprotein detection method to elucidate whether ESAT-6 is glycosylated has not provided any evidence for the glycosylation of ESAT-6 [15], and no evidence of other modifications, such as phosphorylation, acetylation, oxidation or deamidation have been reported. Nevertheless, based on bioinformatic analysis some analogy between the ESAT-6 secretion pathway and type III/IV secretion apparatus of Gram-negative bacteria has been revealed, and post-translational modification of ESAT-6 has been proposed [11].

In the present study using improved conditions for 2-DE, we achieved an excellent separation of different species of ESAT-6 protein present in ST-CF of *M. tuberculosis*, by using a narrow range pH gradient. With MALDI-MS and ESI-MS/MS we attained 100% sequence coverage for several ESAT-6 spots, which enabled us to identify N-terminally acetylated as well as proteolytically truncated species of ESAT-6. In addition, applying 2-DE blot overlay assays we demonstrate differential binding of 10 kDa culture filtrate protein (CFP10) to the acetylated and nonacetylated ESAT-6 species.

8.2
Materials and methods

8.2.1
Protein samples

M. tuberculosis H37Rv (ATCC 27294) was cultured in modified Sauton medium for seven days and short-term culture filtrate (ST-CF) was prepared from culture supernatant by ultrafiltration followed by ammonium sulphate precipitation as described previously [16]. Full length ESAT-6 and CFP10 were synthesized as N-terminal His-tagged fusion proteins in *Escherichia coli* [17, 18]. Recombinant CFP10 (rCFP10) was purified by metal chelating chromatography on Talon® resin (Clontech, Palo Alto,

CA, USA) followed by ion-exchange chromatography through a Hitrap® Q column (Amersham Biosciences, Uppsala, Sweden). Conditions of metal chelating chromatography have been described elsewhere [18]. The Hitrap® Q column was in 25 mM HEPES (pH 6.5), 10% glycerol, 2 M urea, and 0.01% Tween 20. A 0–0.5 M linear gradient of NaCl was used for elution of bound proteins. Fractions were analysed by silver stained SDS-PAGE, and by immunoblots using anti-*E. coli* antibodies [19]. Only fractions with no detectable *E. coli* contaminants were pooled. For purification of rESAT-6, inclusion bodies were isolated [19]; the protein was purified first through a Talon® column, then through a Hitrap® Q column, as described above. Thereafter, the protein was purified further through a Hitrap® SP column in 25 mM HEPES, pH 6.5, 10% glycerol, 3 M urea, and 0.05% Tween 20. Protein concentrations were determined by the bicinchoninic acid test (Micro BCA Protein Assay Reagent kit; Pierce, Oud-Beijerlan, the Netherlands). The purified rCFP10 and rESAT-6 were stored at $-20°C$ in PBS (pH 7.4), 10% glycerol and 0.01% Tween 20, and in 25 mM Tris-HCl, pH 8.4, 10% glycerol and 0.05% Tween 20, respectively.

8.2.2
2-DE blot overlay

ST-CF proteins were separated by 2-DE under similar conditions as described previously [16], except that 18 cm narrow pH range (4–5) immobilised gradient strips were used for the first dimension of 2-DE, and after the completion of isoelectric focusing, 13 cm of the strips' acidic end were applied to 15% SDS-PAGE gels for separation in the second dimension. The proteins were subsequently blotted onto nitrocellulose membranes as described elsewhere [20]. After blocking the membranes in PBS (pH 7.4) containing 0.1% Tween 20 (PBST) and 3% skimmed milk for 1 h at room temperature, or overnight at 4°C, 10 µg/mL of rCFP10 or rESAT-6 in PBST containing 1% skimmed milk and 0.37 M NaCl (binding buffer) were overlaid onto the membranes and incubated overnight at 4°C. Protein-protein interaction complexes were detected by anti-His antibody (Qiagen, Hilden, Germany) using horseradish peroxidase-conjugated rabbit antimouse antibody (Dako, Glostrup, Denmark) in combination with the ECL visualisation system (Amersham Biosciences).

8.2.3
Mass spectrometry

A data set of eight different 2-DE protein spots from pH 4–5 gradient gels of *M. tuberculosis* H37Rv ST-CF was analysed by MALDI-MS peptide mass fingerprinting starting with in-gel digestion by trypsin (porcine sequencing grade modified; Promega, Madison, WI, USA). Using the volatile buffer ammonium carbonate the peptides were directly measured after trypsination without a concentration procedure [21]. Dehydroxy benzoic acid (DHB) was used as a matrix and mass spectra were recorded in the reflection mode of a TOF MALDI mass spectrometer (Voyager Elite; PerSeptive Biosystems, Framingham, MS, USA) with delayed extraction. Spectra were obtained by the summation of 256 laser shots. Internal calibration of the spectra resulted in a mass accuracy of about 30 ppm. Peak detection was evaluated manually and we aspired for complete sequence coverage. Searches with MASCOT (Matrix Science, http://www.matrixscience.com) in the NCBInr or Swiss-Prot database were performed with a mass tolerance of 100 ppm.

Sequence support was obtained by nano-ESI-MS/MS [22] performed as described [23] using the Q-TOF mass spectrometer from Micromass (Manchester, UK). The MS/MS spectra were deconvoluted from m/z values to mass using the software program MAXENT3 (part of MassLynx 4.0; Micromass) and saved as .dta files to identify the protein using MASCOT [24]. Modifications such as N-terminal acetylation, methionine oxidation and unspecific digestion were considered in a second search, if the first one was not successful.

8.3
Results

8.3.1
High resolution separation of acidic ST-CF proteins

Previous proteome analysis used 100–600 µg of ST-CF proteins for 2-DE and MALDI-MS PMF [14, 16]. In the first dimension the proteins were separated either in an immobilised pH gradient 4–7 or in a carrier ampholyte gradient 2–11, and the second dimensional separations were performed in 15% SDS-PAGE gels or 10–20% gradient gels. Under these conditions ESAT-6 focused mainly as two clusters of spots with similar molecular mass, but different pI values. The GroES and CFP10 spots in the low mass acidic region were poorly separated (Fig. 1A). To improve the resolution of these proteins, in the present study we used an 18 cm pH 4–5 IPGphor Strip (Amersham Biosciences) for the first dimension, and the 13 cm acidic end of the strip was subsequently subjected to SDS-PAGE on a 16 × 16 × 10.1 cm gel containing 15% acrylamide. When 1 mg of ST-CF was analysed under these conditions, good resolution was obtained for proteins with molecular mass of 45 kDa or less, and 226 spots were visualised by CBB G-250, with 138 spots below 20 kDa. The resolution of the clustered GroES and CFP10 spots was improved (Fig. 1B). However, some spots still contain a mixture of CFP10 and GroES.

Using ESAT-6 specific monoclonal antibody Hyb76-8, we detected eight ESAT-6 spots by immunoblotting (data not shown). These spots appeared as two triplets and two spots with lower molecular mass as indicated in Fig. 1B. The observed differences in mass and pI suggest multiple PTMs and/or proteolytic cleavages of ESAT-6.

8.3.2
Mass analysis of ESAT-6 spots

Three 2-DE gels of ST-CF were produced as described in Section 3.1, and the eight Hyb76-8 reactive spots from each gel were applied to MALDI-MS PMF and ESI-MS/MS analysis after in-gel digestion by trypsin. Sequence coverage of 100% was

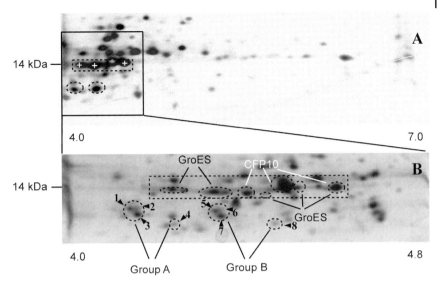

Fig. 1 Comparison of 2-DE separation of low mass acidic ST-CF proteins by different pH intervals. IPG strips were used for the first dimenssional separation, and the second dimension was run in 15% SDS-PAGE gels. Only the region below 21 kDa is shown. A, Silver stained gel containing 85 µg of ST-CF; pH 4–7. The ESAT-6 spots are circled, the GroES spots are indicated by a white + and CFP10 by a white dot; B, CBB G-250 stained gel containing 1 mg ST-CF; pH 4–4.8. The ESAT-6 spots are circled and numbered. The GroES and CFP10 spots are indicated.

obtained for six spots (Tab. 1). Within the triplet spots nos. 5-6-7, 100% sequence coverage was obtained for spots 6 and 7. For spot 5, we identified peptides derived from residues 1–73, but the mass peaks at 2072.93, 2088.92, and 2104.92, representing peptide Thr74-Ala94 were missing. Spots from three separated gels were analysed independently, with several measurements at each time point, but this peptide was not detected in any of the measurements. Attempts to detect this peptide by using CHCA as the matrix also gave negative results. We therefore suggest that the C-terminal peptide of spot 5 is modified.

For triplet 1-2-3, each of the spots revealed a MALDI-MS PMF similar to that of their counterparts in triplet 5-6-7, except that the peaks at 3385.66 nd 3399.69 Da, representing peptide Thr1-Lys32, were missing in all three spots (Tab. 1). Instead, peaks at 3427.67, 3441.66 Da were detected (Fig. 2), and a peptide of this mass was absent in all the spectra of triplet 5-6-7. This suggests that each spot in triplet 1-2-3 differs from their counterpart in triplet 5-6-7 by the same modification on peptide Thr1-Lys32. A mass increase of 42 Da indicates acetylation. ESI-MS/MS analysis of this peptide revealed that the modification site was on N-terminal Thr (Fig. 3) evidenced by the b1 ion which represents acetylated threonine. The b3, b4 and b5 ions further confirm the mass shift of 42. The observed decrease in p*I* of triplet 1-2-3 is consistent with the blocking of the free amino group on Thr.

Tab. 1 ESAT-6 MS analysis: Peptides identified by MALDI-MS and ESI-MS/MS

Amino acid residues	Sequence	Theor. Mass $(M+H)^+$	Chem Score[a]	Spot 1	2	3	4	5	6	7	8
1–32	TEQQWNFAGIEAAASAIQGNVTSIHSLLDEGK[b]	3385.67	10	–	–	–	–	×	×	×	×
	acTEQQWNFAGIEAAASAIQGNVTSIHSLLDEGK[b]	3427.68	5	×	×	×	×	–	–	–	–
33–37	QSLTK	576.34	10	×	×	×	×	×	×	×	×
38–56	LAAAWGGSGSEAYQGVQQK[b]	1907.93	10	×	×	×	×	×	×	×	×
57–73	WDATATELNNALQNLAR[b]	1900.95	100	×	×	×	×	×	×	×	×
74–94	TISEAGQAMASTEGNVTGMFA[b]	2072.93	0.5	–	×	×	–	–	×	×	–
	TISEAGQAMASTEGNVTGMFA + 1Met-ox	2088.92	0.5	–	×	×	–	–	×	×	–
	TISEAGQAMASTEGNVTGMFA + 2Met-ox	2104.92	0.5	–	×	×	–	–	×	×	–
74–83	TISEAGQAMA	978.46	0.5	–	–	–	×	–	–	–	×
	TISEAGQAMA + 1Met-ox[b]	994.45	0.5	–	–	–	×	–	–	–	×
Sequence coverage (%)				78	100	100	100	78	100	100	100

– Not detected
× Detected at least in three independent experiments.
a) Chem Score is calcculated by MS-Fit program
b) The sequence has been confirmed by ESI-MS/MS.

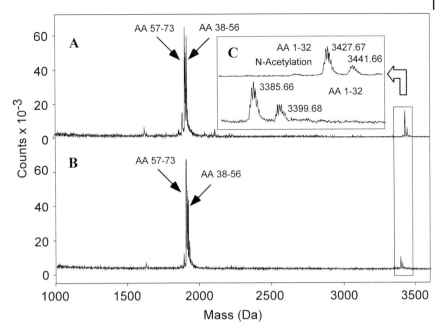

Fig. 2 Peptide AA 1–32 in MALDI-MS in spot 3 and 7. A, Peptide mass fingerprint of spot 3; B, peptide mass fingerprint of spot 7; C, enlarged region of the discriminating region with the acetylated peptide 1–32 in spot 3 and the nonacetylated counterpart in spot 7.

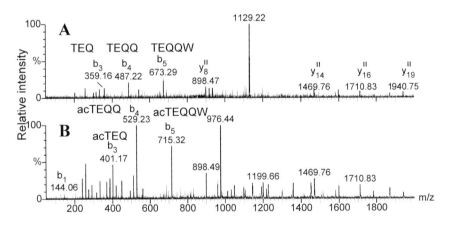

Fig. 3 ESI-MS/MS of the nonacetylated and the acetylated N-terminal peptide. MS/MS spectrum of peptide 1–32 from A, spot 7 and B, spot 3. The differential masses of the b-series are labelled and corresponding sequences are shown.

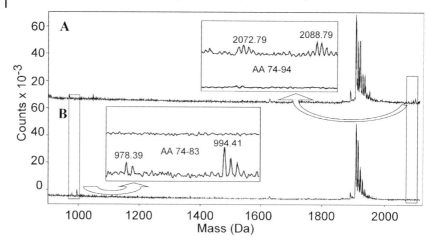

Fig. 4 Mass difference in C-terminal peptide between ESAT-6 spots 7 (A) and 8 (B) determined by MALDI-MS. The enlarged regions of the C-terminal peptides are shown in the insets.

For spots 4 and 8, we obtained 100% sequence coverage of residues 1–73, and spot 4 was N-terminally acetylated. The faster mobility in SDS-PAGE suggests that these two spots have shorter C-terminal peptides. By thorough manual comparison of all the MALDI-MS peptide spectra, we identified a clear peak at 994.41 Da, and a weaker peak at 978.39 Da for both spots 4 and 8. These peaks were absent in all the spectra of the two triplets (Fig. 4). Previously, Renshaw et al. [25] have reported that if proteinase inhibitors were omitted from the refolding buffers, the majority of recombinant ESAT-6 produced in E. coli was truncated at Ala84 [25]. Given that spots 4 and 8 represent this truncation, then the C-terminal peptide resulting from tryptic digestion should be Thr74–Ala83, with a calculated mass of 978.46 Da. The observed peptide masses of 978.39 and 994.41 fit well to this peptide (TISEAGQAMA), without and with the methionine being oxidised, respectively. In addition, the C-terminal truncation would result in the loss of 1 Glu residue, theoretically increasing the pI value by 0.11 (from pI 4.48 to pI 4.59), in agreement with the observed more basic positions of spot 4 in relation to triplet 1-2-3, and of spot 8 in relation to triplet 5-6-7. This evidence strongly suggests that spots 4 and 8 resulted from proteolytic cleavage between Ala83 and Ser84 of N-terminally acetylated and nonacetylated ESAT-6, respectively. ESI-MS/MS analysis was performed for the 994.38 Da peptide, and the sequence identity was confirmed to correspond to peptide Thr74–Ala83 carrying an oxidised Met (results not shown). We have thus also achieved 100% sequence coverage in spots 4 and 8.

In summary, the eight ESAT-6 spots can be divided into two groups (Fig. 1B), each group consisting of four spots, including two full length ESAT-6 (spots 2, 3; and 6, 7). One C-terminally truncated ESAT-6 (spots 4; 8), and a species with unknown modification on peptide Thr74-Ala94 (spots 1, 5). The spots in group A differ from the spots in group B by an N-terminal acetylation The two full-length ESAT-6 spots in each group were indistinguishable from each other by MS analysis.

8.3.3
Interaction of recombinant CFP10 with ESAT-6 spots

Direct interaction between recombinant ESAT-6 and CFP10 has been demonstrated by different techniques [12, 20, 25], however, whether PTM of ESAT-6 has any influence on its interaction with CFP10 was unknown. To address this issue we developed a 2-DE blot overlay assay, in which ST-CF was separated by 2-DE and the separated proteins were transferred onto a nitrocellulose membrane, which was subsequently overlaid with N-terminally His-tagged recombinant CFP10 (rCFP10) or rESAT-6. Binding to endogenous proteins was monitored by anti-His antibody and visualised by ECL-Western detection reagent. When rESAT-6 was overlaid on the membrane, binding to a series of protein spots was obtained (Fig. 5A), and all these spots reacted with the anti-CFP10 antibody in immunoblots (results not shown). When rCFP10 was subsequently overlaid on the membrane, all of the nonacetylated ESAT-6 species, including the C-terminally truncated spot 8, bound to rCFP10, whereas none of the acetylated species interacted (Fig. 5B). A control membrane was prepared in parallel. After electrotransfer, the control membrane was directly incubated with anti-His antibody and no cross-reaction with ESAT-6 and CFP10 spots was observed (data not shown). These results demonstrate that while N-terminally His tagging of ESAT-6 has no effect on its binding to CFP10, N-terminal acetylation clearly has. The biochemical mechanism behind this phenomenon remains to be explored.

Other members of the ESAT-6 protein family have also been identified on the 2-DE map of ST-CF [16]. Under the present experimental conditionss neither rESAT-6 nor rCFP10 bound to any of these proteins, which is in agreement with our previous results that ESAT-6 proteins interact in a pair-wise specific way [20].

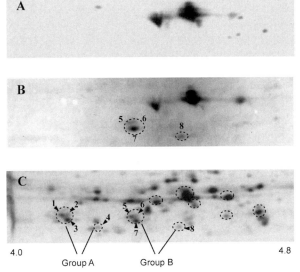

Fig. 5 2-DE blot overlays. A, Low M_r range of a narrow range p*I* 4–4.8 2-DE gel blot of St-CF overlaid with rESAT-6. B, The same membrane was subsequently overlaid with rCFP10. The circled spots correspond to spots 5, 6, 7 and 8. C, The same region of a CBB-stained gel. The anti-CFP10 spots that reacted with rESAT-6 are circled. The ESAT-6 spots are circled and numbered.

8.4 Discussion

Previous studies have suggested that ESAT-6 of M. tuberculosis undergoes PTMs, because secreted endogenous ESAT-6 consistently appears in multiple spots in 2-DE. In the present study, applying MALDI-MS and ESI-MS/MS, we characterised eight ESAT-6 spots in detail (Tab. 1). Sequence coverage of 100% was obtained for six spots, which enabled us to identify four species of full length ESAT-6, of which two were acetylated at the N-terminal Thr and the other two were not. We also identified two species of C-terminally truncated ESAT-6, which differed from each other by N-terminal acetylation. For the remaining two ESAT-6 spots, we have not been able to identify a C-terminal peptide.

N-terminal acetylation is a broad phenomenon in eukaryotes, about 80% mammalian proteins are estimated to bear this modificcation [26]. The biological significance of this modification varies with the particular protein, for example N-terminal acetylation is required for the binding of tropomyosin to actin [27]; and unacetylated rat glycine M-methyltransferase does not exhibit the cooperative behaviour of the acetylated enzyme [28]. On the other hand incorrect N-terminal acetylation of various 20S proteasome subunits causes the loss of specific peptidase activities [29]. Only five proteins of bacterial origin have been reported to carry an acetyl group on the N-terminal residue (reviewed in [26]), and the functional significance of N-terminal acetylation in prokaryotes is not well documented. Wondering whether N-terminal acetylation of ESAT-6 would have any effffect on its interaction with CFP10, we performed 2-DE blot overlay assays. By this technique we were able to investigate the interaction between secreted endogenous ESAT-6 species and recombinant CFP10. Under the experimental conditions N-terminal acetylation abolished ESAT-6 binding activity to CFP10.

Previously, it has been reported that the dissociation constant of ESAT-6:CFP10 complex is less than 10^{-8} M, and that the genes encoding the two proteins are cotranscribed. Consequently, the nascent ESAT-6 and CFP10 proteins would be in close contact with each other. It is tempting to speculate that the formation of ESAT-6: CFP10 complex happened immediately after translation, and N-terminal acetylation of ESAT-6 occurred at a later time point. N-terminal acetylation of ribosome protein L12 of E. coli has also been suggested to occur post-translationally after partial or complete ribosome assembly [26]. If the acetylated ESAT-6 also exhibits a much lower affinity to CFP10 in vivo, then one may speculate that the acetylation event might promote the dissociation of ESAT-6 from CFP10. Whether the acetylated ESAT-6 is able to bind to other M. tuberculosis proteins or host proteins, and thereby differentiates functionally from nonacetylated CFP10 associated ESAT-6 is a challenging question that deserves attention in future studies.

The C-terminal 11 residues of ESAT-6 are not required for interaction with CFP10. Whether the cleavage between Ala83 and Ser84 is a result of specific processing, or is due to nonspecific proteolysis during sample preparation should be investigated systematically in a future study. In this regard, it might be worth mentioning that extracellular subtilisin-like serine protease Mycosin-1 (Rv3883c) is a component of the *esx* gene cluster [30, 31] which encodes the secretion pathway of ESAT-6. The substrates of this protease have not yet been identified.

High diversification into numerous distinct protein species has been shown for human HSP 27 [32], crystallins [33], for mycobacterial HSPX [34] as well as for GroES and CFP10 (Fig. 1B). The structural reasons for this diversity were only partially assigned to PTMs. The analysis of protein composition to verify predictions from transcriptomics and genomics was the major aim for many proteomic studies in recent years. The challenge for the future is to elucidate the chemical structure of the various protein species produced after contact of the nascent protein with its environment. One prerequisite for this goal is to gain 100% sequence coverage for protein species identification. As demonstrated by the current study, a combination of high-resolution 2-DE, complete mass analysis, and 2-DE blot overlays may be key elements in future studies aimed at understanding protein function by proteomics.

We thank Kathryn Wattam and Annette Hansen for excellent technical assistance. This study is part of EU X-TB contract number QLK2-CR-2001-02018.

8.5
References

[1] Andersen, P., Andersen, A. B., Sorensen, A. L., Nagai, S., *J. Immunol.* 1995, *154*, 3359–3372.

[2] Ravn, P., Demissie, A., Eguale, T., Wondwosson, H. *et al.*, *J. Infect. Dis.* 1999, *179*, 637–645.

[3] Brandt, L., Elhay, M., Rosenkrands, I., Lindblad, E. B., Andersen, P., *Infect. Immun.* 2000, *68*, 791–795.

[4] Olsen, A. W., Hansen, P. R., Holm, A., Andersen, P., *Eur. J. Immunol.* 2000, *30*, 1724–1732.

[5] Pym, A. S., Brodin, P., Majlessi, L., Brosch, R. *et al.*, *Nat. Med.* 2000, *9*, 533–539.

[6] Pym, A. S., Brodin, P., Brosch, R., Huerre, M., Cole, S. T., *Mol. Microbiol.* 2002, *46*, 709–717.

[7] Lewis, K. N., Liao, R., Guinn, K. M., Hickey, M. J. *et al.*, *J. Infect. Dis.* 2003, *187*, 117–123.

[8] Hsu, T., Hingley-Wilson, S. M., Chen, B., Chen, M. *et al.*, *Proc. Natl. Acad. Sci. USA* 2003, *100*, 12420–12425.

[9] Tekaia, F., Gordon, S. V., Garnier, T., Brosch, R. *et al.*, *Tuber. Lung Dis.* 1999, *79*, 329–342.

[10] Gey Van Pittius, N. C., Gamieldien, J., Hide, W., Brown, G. D. *et al.*, *Genome Biol.* 2001, 2. Research0044.1–research0044.18, http://genomebiology.com/2001/2/10/research/0044.

[11] Pallen, M. J., *Trends Microbiol.* 2002, *10*, 209–212.

[12] Stanley, S. A., Raghavan, S., Hwang, W. W., Cox, J. S., *Proc. Natl. Acad. Sci. USA* 2003, *100*, 13001–13006.

[13] Guinn, K. M., Hickey, M. J., Mathur, S. K., Zakel, K. L. *et al.*, *Mol. Microbiol.* 2994, *51*, 359–370.

[14] Mattow, J., Schaible, U. E., Schmidt, F., Hagens, K. *et al.*, *Electrophoresis* 2003, *24*, 3405–3420.

[15] Sorensen, A. L., Nagai, S., Houen, G., Andersen, P., Anderson, A. B., *Infect. Immun.* 1995, *63*, 1710–1717.

[16] Rosenkrands, I., King, A., Weldingh, K., Moniatte, M. *et al.*, *Electrophoresis* 2000, *21*, 3740–3756.

[17] Harboe, M., Wiker, H. G., Ulvund, G., Malin, A. S. *et al.*, *Infect. Immun.* 1998, *6*, 717–723.

[18] Skjøt, R. L., Oettinger, T., Rosenkrands, I., Ravn, P. et al., *Infect. Immun.* 2000, *68*, 214–220.

[19] Okkels, L. M., Brock, I., Follmann, F., Agger, E. M. et al., *Infect. Immun.* 2003, *71*, 6116–6123.

[20] Okkels, L. M., Andersen, P., *J. Bacteriol.* 2004, *186*, 2487–2491.

[21] Lamer, S., Jungblut, P. R., *J. Chromatogr. B Biomed. Sci. Appl.* 2001, *752*, 311–322.

[22] Mann, M., Hojrup, P., Roepstorff, P., *Biol. Mass Spectrom.* 1993, *22*, 338–345.

[23] Müller, E. C., Schürmann, M., Rickers, A. Bommert, K. et al., *Electrophoresis* 1999, *20*, 320–330.

[24] Perkins, D. N., Pappin, D. J., Creasy, D. M., Cottrell, J. S., *Electrophoresis* 1999, *20*, 3551–3567.

[25] Renshaw, P. S., Panagiotidou, P., Whelan, A., Gordon, S. V. et al., *J. Biol. Chem.* 2002, *27*, 21598–21603.

[26] Polevoda, B., Sherman, F., *J. Mol. Biol.* 2003, *325*, 595–622.

[27] Urbancikova, M., Hitchcock-DeGregori, S. E., *J. Biol. Chem* 1994, *269*, 24310–24315.

[28] Ogawa, H., Gomi, T., Takata, Y., Dae, T., Fujoka, M., *Biochem. J.* 1997, *327*, 407–412.

[29] Arent, C. S., Hochstrasser, M., *EMBO J.* 1999, *18*, 3575–3585.

[30] Brown, G. D., Dave, J. A., Gey van Pittius, N. C., Stevens, L. et al., *Gene* 2000, *254*, 147–155.

[31] Dave, J. A., Gey van Pittius, N. C., Beyers, A. D., Ehlers, M. R., Brown, G. D., *BMC Microbiol.* 2002, *2*, 30, http://www.biomedcentral.com/1471-2180/2/30.

[32] Scheler, C., Müller, E. C., Stahl, J., Müller-Werdan, U. et al., *Electrophoresis* 1997, *18*, 2823–2831.

[33] Jungblut, P. R., Otto, A., Favor, J., Lowe, M. et al., *FEBS Lett.* 1998, *435*, 131–137.

[34] Mattow, J., Jungblut, P. R., Müller, E. C., Kaufmann, S. H., *Proteomics* 2001, *1*, 494–507.

9
The cell wall subproteome of *Listeria monocytogenes**

Jessica Schaumburg*, Oliver Diekmann*, Petra Hagendorff, Simone Bergmann, Manfred Rohde, Sven Hammerschmidt, Lothar Jänsch, Jürgen Wehland, and Uwe Kärst

The surface subproteome of *Listeria monocytogenes* that includes many proteins already known to be involved in virulence and interaction with host cells has been characterized. A new method for the isolation of a defined surface proteome of low complexity has been established based on serial extraction of proteins by different salts at high concentration, and in all 55 proteins were identified by N-terminal sequencing and mass spectrometry. About 16% of these proteins are of unknown function and three proteins have no orthologue in the nonpathogenic *L. innocua* and might be involved in virulence mechanisms. Remarkably, a relatively high number of proteins with a function in the cytoplasmic compartment was identified in this surface proteome. These proteins had neither predicted or detectable signal peptides nor could any modification be observed except removal of the N-terminal methionine. Enolase (Lmo2455) is one of these proteins. It was shown to be present in the cell wall of the pathogen by immunoelectron microscopy and, along with heat shock factor DnaK (Lmo1473), elongation factor TU (Lmo2653), and glyceraldehyde-3-phosphate dehydrogenase (Lmo2459), it was found to be able to bind human plasminogen in overlay blots and surface plasmon resonance (SPR) experiments. The K_D values of these interactions were determined by SPR measurements. The data indicate a possible role of these proteins as receptors for human plasminogen on the bacterial cell surface. The potential role of this recruitment of a host protease for extracellular invasion mechanisms is discussed.

9.1
Introduction

Listeria monocytogenes is a Gram-positive rod-shaped bacterium that occurs ubiquitously in the environment. As a facultative intracellular pathogen it can cause lis-

---

* Originally published in Proteomics 2004, 10, 2991–3006.

Proteomics of Microbial Pathogens. Edited by Peter R. Jungblut and Michael Hecker
Copyright © 2007 WILEY-VCH Verlag GmbH & Co. KGaA, Weinheim
ISBN: 978-3-527-31759-2

teriosis, a severe food-borne infection in humans and animals leading to syndromes such as meningitis, encephalitis, and sepsis in immunocompromised individuals, or spontaneous abortion in pregnant women. *Listeria* are able to cross the intestinal, blood-brain and placental barrier and gain access to the cytosolic compartment and spread directly from cell to cell thus evading the humoral immune system. * These authors contributed equally.These capabilities are mainly mediated by a number of surface-associated virulence factors that are bound to the thick negatively charged peptidoglycan cell wall. These include the adhesion factors internalin A and B, the pore-forming toxin listeriolysin, the phospholipases A and B, and the actin nucleation factor ActA [1]. Consequently, the direct exposure of bacterial proteins to their environment is supposed to be a key feature of essential virulence factors [2].

The putative surface proteome of *L. monocytogenes* was defined so far only as part of the genome sequencing project [3]. The corresponding surface proteins were organized in groups depending on the retention mechanism to the bacterial surface. These predictions were based on the homology to retention motifs of known surface proteins exhibiting, *e.g.*, GW modules, LPXTG motifs, hydrophobic tails, lipo-boxes, or P60-like domains.

However, a post-genomic validation of proteins that are actually expressed at the bacterial surface is overdue for *L. monocytogenes*. The proteome of Gram-positive bacteria can be fractionated into the four subproteomes of (i) secreted extracellular proteins, (ii) cell wall-associated proteins (this study), (iii) integral and membrane-associated proteins, and (iv) cytoplasmic proteins. Therefore, a functional proteome strategy was developed that comprises methods for both the improved prediction as well as the comprehensive characterization of cell surface-associated proteins resulting in the identification of additional extracellular enzyme functions: serial extraction of surface proteins from intact bacteria was combined with a cytoplasmic enzyme assay in order to generate defined surface-proteomes not contaminated by intracellular proteins. Analyses of qualified fractions resulted in the identification of 55 different proteins. About 50% of all identified proteins exhibited no detectable secretion and retention signal, and many of them are known to be active within the cytoplasm, suggesting that these proteins may have an additional "moonlighting" function [4]. Binding studies with overlay blots and surface plasmon resonance (SPR) analyses demonstrated that four proteins are specific binders for human plasminogen.

9.2
Material and methods

9.2.1
Bacterial strain and growth conditions

Listeria monocytogenes EGDe strain was cultivated in ultrafiltrated (10 kDa cutoff) brain-heart-infusion broth at 37°C and 180 rpm.

9.2.2
Serial extraction of cell wall proteins

400 mL overnight cultures of *L. monocytogenes* were harvested by centrifugation at 4000 × g for 12 min and washed with PBS. Pelleted cells were resuspended in 800 µL 1 M Tris, pH 7.5, and incubated for 30 min at 37°C under gentle shaking. Cells were centrifuged at 4000 × g for 12 min. The supernatant containing solubilized cell wall-associated proteins was removed and stored at −20°C. The pellet was washed with PBS and the incubation of the cells was repeated stepwise with other substances in the following series: 1 M KSCN, 1% CHAPS, 1% CTAB, 1% octylglucoside, and 0.5% SDS. Cells were washed with PBS between each extraction step.

9.2.3
Aminopeptidase C assay

Ten µL freshly prepared cell wall extracts were added to 190 µL 20 mM Tris-HCl, pH 7.4, in microtiter-plates. After mixing with 2 µL 200 mM L-arginine-*p*-nitroanilide (Sigma, St. Louis, MO, USA) in 20 mM Tris, pH 7.4, samples were immediately assayed at 405 nm for 10 min (Spectra-Max250; Molecular Devices, Ismaning, Germany). For a positive control 5 units of aminopeptidase from *Aeromonas proteolytica* (Sigma) in 20 mM Tris, pH 7.4, were used instead of cell wall fractions.

9.2.4
SDS-PAGE Western blotting, and *N*-terminal sequencing

A discontinuous SDS-PAGE was performed according to [5], with 10% gels in the Protean™ II system (Bio-Rad, Hercules, CA, USA) at 70 V overnight. Apparent molecular weights were determined with AIDA 3.51 (Raytest, Straubenhardt, Germany). Blotting was performed with PVDF membranes in a semidry blot chamber as described [6] with 1 mA/cm^2 for 1 h. The membrane was stained with Coomassie Brilliant Blue G-250 and bands were carefully cut out. For *N*-terminal sequencing a gas-phase sequenator (model 494; Applied Biosystems, Palo Alto, CA, USA) with online phenylthiohydantoin amino acid analyzer was used.

9.2.5
2-D-PAGE

Samples containing 500 µg protein were solubilized in rehydration buffer (8 M urea, 2 M thiourea, 1.65 mM Tris, 2% CHAPS, 65 mM DTT, 0.5% IPG buffer, bromophenol blue) to a final volume of 380 µL. Isoelectric focusing was performed at 20°C with the IPGphor system (Amersham Biosciences, Uppsala, Sweden). Samples were applied to the IPGphor strip holders, 18 cm Immobiline 3–10 or 4–7 DryStrips added and covered with silicon oil. The following program was used: 0 V for 4% h, 30 V for 10 h, 30–100 V for 1 h, 100–1000 V for 1 h, 1000–8000 V for 3 h, and 8000 V for 6 h. Afterwards the strips were incubated in equilibration buffer

(6 M urea, 50 mM Tris-HCl, pH 8.8, 30% glycerol, 2% SDS and a trace of bromophenol blue) for 10 min with 2% DTT added and additionally 10 min with 2.5% iodoacetamide added. The second dimension was performed in the Hoefer IsoDalt (Amersham Biosciences) system. The DryStrips were placed on 10–15% gradient polyacrylamide gels. Electrophoresis was carried out at 60 V overnight. Image analysis was carried out using ProteomWeaver 2.1 (Definiens, Munich, Germany).

9.2.6
Gel staining and protein identification by mass spectrometry

2-D gels were stained with RuBPS [7] overnight after fixation in 40% ethanol, 10% acetic acid for 1 h and washing with 20% ethanol twice for 15 min. Detection of the fluorescent staining was performed with a CCD camera (FujiFilm LAS-1000 with ethidium bromide filter, Raytest) and the software Image Reader (FujiFilm). Colloidal Coomassie staining of 1-D and 2-D SDS-PAGE gels was performed according to [8].

Proteins from 1-D and 2-D SDS-PAGE were identified by MALDI mass fingerprinting (Ultraflex; Bruker Daltonics, Bremen, Germany; Mascot, Matrix Science, London, UK) after tryptic in-gel digestion and purification of the resulting peptides. Spots were cut out, reduced, carbamidomethylated and digested with trypsin (sequencing grade; Promega, Madison, WI, USA) with only minor modifications according to published protocols [9, 10]. The digest was dried in vacuum. 30 µL 5% methanol and 0.5% formic acid were added. The peptides were concentrated and desalted using µC18-ZipTip (Millipore, Bedford, MA, USA) and eluted in 1 µL 65% methanol, 0.5% formic acid saturated with 4-hydoxy-α-cyanocinnamic acid as matrix directly onto a MALDI target. Mass fingerprint analyses were carried out with an in-house MASCOT server. Only protein identification results were accepted that satisfied the two conditions of at least 20% sequence coverage and reached twice the default significance criteria from MASCOT.

9.2.7
Immunoelectron microscopy

9.2.7.1 Postembedding labeling studies
Bacteria were fixed with 0.2% glutaraldehyde and 0.5% formaldehyde in cacodylate buffer (0.1 M cacodylate, 0.09 M sucrose, 0.01 M $MgCl_2$, 0.01 M $CaCl_2$, pH 6.9) for 1 h on ice, washed three times with cacodylate buffer containing 10 mM glycine, and processed following the PLT method (progressive lowering of temperature) as follows: dehydration with 10% ethanol on ice, 30% ethanol at $-20°C$, 50, 70, 90, 100% ethanol at $-30°C$, each step for 30 min. Infiltration with the Lowicryl resin K4M and polymerization with UV-light (366 nm) was also performed at $-30°C$. Ultrathin sections were cut with glass knives and mounted onto formvar-coated nickel grids and incubated with polyclonal protein A-purified anti-enolase immunoglobulin G (IgG) antibodies (IgG concentration, 175 µg/mL) for 10 h at 4°C. After several washing steps with PBS sections were incubated with protein A-gold complexes (10 nm in diameter) for 30 min at room temperature for visualization of

the bound antibodies. After several washing steps with PBS containing 0.01% Tween 20 samples were washed with distilled water, air-dried, and counter-stained with 4% aqueous uranyl acetate for 2 min, subsequently washed in distilled water and air-dried. Samples were examined in a Zeiss transmission electron microscope EM910 at an acceleration voltage of 80 kV.

9.2.7.2 Field emission scanning electron microscopic immunolabeling

Bacteria were fixed with 0.5% formaldehyde in PBS for 1 h on ice, washed with PBS containing 10 mM glycine and incubated with polyclonal anti-enolase antibodies (150 µg IgG protein/mL) for 1 h at 30°C. After washing several times with PBS, the bound antibodies were visualized by incubation with protein A-gold complexes (15 nm in diameter) for 30 min at 30°C. After fixation with 2% glutaraldehyde for 10 min at room temperature samples were attached to poly-L-lysine-coated cover slips (12 mm in diameter) and dehydrated with a graded series of acetone (10, 30, 50, 70, 90, 100%) each step for 15 min on ice. After critical-point drying with liquid CO_2 samples were coated with a thin carbon film. Samples were then examined in a Zeiss field emission scanning electron microscope DSM982 Gemini at an acceleration voltage of 5 kV using the Everhardt-Thornley SE-detector and the in-lens detector in a 50:50 ratio.

9.2.8
Overlay blot

Cells were harvested by gentle centrifugation at $1500 \times g$ and the pellets washed twice with PBS. Proteins noncovalently attached to the cell surface were extracted from the cells by treatment with 0.5 M Tris-HCl, pH 7.5. The supernatant was filtered through a 0.22 µM membrane to remove remaining cells. Protein extracts were precipitated by addition of 80% ice-cold acetone and stored at $-20°C$ overnight. The precipitate was solubilized in Laemmli-buffer, subjected to SDS-PAGE with 12% polyacrylamide (PAA) as described [5], and either stained with Coomassie Brilliant Blue (CBB) or subsequently transferred to a nylon membrane (Immobilon-P; Millipore) using a semidry blotting system. The membranes were blocked by incubation in 10% fat-free milk in 10 mM PBS before performing subsequent binding experiments. Incubation with human plasminogen (Sigma) was performed at a concentration of 2 µg/mL. Binding was detected by incubation of the membrane with peroxidase-labeled anti-plasminogen antibodies and subsequent addition of a substrate solution containing 1 mg/mL 4-chlor-1-naphthol and 0.1% H_2O_2 in PBS. Immunoblot analyses with anti-enolase and anti-gapDH antiserum were performed using a peroxidase-conjugated second antibody and substrate solution.

9.2.9
Ligand fishing

Biacore assays were performed to isolate proteins binding to the kringle 1–3 domain (fragment Y^{80}-P^{353}) of human plasminogen from *Listeria* cell lysate. The kringle domain (70 µL, 50 µg/mL in 10 mM acetate, pH 5.5) was coupled covalently at a flow rate of 10 µL/min onto N-hydroxysuccinimide (NHS; 0.05 M)/N-ethyl-N'-(diethylaminopropyl) carbodiimide (EDC, 0.2 M)-activated CM5 sensor chips (Biacore, Freiburg, Germany) essentially as described by the manufacturer. Whole-cell extracts were generated by breaking the cells at 15 000 psi in a French press (SLM-Aminco), and 1 mM Pefabloc-SC (Roth Chemicals, Karlsruhe, Germany) was added to prevent proteolysis. Cell disruption was repeated three times. The suspension was then centrifuged at $10\,000 \times g$ (10 min, 4°C), dialyzed against HBS-EP (10 mM HEPES, 150 mM NaCl, 3 mM EDTA, 0.05% Tween 20, pH 7.4) and centrifuged at $10\,000 \times g$ (10 min, 4°C) again. The clear supernatant was applied to a Biacore optical biosensor (Biacore 3000 system with microrecovery option). Ten µL of the protein solution were injected at a flow rate of 1 µL/min. Binding of proteins to the sensor chip surface was followed by SPR detection. After application of the protein sample an extra clean procedure and a bypass wash of the integrated µ-fluidic cartridge with running buffer HBS-EP was performed to reduce carry over. Microrecovery of the bound proteins was achieved in 4 µL 10 mM NaOH. Incubation time for the recovery solution was 1 min. Recovery was complete after this procedure. The experiment was repeated several times to collect enough material for 1-D gel electrophoresis followed by staining with colloidal Coomassie blue, tryptic digest, and identification by MALDI mass fingerprinting.

9.2.10
Cloning and purification of proteins

After identification of potential plasminogen binders these proteins were overexpressed in *E. coli* and purified to apparent homogeneity. The purified proteins were used in direct binding studies with the kringle domain for determination of the corresponding K_D values. For GAPDH and DnaK, after PCR-amplification the coding DNA sequences were cloned into expression vector pQE30. Expression in M15 cells was induced by addition of 1 mM isopropyl-β-D-thiogalactopyranoside (IPTG). After 4 h of incubation at 37°C the cells were lysed and amino terminally His-tagged proteins were purified using Ni-NTA agarose. Enolase was expressed in a similar way: induction with 1 mM IPTG was performed at 30°C for 5 h and cells broken up by French Press treatment. Purification of the protein was achieved by using a Q-Sepharose followed by a MonoQ column, both running in a gradient of 0–1 M NaCl, 20 mM HEPES, pH 8.0, and a gel filtration on a Superdex75 Column (1.6×60 cm; 20 mM HEPES, 100 mM NaCl, pH 8.0). Expression of elongation factor TU was achieved after cloning of the coding DNA into pGEX-6P1 and transformation into BL21 cells. Induction with 0.1 mM IPTG was followed by shaking for 4 h at 37°C. The GST-tagged protein was purified using glutathione sepharose.

The GST-tag was removed by cleavage with precision protease. All inserts were sequenced in their expression vectors. The purified proteins were identified by MALDI mass fingerprinting.

9.2.11
Kinetic analyses using SPR detection

The association and dissociation reactions of the kringle domain to proteins derived from *Listeria monocytogenes* were analyzed on a Biacore 2000 system using CM 5 sensor chips. The proteins were coupled in 10 mM acetate (pH = $pI_{protein} - 1$) as described above. Binding of analytes was performed in HEPES buffered saline-EDTA polysorbate (HBS-EP) as running buffer at 20°C using flow rates of 20–40 µL/min. The affinity surface was regenerated between subsequent sample injections of analytes with 10 µL 10 mM NaOH. Binding was assayed with immobilized kringle domain as well as with immobilized *Listeria* proteins. Resulting K_D values were calculated using Biaevaluation 3.2 software (Biacore) by fitting a 1:1 Langmuir binding curve to the measured data set.

9.2.12
Bioinformatic analysis

This analysis was carried out as described in Trost et al. (manuscript submitted). Briefly, the tools NN and HMM of SignalP 2.0.b2 [a], SMART [b] and Sosui Signal beta [c] were used to analyze all coding sequences in the genome of *L. monocytogenes* EGDe. All coding sequences (CDS) for which at least one tool indicated a signal sequence and in addition those for which the annotation indicated a function outside the cytoplasmic membrane were further analyzed using PSORT I [d], iPSORT [d], and AnTheProt [e]. If possible, all tools were restricted to Gram-positive/prokaryotic data sets. The CDS were then grouped according to the number of positive signal sequence detections and this list compared to a detailed prediction of transmembrane domains [11, f] to separate CDS with a signal-anchor or N-terminal transmembrane helix (TM) domain from those with a secretion signal. All CDS with more than two predicted TM domains were considered integral membrane or membrane-associated proteins. Those with one or two predicted TM domains were individually checked for the position of these domains. Single TM helices predicted within the first 40 N-terminal amino acids matching a positive signal sequence prediction or a C-terminal TM helix matching the prediction of a membrane retention signal were considered as exported proteins. The plausibility of this classification was checked by a comparison with the annotation [g] and InterPro data for proteins with no known function.

9.3
Results

For the intracellular pathogen *L. monocytogenes* as well as for many, if not all, other pathogens, surface proteins are the prime interaction partners of the bacterial cell with the host. They possess key functions in adaptation to the environment, *e.g.*, stress and in adhesion and invasion processes during infection. Using proteomics for the analysis of defined surface subproteomes, we resolved about 120 and 180 spots from the Tris and KSCN extract, respectively, about 150 of which are unique. Particular protein signals (or spots) resulting in the identification of new surface proteins in *L. monocytogenes* were compared with the predictions from the genome annotation.

9.3.1
Prediction and validation of cell wall-associated proteins

9.3.1.1 Prediction of exported proteins
Since the genome annotation of *Listeria monocytogenes* was already completed in 2001, we started our analysis by reviewing and specifying the number of expected surface-associated proteins. For this analysis we could take into account recent publications related to the topology of membrane proteins and our newly developed integrated strategy for the prediction of secretion signals: the results of the different signal peptide predictions were combined by grouping the proteins into classes according to the number of matching predictions yielding 532 CDS with at least 3 coinciding results. As this analysis largely excluded information on transmembrane domains, this list was compared to the data from a multitool analysis of whole-genome transmembrane protein distribution and topology [11, f] that included *L. monocytogenes* EGDe. This identified 255 proteins as membrane-associated with two or more TMs, leaving 270 as probably exported. For a further division of these proteins into surface-associated and those truly secreted into the environment, the presence of cell surface retention signals was checked based on the available genome annotation [3, g]. 132 proteins were labeled surface-associated and 138 proteins were defined as secreted (Trost *et al.*, manuscript submitted).

9.3.1.2 Validation of the cell wall subproteome
For generating the cell wall protein fractions we established a new procedure to prepare defined subproteomes. By serial extraction surface proteins are solubilized stepwise to yield mainly noncovalent bound proteins and fractions of low complexity. To establish the serial extraction, different substances were tested with different solubilization characteristics. We solubilized surface proteins of intact bacteria stepwise with Tris buffer, KSCN, CHAPS, octylglucoside, CTAB, and SDS. The different serial extracts had to be screened for the absence of cytoplasmic contaminants to exclude cell lysis during treatment. Therefore, *p*-nitroanilide derivatives that can be converted by the strictly cytoplasmic enzyme aminopeptidase C (PepC) in a photometrically measurable reaction at 405 nm were added to the sur-

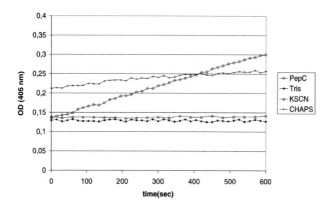

Fig. 1 PepC-assay of serial surface extracts of *L. monocytogenes*. Surface extracts were obtained by incubation with 1 M Tris, pH 7.5, 1 M KSCN or 1 M CHAPS. By adding L-arginine-*para*-nitroanilide to a final concentration of 2 mM and measuring the absorption at 405 nm for 10 min they were screened for aminopeptidase C activity as a marker of cytoplasmic contamination. As a positive control 5 units of purified aminopeptidas C from *Aeromonas proteolytica* were used. (○) PepC; (■) Tris; (△) KSCN; (◇) CHAPS.

face fractions. In this newly established PepC assay only the Tris- and KSCN-fractions showed no PepC-activity and therefore were assumed to be free of cytoplasmic contaminations whereas the CHAPS- and following fractions revealed cytoplasmic contamination by PepC-activity (Fig. 1).

The serial extracts were first separated by 1-D gel electrophoresis (Fig. 2). The protein patterns confirmed the PepC results as the Tris- and KSCN-fraction displayed varying band patterns. From the CHAPS-fraction onwards the protein pattern showed no significant differences but a higher complexity, suggesting a disruption of the cells during these treatments. Twenty-three protein bands representing 21 different proteins from the Tris- and KSCN-fractions were identified by *N*-terminal sequencing after Western blotting and Coomassie staining (Fig. 2, Tab. 1). The identification revealed seven proteins (Nos. 3, 9, 11, 34, 50, 52, 55) with a predictable signal sequence and surface association motifs containing known cell wall components like Ami or the virulence factor Internalin B that was identified twice with identical *N*-termini but different molecular weights. Fourteen identified proteins showed no predictable signal sequence or had predicted cytoplasmic functions like enolase or phosphoglycerate mutase. The most abundant proteins in these extracts were internalin B, Lmo2691 (an autolysin), glyceraldehyde-3 phosphatase, Lmo1847 (a lipoprotein), Ami, and P60/Iap.

To improve the separation and resolution of the protein fractions, 2-D gels of the Tris- and KSCN-fraction were generated (Figs. 3a and b). Protein spots were stained with ruthenium(II) tris-(bathophenanthroline disulfonate) (RuBPS) and identified by MALDI- or Q-TOF analysis. Altogether, 23 and 44 protein spots representing 19 and 37 proteins, respectively, were identified from the 2-D gels of the Tris and KSCN fraction, respectively, which together with the data from 1-D PAGE resulted

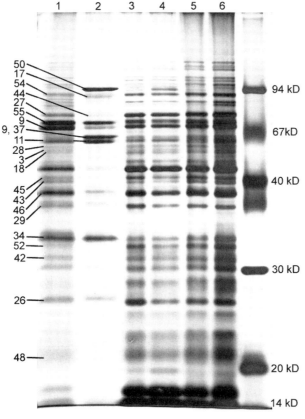

Fig. 2 Silver-stained SDS-PAGE of serial extracts from L. monocytogenes. Surface proteins were solubilized by 1 M Tris, pH 7.5 (lane 1), 1 M KSCN (lane 2), 1% CHAPS (lane 3), 1% CTAB (lane 4), 1% octylglucoside (lane 5), and 0.5% SDS (lane 6) and separated in 10% PAA gels. Identification was performed by transferring unstained proteins to a PVDF-membrane followed by Coomassie Brilliant Blue staining and N-terminal sequencing. Identified proteins are indicated by numbers and listed in Tab. 1.

in the identification of 55 different proteins (Tab. 1). Twenty-seven of these proteins possessed a predicted signal sequence, among them the known surface protein TcsA. Twenty-eight proteins had no predicted signal sequence and are so far known as cytoplasmic enzymes, like enolase, glyceraldehyde-3-phosphate dehydrogenase, and DnaK. The most abundant proteins in the 2-D gels were again Lmo1847, P45, TcsA, and two proteins of unknown function, Lmo0791 and Lmo1068.

Nine proteins were identified based on both 1-D PAGE and 2-D PAGE analyses, respectively: the invasion-associated protein Iap (Lmo0582), superoxide dismutase (Lmo1439), Lmo1847, the two chaperons DnaK (Lmo1473) and GroEL (Lmo2068), enolase (Lmo2455), glyceraldehyde-3-phosphate dehydrogenase (Lmo2459), the lipoprotein Lmo2637, and the translation elongation factor Fus (Lmo2654). The 2-D separation enabled the identification of 33 additional proteins from surface protein extracts, whereas 12 proteins were only identified in 1-D SDS-PAGE, among them the virulence factor InlB and two autolysins (Lmo2558 and Lmo2691) with a predictable signal peptide.

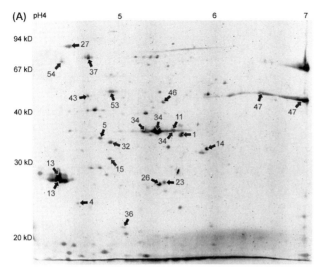

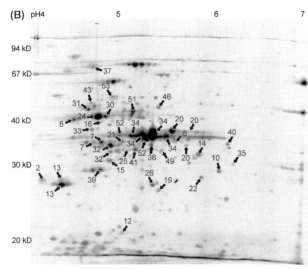

Fig. 3 2-D map of surface proteins of L. monocytogenes. (A) Bacteria cells from overnight cultures in brain-heart-infusion (BHI) were harvested and incubated with 1 M Tris, pH 7.5, for 30 min. The supernatant was precipitated with TCA-acetone and separated by 2-D gel electrophoresis. After staining with RuBPS, indicated protein spots were identified by MALDI-MS and are listed in Tab. 1. (B) After Tris extraction and PBS washing bacteria cells were incubated with 1 M KSCN for 30 min. The supernatant was precipitated with TCA-acetone and separated by 2-D gel electrophoresis. After staining with RuBPS, indicated protein spots were identified by MALDI-MS and are listed in Tab. 1.

Taken together, the identified proteins from serial cell wall extractions included two surface proteins with a GW-module (InlB and Ami), two P60-like proteins (Iap and Spl), 20 lipoproteins, most of them substrate-binding components of transporters, and three further proteins with a signal peptide that have so far no predictable surface association (Lmo0443, Lmo0950, Lmo2691). Interestingly, the genes coding for the two lipoproteins, Lmo1068 with unknown function, and Lmo2331 with homology to a bacteriophage protein, showed no orthologue in the apathogenic strain Listeria innocua. The identification revealed nine proteins of unknown function, seven of which showed homology to other genomes, among them four putative lipoproteins (Lmo0047, Lmo0791, Lmo1757, and Lmo2636) and two proteins

Tab. 1 Proteins identified in serial extracts of L. monocytogenes from 1-D PAGE by N-terminal sequencing and from 2-D PAGE by MALDI-MS

Spot	Gene	Description	pI	(kDa)	Cat.[a]	SP[a]	Class
1	Lmo0013	qoxA, quinol oxidase subunit II, putative lipoprotein	6	33	1.4	Yes	5
2	Lmo0047	Unknown, lipoprotein	3.7	23	5.2	Yes	6
3	Lmo0135	Similar to oligopeptide ABC transport system substrate-binding proteins, lipoprotein	–	56	1.2	Yes	7
4	Lmo0197	Similar to B. subtilis SpoVG protein	4.1	20	1.7	No	–
5	Lmo0214	mfd, transcription-repair coupling factor	4.7	32	3.2	No	–
6	Lmo0223	cysK, highly similar to cysteine synthase	5.7	33	2.2	No	–
7	Lmo0285	Putative lipoprotein	4.6	33	1.2	Yes	6
8	Lmo0292	Similar to heat-shock protein htrA serine protease	4.3	38	4.1	No	–
9	Lmo0434	inlB, Internalin B	–	70/64	1.8	Yes	4
10	Lmo0443	Similar to B. subtilis transcription regulator LytR	6.2	25	3.5.2	No	4
11	Lmo0582	iap, P60 extracellular protein, invasion-associated protein Iap	5.8	33	1.2	Yes	6
12	Lmo0739	Similar to 6-phospho-β-glucosidase	5	8	2.1.1	No	–
13	Lmo0791	Unknown, lipoprotein	3.8	24	5.2	Yes	7
14	Lmo0950	Unknown	6.4	29	5.2	No	6
15	Lmo1041	Similar to molybdate ABC transporter-binding protein, lipoprotein	4.7	27	1.2	Yes	5
16	Lmo1053	pdhB, highly similar to pyruvate dehydrogenase (E1 β-subunit)	4.6	37	2.1.2	No	–
17	Lmo1054	pdhC, highly similar to pyruvate dehydrogenase (dihydrolipoamide acetyltransferase E2 subunit)	–	89	2.1.2	No	–
18	Lmo1055	pdhD, highly similar to dihydrolipoamide dehydrogenase, E3 subunit of pyruvate dehydrogenase complex	–	54	2.1.2	No	–
19	Lmo1059	Unknown	5.4	20	6	No	–
20	Lmo1068	Unknown, lipoprotein	5.6	35	6	Yes	6
21	Lmo1073	Similar to metal-binding protein (ABC transporter), lipoprotein	45	32	1.2	Yes	7
22	Lmo1291	Similar to acyltransferase (to B. subtilis YrhL protein)	6	23	1.1	No	–
23	Lmo1314	frr, highly similar to ribosome recycling factors	5.6	22	3.7.5	No	–

Tab. 1 Continued

Spot	Gene	Description	p/	(kDa)	Cat.[a]	SP[a]	Class
24	Lmo1388	tcsA, T cell-stimulating antigen, lipoprotein	4.5	39	1.2	Yes	6
25	Lmo1426	Similar to glycine betaine/carnitine/choline ABC transporter (osmoprotectant-binding protein), lipoprotein	5	30	1.2	Yes	6
26	Lmo1439	sod, superoxide dismutase	5.6	23	4.2	No	–
27	Lmo1473	dnaK, class I heat-shock protein (molecular chaperone) DnaK	4.4	55	3.9	No	–
28	Lmo1634	Similar to alcohol-acetaldehyde dehydrogenase	–	58	2.1.1	No	–
29	Lmo1657	tsf, translation elongation factor	–	37	3.5.3	No	–
30	Lmo1671	Similar to ABC transporter and adhesion proteins, lipoprotein	4.8	39	1.2	Yes	6
31	Lmo1730	Similar to sugar ABC transporter-binding protein, lipoprotein	4.4	42	1.2	Yes	7
32	Lmo1738	Similar to amino acid ABC transporter (binding protein), lipoprotein	4.8	32	1.2	Yes	7
33	Lmo1757	Similar to unknown protein, lipoprotein	4.6	36	5.2	Yes	7
34	Lmo1847	Similar to adhesion binding proteins and lipoproteins with multiple specificity for metal cations (ABC transporter)	5.3	33	1.2	Yes	6
35	Lmo1967	Similar to toxic ion resistance proteins	6.6	27	4.2	No	–
36	Lmo2021	Similar to unknown protein	5	13	5.2	No	–
37	Lmo2068	groEL, class I heat-shock protein (chaperonin) GroEL	4.6	52	3.9	No	–
38	Lmo2219	Similar to post-translocation molecular chaperone, lipoprotein	5.4	32	1.6	Yes	7
39	Lmo2331	Weakly similar to gp32_bacteriophage A118 protein, lipoprotein	4.6	25	4.4	Yes	7
40	Lmo2415	Similar to ABC transporter, ATP-binding protein	6.5	32	1.2	No	–
41	Lmo2417	Conserved lipoprotein (putative ABC transporter-binding protein)	5.1	30	1.2	Yes	7
42	Lmo2425	Similar to glycine cleavage system protein H	8.1	42	2.2	No	–
43	Lmo2455	eno, highly similar to enolase	4.6	43	2.1.2	No	–
44	Lmo2456	pgm, highly similar to phosphoglycerate mutase	–	76	2.1.2	No	–
45	Lmo2458	pgk, highly similar to phosphoglycerate kinase	–	42	2.1.2	No	–
46	Lmo2459	Highly similar to glyceraldehyde-3-phosphate dehydrogenase	5.4	42	2.1.2	No	–
47	Lmo2505	spl, peptidoglycan lytic protein P45	8.1	42	1.1	Yes	6

Tab. 1 Continued

Spot	Gene	Description	p*I*	(kDa)	Cat.[a]	SP[a]	Class
48	Lmo2511	Similar to conserved hypothetical proteins like to *B. subtilis* YvyD protein	–	21	5.2	No	–
49	Lmo2556	fbaA, highly similar to fructose-1,6-bisphosphate aldolase type II	5.5	30	2.1.1	No	–
50	Lmo2558	ami, autolysin; amidase	–	95	1.8	Yes	6
51	Lmo2636	Conserved hypothetical lipoprotein	5.1	41	5.2	Yes	7
52	Lmo2637	Conserved lipoprotein	4.9	34	1.8	Yes	7
53	Lmo2653	tufA, highly similar to translation elongation factor EF-Tu	4.8	45	3.7.4	No	–
54	Lmo2654	fus, highly similar to translation elongation factor G	4.3	51	3.7.4	No	–
55	Lmo2691	Similar to autolysin, N-acetylmuramidase	–	73	1.1	No	3

p*I* and kDa refer to the apparent isoelectric points and molecular masses determined form the gels. Proteins without p*I* value were identified by 1-D PAGE. Cat and SP refer to the functional category and signal peptide prediction, respectively, as given by [3]. Class refers to the classification from 7 to 3 according to the number of positive signal peptide predictions from this study; proteins with less than 3 positive results were considered nonsecretory.

a) Functional category; SP prediction according to [3] and its supplementary material with no homology to proteins from any other organism sequenced so far (Lmo1059 and Lmo1068). Altogether, six of the genes of the identified proteins have a putative binding site for the virulence regulator PrfA upstream the coding region: the lipoprotein Lmo0013, the cysteine synthase Lmo0223, the pyruvate dehydrogenase Lmo1053, and the lipoproteins Lmo1730, Lmo1847, and Lmo2219. In conclusion, our prediction and validation strategy revealed three further proteins (Lmo0443 in class 4, Lmo0950 in class 6, Lmo2691 in class 3) that were not considered to be secreted based on previous genome studies.

9.3.2
Analysis of protein processing and localization

The identification of 28 proteins that have no predicted signal sequence and are assumed to have cytoplasmic functions in cell wall extracts was surprising and led us to further investigations:
(i) If cytoplasmic proteins are recruited to the cell surface by chance subsequent to cell lysis during bacterial cultivation, proteolytic fragments of the corresponding proteins should be present. Therefore, we first checked the positions of the proteins identified in 2-D gels and compared the apparent molecular weights and isoelectric points of these proteins with the calculated values but did not observe significant differences.
(ii) In order to reveal putative post-translational processing of secretion signals not predictable presently N-terminal sequencing was done with sets of secretory and cytoplasmic proteins, respectively. In this study, 21 different N-terminal sequences were obtained among them 7 of secretory and 14 of the putative cytoplasmic proteins (Tab. 2). The secretory proteins were found to be processed. However, the determined N-termini were all different from those predicted in the annotation of the genome sequence. The N-terminal sequences of the non-secretory proteins corresponded to the sequences predicted from the genome and revealed no cleavage of a transport signal. As an interesting observation, in 12 cases the N-terminal methionine was missing. Only two N-termini contained the first amino acid methionine encoded by the start codon ATG in the genome sequence. A cleavage of the N-terminal methionine by a methionine aminopeptidase can take place and is assumed to influence the half time of a protein which is important for the turn-over within the cell. These analyses indicated that the proteins were present in the extracellular compartment in an active form and led to the assumption that they may have an additional function at this place.
(iii) To substantiate the reliability of the developed extraction method we selected enolase as a candidate protein and performed immunoelectron microscopic labeling studies with anti-enolase antibodies visualized by protein A-gold complexes to reveal its localization. Postembedding labeling of ultrathin sections revealed the presence of enolase both within the bacterial cytoplasm and the cell wall region (Figs. 4A–E). In addition, immunoscanning electron microscopy clearly showed that enolase is present and accessible on the bacterial cell surface (Fig. 4F).

Tab. 2 N-terminal sequences of proteins from serial extracts of L. monocytogenes

N-Terminus	Gene	Description	M_r (kDa)
CGGSGASSDKANG	Lmo0135	Oligopeptide-binding protein (ABC transporter)	56
ETITVPTPIKQI	Lmo0434	InlB, internalin B	70
ETITVPTPIKQI	Lmo0434	InlB, internalin B	64
TVVVEAGDTLWGIAQ	Lmo0582	Iap (P60), invasion-associated protein	61
AYSFKLPDIGEPI	Lmo1054	PdhC, pyruvate dehydrogenase (E2)	89
VVGDFPSERDTIVIG	Lmo1055	PdhD, pyruvatde hydrogenase (E3)	54
MTYELPKLPYTYDALE	Lmo1439	Sod, superoxide dismutase	27
SKIITGIDLGTTN	Lmo1473	DnaK, molecular chaperone	77
AIKENAAQEVLEVQK	Lmo1634	Alcohol acetaldehyde dehydrogenase	58
ANITAQMVKELREKT	Lmo1657	Tsf, translation elongation factor	37
CSSQNSDSKKTDG	Lmo1847	Adhesion binding protein (ABC transporter)	34
AKDIKFSEDA	Lmo2068	GroEL, chaperonin	64
SLPKDLLYTEEH	Lmo2425	Glycin cleavage system protein H	32
SIITEVYAREVL	Lmo2455	Eno, enolase	40
SKSPVAIIILGDGF	Lmo2456	Pgm, phosphoglycerate mutase	76
AKKVVTDLDLKD	Lmo2458	Pgk, phosphoglycerate kinase	42
TVKVGINGFGR	Lmo2459	Gap, glyceraldehyde-3-phosphate dehydrogenase	39
MLKYNIRGENIEV	Lmo2511	Conserved hypothetical protein	21
SIDPVQKTD	Lmo2558	Ami, autolysin, amidase	95
CGSSDDSSKDK	Lmo2637	Conserved lipoprotein	33
AREFSLEKTRNIG	Lmo2654	Fus, translation elongation factor G	91
DETAPADEASKSAEA	Lmo2691	Autolysin, N-acetylmuramidase	73

9.3.3
Analysis of plasminogen-binding proteins

Since enolase from other pathogens was already demonstrated to bind specifically plasminogen [12], we decided to screen systematically for plasminogen-binding listerial surface proteins in the Tris fraction (Figs. 2 and 3a) to find out whether similar mechanisms could also be true for L. monocytogenes and may be a reason for the presence of cytosolic proteins in the cell wall. In an overlay blot, binding of human plasminogen was observed for four bacterial components with apparent molecular masses of 66, 47, 44, and 33 kDa within Tris extracts. The most intense binding signals were generated by proteins with molecular masses of 44 and 33 kDa (Fig. 5, lane B). Unambiguous identification of the four bands by tryptic digest and MALDI mass fingerprinting was not possible. However, staining of the blot with anti-enolase antibodies revealed a protein band of 46 kDa (Fig. 5, lane C).

A Biacore experiment was set up to enrich and isolate proteins binding to plasminogen. The kringle 1–3 domain (fragment Y^{80}-P^{353}) of human plasminogen was chosen as a ligand because it lacks proteolytic activity, thus facilitating the isolation of intact proteins. This fragment has been shown previously to be sufficient for binding experiments with α-enolase of *Streptococcus pneumoniae* [12]. Proteins

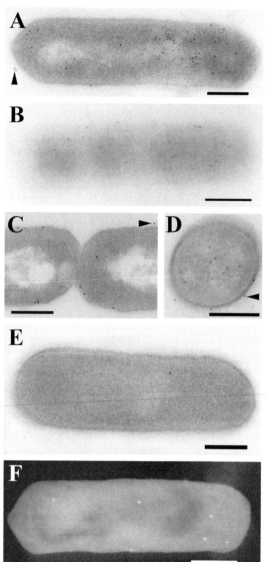

Fig. 4 Immunoelectron microscopic labeling studies of enolase in *L. monocytogenes* EGD. (A)–(D) Enolase label (black dots) is localized in the cytoplasm of *L. monocytogenes* EGD (A and D); furthermore, some labeling was detectable at the cell wall region (C and D) and on the outside of the cell wall (marked with arrow heads in A and C). (B) Section through the cell wall region, exhibiting labeling for enolase over the entire length of the bacterium. (E) Control experiment where incubation with the polyclonal anti-enolase antibody was omitted; no label is detectable. (F) Scanning immunolabeling of enolase on the surface of *L. monocytogenes* EGD. The white dots represent gold-particles which are bound to anti-enolase antibodies demonstrating that enolase is localized on the cell surface of the bacterium since the antibody has access to enolase from the outside. Bars represent 0.25 μm.

binding to the covalently coupled kringle 1–3 domain were eluted with 10 mM NaOH and separated by SDS-PAGE (Fig. 6). After staining with Coomassie blue, the bands were identified by MALDI mass fingerprinting. The identified proteins are: heat shock factors GroEL (Lmo2068, M_r 57 367 Da) and DnaK (Lmo1473, M_r 66 144 Da), elongation factor TU (Lmo2653, M_r 43 342 Da), enolase (Lmo2455, M_r 46 472 Da), and glyceraldehyde-3-phosphate dehydrogenase (Lmo2459, M_r 36 286 Da) (Fig. 6B). Under identical conditions no binding on a deactivated CM5 reference cell was observed in absence of kringle 1–3 (Fig. 6A).

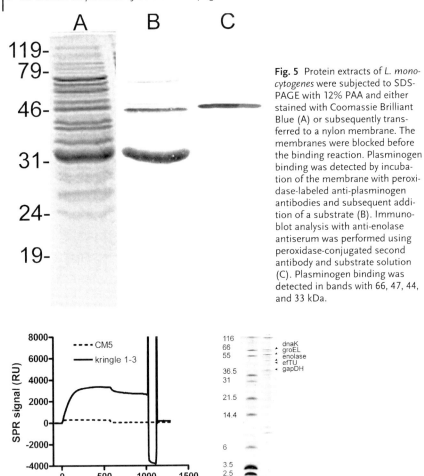

Fig. 5 Protein extracts of *L. monocytogenes* were subjected to SDS-PAGE with 12% PAA and either stained with Coomassie Brilliant Blue (A) or subsequently transferred to a nylon membrane. The membranes were blocked before the binding reaction. Plasminogen binding was detected by incubation of the membrane with peroxidase-labeled anti-plasminogen antibodies and subsequent addition of a substrate (B). Immunoblot analysis with anti-enolase antiserum was performed using peroxidase-conjugated second antibody and substrate solution (C). Plasminogen binding was detected in bands with 66, 47, 44, and 33 kDa.

Fig. 6 *Listeria monocytogenes* were lysed with a french pressure cell in HBS-EP and centrifuged at 20 000 × g to remove debris. The lysate was tested for proteins binding to human plasminogen using Biacore 3000 SPR-detection. The proteolytically inactive kringle domain of plasminogen was immobilized to a CM5-sensorchip using EDC/NHS coupling chemistry. (A) Incubation with the lysate was performed at a flow rate of 1 µL/min. The chip was washed and the bound proteins were eluted in 4 µL 10 mM NaOH using the microrecovery option of the instrument. 1000 RU correspond to approximately 1 ng of bound protein. The experiment was repeated several times to collect material for subsequent analysis by SDS-PAGE. (B) The eluted sample was applied to SDS-PAGE and stained with colloidal Coomassie blue. Proteins were identified by MALDI mass fingerprinting.

To make sure that the occurrence of these five high-abundant cytosolic proteins in the eluted sample was not due to contamination within the instrument, we decided to express them in E. coli and to determine the affinities of the purified proteins to human plasminogen.

Expression of GAPDH and DnaK using pQE30 in M15 cells as expression system resulted in N-terminally His-tagged proteins that could easily be purified with Ni-NTA agarose. Concentrations of 500 µg/mL GAPDH and 240 µg/mL DnaK were achieved. Enolase could not be purified by use of Ni-NTA agarose, probably because of a nonfunctional or unaccessible His-tag. Purification by classical chromatography resulted in a protein concentration of 400 µg/mL. EF-TU could only be expressed when fused with GST. Purification and cleavage with precision protease yielded a concentration of 120 µg/mL. Expression of GroEL failed with every expression system mentioned above. Attempts to purify the protein fused with an amino terminal His-tag resulted in the isolation of a 25 kDa fragment, indicating degradation of the expression product.

First binding experiments using human plasminogen as analyte in solution failed due to the presence of activated plasmin in solution which digested the immobilized ligands during incubation (data not shown). Even plasmin inhibited by PefablocSC showed significant proteolytic activity towards the immobilized ligands. Again, the plasminogen fragment kringle 1–3 was chosen which lacks the proteolytic activity of the intact protein. GAPDH was immobilized onto a CM5 sensor chip and the binding of kringle 1–3 in solution was monitored. 2.5–625 nM kringle 1–3 was applied. Affinity was calculated by association and dissociation curves to a 1:1 Langmuir binding model. The K_D value of the interaction is 56 nM (Fig. 7a). In the same way DnaK was immobilized and kringle 1–3 was applied in a concentration range from 1.5–375 nM. Fitting these curves to a 1:1 Langmuir model results in a calculated K_D value of 25 nM (Fig. 7b). Only minor deviations of the binding curves from a 1:1 model were observed in the concentration range used for determination of K_D values of both interactions.

However, binding curves obtained by application of high concentrations of kringle 1–3 differed significantly from a 1:1 binding model. This can be explained by an oligomerization of kringle 1–3. This effect contributes increasingly to the binding signal in concentrations above 100 nM kringle 1–3 (Biacore binding curves of kringle 1–3 in solution binding to immobilized kringle 1–3 are not shown). Therefore, interactions of enolase and EF-TU to kringle 1–3 were monitored *vice versa* by immobilizing the kringle domain onto a CM5 chip and application of *Listeria* proteins in solution. Enolase was applied to a surface with immobilized kringle domain in a concentration range from 1.1 nM–2.2 µM. Assuming a Langmuir 1:1 binding model, the calculated K_D value is 60 nM (Fig. 7c). Binding of EF-TU was measured the same way. The K_D value was calculated to be 24 nM (Fig. 7d). All K_D values are summarized in Tab. 3.

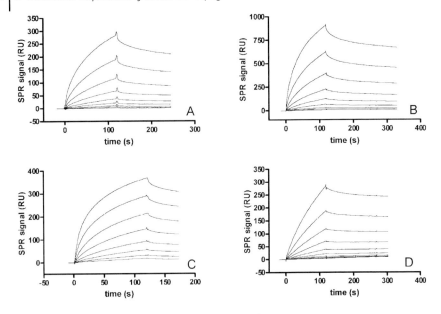

Fig. 7 (A) GAPDH was immolilized on a CM5 chip. Binding was measured in a concentration range from 2.5 to 625 nM kringle 1–3. The analyte was applied with a flow rate of 20 μL/min. The K_D value of the interaction is 56 nM. (B) DnaK was immobilized on a CM5 chip. Binding was measured in a concentration range from 1.5 to 375 nM kringle 1–3. The analyte was applied with a flow rate of 20 μL/min. The K_D value of the interaction is 25 nM. (C) Kringle 1–3 was immobilized on a CM5 chip. Binding was measured in a concentration range from 1.1 to 2.2 μM enolase. The analyte was applied with a flow rate of 30 μL/min. The K_D value of the interaction is 60 nM. (D) Kringle 1–3 was immobilized on a CM5 chip. Interaction was measured in a concentration range from 1.4 to 700 nM EF-TU. The analyte was applied with a flow rate of 40 μL/min. The K_D value of the interaction is 24 nM. All Biacore experiments were performed in HBS-EP at 20°C.

Tab. 3 K_D values for the interactions of the kringle domain 1–3 (fragment Y^{80}-P^{353}) of human plasminogen to various *Listeria* proteins present on the cell wall of *L. monocytogenes*

Binding to kringle 1–3	DnaK Lmo1473	EF-TU Lmo2653	Enolase Lmo2455	GAPDH Lmo2459
K_D value (M)	2.5e-8	2.4e-8	6.0e-8	5.6e-8

All values were determined by direct Biacore binding studies at 20°C. K_D values were calculated by Biaevaluation 3.2 software (Biacore) by fitting a Langmuir 1:1 binding function to the measured data set.

9.4 Discussion

9.4.1 Analysis of proteins identified from surface extracts

Surface proteins are of high interest for functional investigations of bacterial pathogens. They mediate the direct interactions between the bacteria and the host cell. In *L. monocytogenes* surface proteins have to fulfil an exceptionally high number of different functions due to the facultative intracellular lifestyle of this pathogen and its ability to adhere to and invade different types of host cells of different species. We performed an improved prediction of secretion signals. The sequence of the genome revealed the existence of 132 proteins with a predictable signal sequence and surface association modules. Only very few of these have been identified and characterized like the family of internalins that include essential virulence factors [13–15]. The number of predicted lipoproteins which are noncovalently linked to the cell membrane – 68 in *L. monocytogenes* – is higher than in any other sequenced genome of Gram-positive pathogenic bacteria [2].

In this study, we developed a new method to obtain defined surface protein fractions of low complexity based on an earlier observation that InlB can be extracted almost quantitatively by Tris alone [16]. By the use of Tris, KSCN, CHAPS, octylglucoside, CTAB, and SDS one after the other in a serial extraction, we produced partial surface protein fractions. We established the PepC assay to exclude a contamination by cell lysis during extraction. By this criterion, the Tris and KSCN extractions were done without the loss of membrane integrity and the uncontrolled release of cytoplasmic components. Altogether, 55 different proteins were identified by N-terminal sequencing and mass spectrometry after serial extraction. Among them were two proteins containing a GW module for surface association, of which nine were predicted from the genome sequence [2, 3]. These two GW-proteins, InlB and Ami, were already known and characterized. InlB was highly abundant in our extracts and represents a major surface protein responsible for invasion of *L. monocytogenes* into hepatocytes and some epithelial cell lines [13, 17]. Ami is an autolysin that contributes to the adhesion of *L. monocytogenes* to eukaryotic cells [18] and needs to be further characterized.

We also identified two P60-like proteins of which four proteins are predicted in the genome that are noncovalently linked to the cell wall [2, 3]. These proteins, Iap/P60 and P45/Spl, have been roughly characterized before [19, 20] and a more detailed investigation of their role in virulence has to be done. Out of 68 putative lipoproteins [2, 3] that are anchored in the membrane *via* an N-terminal diacylglyceryl moiety we identified 20. Finding such a high number of lipoproteins in the extracts obtained under mild solubilizing conditions with Tris buffer was rather unexpected. Unlike GW- and P60-proteins they are not associated with structures in the peptidoglycan on the very surface of the bacteria but are inserted into the cytoplasmic membrane. As lipoproteins were also found in the secretome of Gram-positive bacteria ([21] and Trost *et al.*, manuscript submitted), it seems likely that they are easily released, maybe even by an active process like proteolytic shaving or a hypothetical release factor. Among these are four lipoproteins of unknown function, Lmo0047, Lmo0791, Lmo1757, and Lmo2636, that showed homology to hypothetical proteins of other genomes only. One identified lipoprotein, Lmo1068, had no homologues in other genomes and is thus specific for *Listeria*. Similar results were obtained in an analogous study with *Bacillus subtilis* that used a single salt extraction step with 1.5 M LiCl and did not include a marker enzyme assay [22]. This approach yielded a comparatively simple subproteome that in accordance with our approach included both predicted and unexpected cell wall-bound proteins. From the group of proteins that are covalently cell wall-bound *via* an LPXTG motif, of which 41 are predicted from the genome sequence, as expected, none could be identified in the serial extracts. Covalently linked proteins represent a problematic group of surface proteins that are difficult to isolate from bacterial cell walls [23, 24]. For *Listeria*, Bierne and colleagues [25] identified 3 of 41 predicted LPXTG-proteins from purified peptidoglycan extracts by a comparative ESI-MS/MS analysis of direct tryptic digests. Very likely it is the low expression of these proteins or/and the tight linkage to the peptidoglycan of the cell wall that makes it difficult to solubilize and separate these proteins in gels.

We also identified three proteins that have a predictable signal peptide but were not grouped into the secretory proteins [3] or possess any of the known surface association mechanisms. Lmo0443 shows homology with the transcriptional regulator LytR of *B. subtilis*, Lmo0950 is a protein of unknown function, and Lmo2691 was recently described as muramidase MurA [26].

As the genome of the apathogenic species *Listeria innocua* was also sequenced [3], a comparison can indicate new virulence-related proteins. In this study, besides the known virulence factors, three proteins were identified that have no orthologues in the *L. innocua* genome and are *L. monocytogenes*-specific: Lmo0739, Lmo1068, and Lmo2331. To corroborate and extend this analysis the newly established procedure was also used for a comparative proteome analysis of the wild-type strain with a deletion mutant of PrfA, the central transcription regulator of virulence proteins in *Listeria* (Schaumburg *et al.*, in preparation). All the unknown proteins and proteins with no orthologous gene in the apathogenic *L. innocua* might be involved in virulence and their functions have to be characterized to get a complete understanding of the pathogenesis of *L. monocytogenes*.

9.4.2
Analysis of protein processing and localization in surface extracts

Interestingly, 28 of the identified proteins from serial extracts possess no predictable signal peptide but a putative cytoplasmic function. These include, *e.g.*, the glycolytic enzymes enolase (Eno), glyceraldehyde-3-phosphate dehydrogenase (GAP), the elongation factor TU (TufA), and the stress proteins GroEL and DnaK. The available *N*-terminal sequence data comprising both proteins with and without a predicted secretion signal peptide were reanalyzed for clarification. The seven proteins with a predicted secretion signal were indeed found to be processed, however, the determined cleavage sites differed from the predictions of the software tools used. In contrast, the N-terminal sequencing of the proteins without predicted secretion signal revealed no aberration compared to the predicted genome sequence and no cleavage of an unknown transport signal. Furthermore, the apparent molecular weights and isoelectric points determined from 2-D gels did not indicate any degradative processing. However, 12 of 14 sequences of uncleaved cytoplasmic proteins lacked the N-terminal methionine indicating a cleavage by methionine aminopeptidase that – if these proteins really do have a "moonlighting" function outside the *Listeria* cell – might influence the half-life of these proteins as described in the N-end rule [27] and shown for the listerial virulence factors ActA and P60 in the mammalian cytosol [28, 29]. As a support for this assumption, we decided to check for the presence of enolase on the surface of *L. monocytogenes* by immunoelectron microscopy. Here, we observed that enolase remained bound to the listerial surface although it lacks a known cell wall retention signal as well as a secretion signal. This matches the results obtained for enolase of *S. pneumoniae* [30].

9.4.3
Possible function of plasminogen binding in virulence

Along with the glycolytic enzymes enolase (Eno, Lmo2455) and glyceraldehyde-3-phosphate dehydrogenase (GAPDH, Lmo2459) the heat-shock factors GroEL (Lmo2068) and DnaK (Lmo1473) as well as elongation factor TU (Lmo2653) belong to the "unexpected" proteins within the *Listeria* cell wall proteome. Our findings raise three questions: (i) how do these proteins leave the bacterial cell, (ii) how are they retained on the cell surface, and (iii) do they have an additional function when associated with the cell wall different from their known function inside the bacterial cell?

The secretion mechanism of "moonlighting" proteins is subject to many speculations. Export by ABC-transporters and passive release of proteins upon osmotic stress mediated by the mechanosensitive channel MscL or membrane permeabilizing bacteriophage holins were postulated to be involved in the release of proteins [31–33] – phages A118 and PSA were described in *L. monocytogenes* [34, 35]. Additionally, an acylation-dependent protein export was shown for *Leishmania* by Denny et al. [36].

However, the release of cytosolic proteins by a yet unknown mechanism or even by lysis of bacteria cannot be excluded. Very interestingly, Boël et al. [37] recently demonstrated that enolase of *E. coli* is automodified by 2-phosphoglycerate on Lysin 341. A change to glutamic acid at this position hardly influenced the catalytic activity of the enzyme but completely abolished its export thus indicating that its export is a cellular process. Anyway, lysis of bacteria is not likely in our experiments as controls did not point to any substantial contamination of the cell wall extracts by aminopeptidase C activity. We focused on the third question regarding an additional function of these proteins when bound to the cell wall. Cytosolic proteins were found to be released by many prokaryotic and eukaryotic organisms. Heat shock proteins and chaperones have classically been considered to be induced when the cell is under stress, particularly heat stress, and are involved in preserving normal folding and hence function of proteins. Today there is evidence that these proteins can be released from cells to interact with specific receptors presented by other cells [38]. Glyceraldehyde-3-phosphate dehydrogenase of *Staphylococcus* and *Streptococcus* was reported to bind to fibronectin and laminin, to cause membrane fusion and to have a phosphotransferase/phosphokinase activity [39]. Moreover, prokaryotic glyceraldehyde-3-phosphate dehydrogenase binds to several mammalian proteins including lysozyme, laminin, fibronectin, plasmin, actin, and transferrin [40–42].

Especially interesting is the finding that, in addition to glyceraldehyde-3-phosphate dehydrogenase, surface-displayed enolase of *Streptococcus pneumoniae* also recruits plasminogen from the host to the surface of the pathogen and facilitates subsequent activation of the protease by tissue-type plasminogen activator tPA [43].

Plasminogen is a 92 kDa glycoprotein which is present in human plasma in a concentration of 2.0–2.4 µM [44, 45]. Its active form, the protease plasmin, is composed of five triple disulfide-bonded kringle domains and a catalytic domain with serine protease activity [46]. Besides its function in fibrinolysis, it has a broad substrate specificity against extracellular matrix components like fibronectin, laminin, and vitronectin [47, 48]. Additionally, it induces activation of other matrix-degrading proteases, such as collagenases [49]. The binding of plasminogen by surface-displayed enolase suggests that the glycolytic enzyme may help *S. pneumoniae* to pass through the extracellular matrix (ECM) of its mammalian host by misusing the broad substrate specificity of plasmin [12, 30]. The finding of surface-displayed enolase is further supported by a clinical study that correlates infection of patients with *S. pneumoniae* with development of antibodies against streptococcal enolase. These antibodies cross-react with human enolase and subsequently lead to autoimmune reactions of the patients [50].

Many reports are available demonstrating the direct correlation between increased expression of enolase both at the DNA level and at the protein level, and progression of tumors, such as neuroendocrine tumors, neuroblastoma, and lung cancers, including small-cell carcinoma [51–56]. It is likely that the overexpression of enolase on the surface of such cells in various cancerous conditions may result in acquiring more plasminogen and hence in the spread of tumor cells [57]. In addition to cancer cells, enolase has been shown to be expressed on the surface of neuronal [58] and some hematopoietic cells as a novel plasmin(ogen) receptor [59, 60].

An overlay blot with proteins extracted from the surface of *L. monocytogenes* incubated with human plasminogen indicated that this unusual "moonlighting" function as plasminogen receptor may be true not only for listerial glyceraldehyde-3-phosphate dehydrogenase and enolase, as described for other organisms, but binding was also observed for at least two more bacterial components detected with estimated molecular masses of 66 and 44 kDa. The shortcoming of this method is the use of denaturing conditions in SDS-PAGE and electroblotting, which means that binding of plasminogen can only be tested with unfolded listerial proteins on the blot membrane. Additionally, the blotted proteins are not easily identified. In order to verify the binding of listerial proteins to plasminogen under native conditions and to identify binding proteins we designed a "fishing" experiment using the biosensor Biacore 3000. A proteolytically inactive fragment of human plasminogen comprising the first three kringle motifs was chosen as a bait for fishing interaction partners of plasmin(ogen) out of a *Listeria* lysate containing intra- and extracellular proteins. Even though the sample was derived from a whole-cell lysate, the identified binding proteins strongly correlated to those found in surface extracts of *L. monocytogenes*. Five major bands in the PAA gel were identified as DnaK, GroEL, EF-TU, enolase, and GAPDH. The binding was shown to be specific to the kringle 1–3 motif of human plasminogen, because there was no binding to a reference cell in absence of the protein. Additionally, the molecular masses of DnaK, EF-TU, enolase, and GAPDH correlate to plasminogen-binding bands of 66, 47, 44, and 33 kDa in a plasminogen overlay blot of surface proteins extracted by 500 mM Tris buffer (Fig. 5, lane B).

All five proteins found to bind to the kringle motifs of human plasminogen are highly abundant. To confirm that these findings are not due to contamination, affinities of all proteins except GroEL, which was not expressed by *E. coli* under different conditions as described above, towards the kringle 1–3 motif were quantified in direct Biacore binding studies. These direct binding studies confirmed that DnaK, EF-TU, enolase, and GAPDH are strong plasminogen binders. GroEL may also belong to this group of "moonlighting" proteins. It was identified in the eluate of a kringle 1–3 surface in a Biacore "fishing" experiment. However, it was not detected in a plasminogen overlay blot. Additionally, a final proof by direct interaction studies of the purified protein cannot be provided. The nanomolar K_D values of the measured interactions indicate a possible role as plasminogen receptors for these cell wall-bound "cytosolic" proteins because plasma concentration of plasminogen in humans is 40- to 80-fold higher. This means, that the receptors presented on the cell surface will probably be saturated with human plasminogen under *in vivo* conditions. In analogy to *Streptococcus pneumoniae* and various tumors the broad substrate specificity of the protease against ECM molecules may be misused by the pathogen to invade the host tissue by penetrating through the extracellular matrix. This suggestion is further supported by the finding that *L. monocytogenes* possesses adhesins for fibronectin [61] and by electron microscopy pictures published by Daniels *et al.* [62], revealing the adherence of *L. monocytogenes* to the basal lamina of the small intestine in areas of damaged epithelium, thus indicating a novel invasion mechanism of the pathogen by passing through the ECM of its host.

9.5
Concluding remarks

In conclusion, surface proteins of pathogenic bacteria are important for adhesion to and invasion of mammalian host cells. *Listeria monocytogenes* uses a number of known proteins mainly identified by genetic screens to breach and evade host defences. In this study we mapped the cell wall proteome and identified many proteins of unknown function, three of which have no orthologue in the non-pathogenic *L. innocua* and might be involved in virulence. Although we could largely exclude contamination by cytoplasmic proteins, a substantial fraction of the identified proteins had neither a predicted secretion signal nor a known surface-binding domain. For one of these proteins, enolase, the gene of which is present as a single copy without a paralogue, we could demonstrate its retention to the cell wall and, together with three more proteins, a highly specific binding of human plasminogen. While it has to remain open how these proteins pass the cytoplasmic membrane and associate with the cell wall, the binding assays point to an additional function and involvement of these proteins in the crossing of the intestinal barrier by the pathogen. These results also demonstrate that proteome analyses are essential to verify and extend theoretical predictions of protein localization and function from genome sequences.

We are grateful for the excellent technical assistance of Steffen Pahlich, Jaqueline Majewski, Reiner Munder, and P. Matzander. We would like to thank Prof. Dr. Sebald from University of Bielefeld for providing access to a Biacore 3000 with microrecovery option. This work was supported by grants from the "Verbundprojekt: Neue Methoden zur Erfassung des Gesamtproteoms von Bakterien" (031U207D) of the Federal Ministry of Education and Research and the European Union (Realis/QLG2-CT-1999-00932).

9.6
References

[1] Vazquez-Boland, J.-A., Kuhn, M., Berche, P., Chakraborty, T. et al., *Clin. Microbiol. Rev.* 2001, *14*, 584–640.

[2] Cabanes, D., Dehoux, P., Dussurget, O., Frangeul, L., Cossart, P., *Trends Microbiol.* 2002, *10*, 238–245.

[3] Glaser, P., Frangeul, L., Buchrieser, C., Amend, A. et al., *Science* 2001, *294*, 849–852.

[4] Jeffery, C. J., *TIBS* 1999, *24*, 8–11.

[5] Laemmli, U. K., *Nature* 1970, *227*, 680–685.

[6] Towbin, H., Staehlin, T., Gordon, J., *Proc. Natl. Acad. Sci. USA* 1979, *76*, 4350–4354.

[7] Rabilloud, T., Strub, J.-M., Luche, S., van Dorsselaer, A., Lunardi, J., *Proteomics* 2001, *1*, 699–704.

[8] Neuhoff, V., Arold, N., Taube, D., Ehrhardt, W., *Electrophoresis* 1988, *9*, 255–262.

[9] Shevchenko, A., Wilm, M., Vorm, O., Mann, M., *Anal. Chem.* 1996, *68*, 850–858.

[10] Wilkins, M. R., Gasteiger, E., Bairoch, A., Sanchez, J. C. et al., *Methods Mol. Biol.* 1999, *112*, 531–552.

[11] Arai, M., Ikeda, M., Shimizu, T., *Gene* 2003, *304*, 77–86.

[12] Bergmann, S., Wild, D., Diekmann, O., Frank, R. et al., *Mol. Microbiol.* 2003, *49*, 411–423.

[13] Dramsi, S., Biswas, I., Maguin, E., Braun, L. et al., *Mol. Microbiol.* 1995, *16*, 251–261.

[14] Gaillard, J.-L., Berche, P., Frehel, C., Gouin, E., Cossart, P., *Cell* 1991, *65*, 1127–1141.

[15] Raffelsbauer, D., Bubert, A., Engelbrecht, F., Scheinpflug, J. et al., *Mol. Gen. Genet.* 1998, *260*, 144–158.

[16] Müller, S., Hain, T., Pashalidis, P., Lingnau, A. et al., *Infect. Immun.* 1998, *66*, 3128–3133.

[17] Jonquieres, R., Pizarro-Cerda, J., Cossart, P., *Mol. Microbiol.* 2001, *42*, 955–965.

[18] Milohanic, E., Jonquieres, R., Cossart, P., Berche, P., Gaillard, J. L., *Mol. Microbiol.* 2001, *39*, 1212–1224.

[19] Kuhn, M., Goebel, W., *Infect. Immun.* 1989, *57*, 55–561.

[20] Schubert, K., Bichlmaier, A. M., Mager, E., Wolff, K. et al., *Arch. Microbiol.* 2000, *173*, 21–28.

[21] Antelmann, H., Tjalsma, H., Voigt, B., Ohlmeier, S. et al., *Genome Res.* 2001, *11*, 1484–1502.

[22] Antelmann, H., Yamamoto, H., Sekiguchi, J., Hecker, M., *Proteomics* 2002, *2*, 591–602.

[23] Navarre, W. W., Schneewind, O., *Microbiol. Mol. Biol. Rev.* 1999, *63*, 174–229.

[24] Osaki, M., Takamatsu, D., Shimoji, Y., Sekizaki, T., *J. Bacteriol.* 2002, *184*, 971–982.

[25] Bierne, H., Mazmanian, S. K., Trost, M., Pucciarelli, M. G. et al., *Mol. Microbiol.* 2002, *43*, 869–881.

[26] Carroll, S. A., Hain, T., Technow, U., Darji, A. et al., *J. Bacteriol.* 2003, *185*, 6801–6808.

[27] Varshavsky, A., *Proc. Natl. Acad. Sci. USA* 1996, *93*, 12142–12149.

[28] Moors, M. A., Auerbuch, V., Portnoy, D. A., *Cell. Microbiol.* 1999, *1*, 249–257.

[29] Sijts, A. J., Pilip, I., Pamer, E. G., *J. Biol. Chem.* 1997, *272*, 19261–19268.

[30] Bergmann, S., Rohde, M., Chhatwal, G. S., Hammerschmidt, S., *Mol. Microbiol.* 2001, *40*, 1273–1287.

[31] Kuchler, K., *Trends Cell Biol.* 1993, *3*, 421–425.

[32] Berrier, C., Garrigues, A., Richarme, G., Ghazi, A., *J. Bacteriol.* 2000, *182*, 248–251.

[33] Wang, I. N., Smith, D. L., Young, R., *Annu. Rev. Microbiol.* 2000, *54*, 799–825.

[34] Vukov, N., Moll, I., Blasi, U., Scherer, S., Loessner, M. J., *Mol. Microbiol.* 2003, *48*, 173–186.

[35] Zimmer, M., Sattelberger, E., Inman, R. B., Calendar, R., Loessner, M. J., *Mol. Microbiol.* 2003, *50*, 303–317.

[36] Denny, P. W., Gokool, S., Russell, D. G., Field, M. C., Smith, D. F., *J. Biol. Chem.* 2000, *275*, 11017–11025.

[37] Boël, G., Pichereau, V., Mijakovic, I., Mazé, A. et al., *J. Mol. Biol.* 2004, *337*, 485–496.

[38] Maguire, M., Coates, A. R., Henderson, B., *Cell Stress Chaperones* 2002, *7*, 317–329.

[39] Sirover, M. A., *Biochim. Biophys. Acta* 1999, *1432*, 159–184.

[40] Pancholi, V., Fischetti, V. A., *J. Exp. Med.* 1992, *176*, 415–426.

[41] Lottenberg, R., Broder, C. C., Boyle, M. D., Kain, S. et. al., *J. Bacteriol.* 1992, *174*, 5204–5210.

[42] Modun, B., Williams, P., *Infect. Immun.* 1999, *67*, 1086–1092.

[43] Eberhard, T., Kronvall, G., Ullberg, M., *Microb. Pathog.* 1999, *26*, 175–181.

[44] Robbins, K. C., Summaria, L., *Methods Enzymol.* 1976, *45*, 257–273.

[45] Pollanen, J., Stephens, R. W., Vaheri, A., *Adv. Cancer Res.* 1991, *57*, 273–328.

[46] Miyashita, C., Wenzel, E., Heiden, M., *Haemostasis* 1988, *18*, 7–13.

[47] Gonzales-Gronow, M., Stack, S., Pizzo, S. V., *Arch. Biochem. Biophys.* 1991, *286*, 625–628.

[48] Duval-Jobe, C., Parmely, M. J., *J. Biol. Chem.* 1994, *269*, 21353–21357.

[49] Wong, A. P., Cortez, S. L., Baricos, W. H., *Am. J. Physiol.* 1992, *263*, 1112–1118.

[50] Fontan, P. A., Pancholi, V., Nociari, M. M., Fischetti, V. A., *J. Infect. Dis.* 2000, *182*, 1712–1721.

[51] Kato, K., Asai, R., Shimizu, A., Suzuki, F., Ariyoshi, Y., *Clin. Chim. Acta* 1983, *127*, 353–363.

[52] Royds, J. A., Parsons, M. A., Taylor, C. B., Timperley, W. R., *J. Pathol.* 1982, *137*, 37–49.

[53] Niklinski, J., Furman, M., *Eur. J. Cancer Prev.* 1995, *4*, 129–138.

[54] Ledermann, J. A., *Eur. J. Cancer* 1994, *30A*, 574–576.

[55] Ebert, W., Muley, T., Drings, P., *Anticancer Res.* 1996, *16*, 2161–2168.

[56] Kaiser, E., Kuzmits, R., Pregant, P., Burghuber, O., Worofka, W., *Clin. Chim. Acta* 1989, *183*, 13–31.

[57] Lopez-Alemany, R., Correc, P., Camoin, L., Burtin, P., *Thromb. Res.* 1994, *75*, 371–381.

[58] Nakajima, K., Hamanoue, M., Takemoto, N., Hattori, T. et al., *J. Neurochem.* 1994, *63*, 2048–2057.

[59] Miles, L. A., Dahlberg, C. M., Plescia, J., Felez, J. et al., *Biochemistry* 1991, *30*, 1682–1691.

[60] Redlitz, A., Fowler, B. J., Plow, E. F., Miles, L. A., *Eur. J. Biochem.* 1995, *227*, 407–415.

[61] Gilot, P., Andre, P., Content, J., *Infect. Immun.* 1999, *67*, 6698–6701.

[62] Daniels, J. J., Autenrieth, I. B., Goebel, W., *FEMS Microbiol. Lett.* 2000, *190*, 323–328.

10
Low virulent strains of *Candida albicans*: Unravelling the antigens for a future vaccine*

Elena Fernández-Arenas, Gloria Molero, César Nombela, Rosalía Diez-Orejas and Concha Gil

Several low virulent *Candida albicans* mutant strains: CM1613 (deleted in the Mitogen Activated Protein (MAP) Kinase *MKC1*), CNC13 (deleted in the MAP-kinase *HOG1*) and the morphological mutant 92' were used as vaccines employing a murine model of systemic candidiasis. In this vaccination trial, only the CNC13 strain was able to induce protection against a subsequent infection with a lethal dose of the wild-type strain. The protection induced by CNC13 vaccinated animals resulted in 60–70% percent of survival. These results demonstrate that collaboration between cellular and humoral responses, induced by the CNC13 mutant, elicited a long lasting and effective protection. Using a proteomic approach (two-dimensional gel electrophoresis followed by Western blotting), twenty-five *C. albicans* immunogenic proteins were detected and identified by matrix-assisted laser desorption/ionization and/or tandem mass spectrometry. We were able to define an antibody pattern in the sera from the nonvaccinating strains (92' and CM1613), which was different from the profile detected in the sera from surviving animals (vaccinated with the CNC13 mutant). The utility of this proteomic approach has allowed us to identify antigens that induce protective IgG2a antibody isotype in the sera from vaccinated animals: enolase (Eno1p), pyruvate kinase (Cdc19p), pyruvate decarboxylase (Pdc11p), a component from the 40S ribosomal subunit (Bel1p), triosephosphate isomerase (Tpi1p), DL-glycerol phosphatase (Rhr2p), fructose-bisphosphate aldolase (Fba1p) and two new protective antigens: IMP dehydrogenase (Imh3p), and acetyl-CoA synthetase (Acs2p). The antigenic proteins that promote protective antibodies described in this work are excellent candidates for a future fungal vaccine; their heterologous expression and vaccine design is currently underway.

* Originally published in Proteomics 2004, 10, 3007–3020

10.1
Introduction

Candida albicans is part of the microbial flora that colonizes the mucocutaneous surfaces of the oral cavity, gastrointestinal tract and vagina. The high levels of morbidity and mortality induced by *C. albicans* in hospitalized patients mean that this species is now one of the most prominent human pathogens [1], especially in immunocompromised hosts. Diagnosis of fungal diseases is not always possible because of difficulties in the isolation of the fungus from clinical samples. Current antifungal treatments are hampered by not being rapidly fungicidal, and by having a limited spectrum, toxicity concerns, and emerging resistance. Hence, there is considerable interest in preventing fungal diseases through the development of vaccines based on overcoming host immune deficits *via* stimulation of cellular and/or humoral immune responses [2]. After 200 years of experience, vaccinology has proven to be very effective in preventing many infectious diseases [3]. A large proportion of the currently available vaccines falls into three major categories; attenuated or killed microorganisms; bacterial toxins; and polysaccharide-carrier protein conjugates. However, new approaches to vaccine development are needed because, firstly, for several infectious diseases traditional approaches have failed and secondly, today, regulatory authorities require extremely safe vaccines. At present, there are no licensed vaccines for the prevention or treatment of human fungal infections [4].

As for other fields in the medical sciences, it is expected that vaccinology will greatly benefit from emerging technologies such as DNA microarrays, bioinformatics and proteomics [3]. Proteomics is a very useful tool to examine host response to microbial infections [5]. For example, the identification of immunogenic proteins by a combination of 2-DE and immunoblotting has been used with a number of bacteria [6–9]. In *C. albicans*, several studies by our group have demonstrated the utility of this proteomic approach for the detection and identification of antigenic proteins [10–12].

In experimental candidiasis, the best protective effects have been achieved by immunization with viable cells from virulent strains of *C. albicans* [13–15], while immunization with killed cells or subcellular components of the organism has not always been successful [16–22]. Other approaches for searching for anti-*Candida* protection have been vaccination with a low virulent strain in a murine model with systemic candidiasis [23, 24]; or the study of the protection spontaneously acquired after recovery from a vaginal infection [13, 25].

It is generally believed that the major function of T lymphocytes in *Candida* infections is the production of cytokines with activating and deactivating signals to fungicidal effector phagocytes. Protective immunity to *C. albicans* is mediated by Th1 responses [26–28], but some Th2 cytokines are necessary to maintain the Th1-dependent immune resistance against fungi [29]. The necessity for this collaboration in the development of a future vaccine has already been described for other microorganisms [30, 31]. The employment of an attenuated *C. albicans* strain (CNC13) to selectively induce high levels of protective antibodies has demonstrated the importance of the humoral response in the resolution of systemic candidiasis [32]. In the present work, we describe the vaccination capacity of this mutant strain, as compared with other low virulent mutant strains. The collaboration of both humoral and cellular immune responses in the induction of a long lasting protection of the CNC13 mutant strain, has been also described in a murine model of systemic candidiasis. The use of proteomic techniques has led us to identify protective antibodies selectively induced by the CNC13 strain that are not produced by the other attenuated mutant strains assayed for protection, and these results are in agreement with their nonvaccination capacity.

10.2
Materials and methods

10.2.1
Microorganism and culture conditions

C. albicans SC5314 (wild-type) [33] was employed for the generation of the following mutant strains: CNC13 strain: *ura3Δ::imm^{434}/ura3Δ::imm^{434} his1Δ::hisG/his1Δ::hisG hog1::hisG-URA3-hisG/hog1::hisG* [34] and CM1613 strain: *mkc1Δ::hisG-CaURA3-hisG/mkc1Δ::hisG ura3Δ::imm434 ura3Δ::imm434* [35]. *C. albicans* 1001 (ATCC 64385) was obtained from the Spanish Type Culture Collection (Departamento de Microbiología, Universidad de Valencia, Spain); the morphological mutant *C. albicans* 92' derived from the parental 1001 strain after UV-light treatment [36] was unable to give rise to mycelia under the conditions usually employed to induce hyphal morphogenesis. All strains were grown on yeast extract dextrose (YED) agar plates (1% Difco yeast extract, 2% glucose, 2% agar) and incubated at 30°C for at least 2 d.

10.2.2
Mice

Inbred BALB/c mice, ranging in age from 6–8 weeks, with a weight of about 20 g, were obtained from Harlan France (Sarl, France). IFN-γR$^{0/0}$, IL-4 knock out (KO) and the parental mice strains [37] were provided by Dr. M. Fresno (Centro de Biología Molecular, Madrid, Spain). All animal studies were carried out at the Animal Facilities of the School of Medicine at the Complutense University in Madrid, Spain. All animal handling procedures were performed according to the European Community's ethical rules for animal experimentation (According to specifications of Directive 86/609/EEC).

10.2.3
Systemic infection conditions and generation of immune sera

For the infection conditions, *Candida albicans* cells were harvested from YED agar plates, washed twice with PBS and diluted to the desired density in the same buffer prior to injection into the lateral tail vein of mice at a volume of 0.5 mL [38, 39].

Survival experiments were carried out in groups of 10 mice and mortality was monitored for 30 days in order to calculate both the median survival time (MST) and the percentage of mortality of each strategy of vaccination. Pooled sera, from at least 10 mice each, were obtained on different days after the challenge by cardiac puncture. Serum from nonvaccinated mice was obtained as a negative control. All sera were stored at −20°C. In all the cases, at least two different batches of sera were obtained and analyzed separately.

10.2.4
T cell purification and passive immunization

Balb/c mice were inoculated intravenously with 1×10^6 cells from *C. albicans* CNC13. Spleens from mice were collected aseptically in a laminar flow hood 15 d after infection. Cell suspensions were obtained by mechanical disruption of freshly isolated spleens, in a complete culture medium (MCC) containing RPMI 1640, 10% foetal calf serum, 2 mM L-glutamine, 100 U of penicillin *per* mL and 100 µg of streptomycin *per* mL. To remove macrophages the cell suspension was placed in petri culture dishes for 1 h at 37°C in an atmosphere containing 5% CO_2. After the incubation, nonadherent cells were removed by gentle washing with a Pasteur pipette, and erythrocytes were removed by incubating in lysing buffer (0.15 M ammonium chloride, 0.01 M potassium hydrogen carbonate, 0.001 M EDTA) for 10 min at 4°C. The cells were then washed twice by centrifugation (1000 rpm) for 10 min at 4°C and suspended in MCC buffer. Cells were incubated for 1 h at room temperature on anti-Ig-coated plates (polyclonal goat anti-mouse Ig; Sigma, St. Louis, MO, USA) in order to remove B cells. Finally, the T cell suspension was collected and counted in a Neubauer chamber and viability was assessed by trypan blue exclusion. For passive immunization, T cells were washed in saline buffer and diluted to the desired density (8×10^6 cells) to be injected intravenously in a volume of 0.5 mL/mouse.

10.2.5
2-DE

10.2.5.1 Protoplast lysate preparation
Cytoplasmic extracts from protoplasts were prepared as previously described [10] with minor modifications. Briefly, *C. albicans* SC5314 cells were grown in liquid YED medium at 28°C up to an optical density of 2 at 600 nm, harvested and washed once with water. Cells were resuspended at a density of $1-2 \times 10^9$ cells/mL and incubated in a pretreatment solution (10 mM Tris-HCl pH 9, 5 mM EDTA, 1% v/v 2-mercapto-ethanol) at 28°C by shaking at 80 rpm for 30 min. After washing with 1 M sorbitol, they were resuspended in the same solution at a density of 5×10^8 cells/mL and 30 µL/mL glusulase (Perkin Elmer, Boston, MA, USA) was added. Cells were then incubated with gentle shaking until more than 90% protoplasts were obtained. Protoplasts were gently washed three times with 1 M sorbitol to eliminate any trace of glusulase. The pellet was resuspended in 2 mL of cold lysis buffer (50 mM Tris-HCl, pH 7.5, 150 mM NaCl, 1 mM DTT, 0.5 mM PMSF and

5 µg/mL of leupeptin, pepstatin and antipain (Sigma)), vortexed for 1 min, and cooled on ice for 2 min (this procedure was repeated three times). The suspension was centrifuged for 15 min at 16 060 × g at 4°C. Supernatants were again transferred to new tubes and centrifuged. The supernatants were stored at −80°C. Protein quantification was performed using the Bradford assay.

10.2.5.2 Analytical and micropreparative 2-DE

2-DE was performed as previously reported [10] with some modifications. Approximately 250 µg of protein from protoplast lysates were loaded for analytical gels and 2 mg for micropreparative gels. IEF was carried out on the IPGphor unit (Amersham Biosciences, Uppsala, Sweden) at 15°C with the following programs: for analytical gels 500 V for 1 h, 500–2000 V in 1 h and 8000 V for 5.5 h; for micropreparative gels 30 V (active rehydration) for at least 13 h, 500 V for 1 h, 1000 V for 1 h, 2000 for 1 h, 2000–5000 V in 3 h and 8000 V for 11 h. IPG strips providing a nonlinear pH 3–10 gradient (18 cm long, Amersham Biosciences) were used. After equilibration of the strips, the second dimension separation by M_r was carried out on homogenous 10% T, 1.6% C (piperazine diacrylamide was used as a cross-linker) polyacrylamide gels (1.5 mm thick). Electrophoresis was conducted at 40 mA/gel constant current for 6 h in a Protean II cell (Bio-Rad, Richmond, CA, USA). Analytical gels were silver-stained as described by Bjellqvist *et al.* [40] with a few modifications, and preparative gels were stained using R-250 CBB and silver-staining compatible with MS analyses [41]. Silver-stained gels were scanned with a GS-690 Imaging Densitometer (Bio-Rad) and analyzed with MELANIE 3.0 computer software (Bio-Rad). Using MELANIE 3.0 tools, protein spots were detected, enumerated, quantified (optical density, area, volume, optical density % and volume %) and characterized with respect to their M_r and pI by bilinear interpolation between landmark features on each image previously calibrated with respect to internal standards (Bio-Rad). For each experiment, at least three gels were analyzed to guarantee representative results.

10.2.6
Immunoblot analyses

Analytical gels were electroblotted onto NC membranes in Towbin buffer at 50 mA overnight as previously described [12]. Sera were diluted 1:100. Immunoreactive spots were detected using the polyclonal antibodies horseradish peroxidase-labeled anti-mouse Igs (Amersham Biosciences), IgGs (Stressgen, San Diego, CA, USA) and IgG2a (Cultek, Madrid, Spain) at dilutions of 1:2000 and a chemiluminescence-based kit (ECL; Amersham Biosciences). The membranes were wrapped in plastic foil and exposed to Kodak BioMax MR1 film for up to 2–4 min (Amersham Biosciences). At least three immunoblots were performed for each serum. The same membrane was stripped and reused with different sera to correlate the differences in antigen expression. The membranes were stripped using the following buffer: 7.5 g glycine in 700 mL of MilliQ water (Millipore, France), pH 2.2 (adjusted with HCl),

10% NP-40, 1% SDS final concentration and addition of MilliQ water up to 1 L. The membranes were incubated in this solution for 1.5 h. Immobilized proteins on NC membranes were stained with SYPRO Ruby (Bio-Rad), a novel ruthenium-based fluorescent dye, following the manufacturer's instructions. In short, each membrane was completely immersed in 7% acetic acid, 10% methanol and incubated at room temperature for 10 min in a polystyrene dish with continuous, gentle agitation. Subsequently, the membrane was washed four times with deionized water for 5 min each. Next, it was completely immersed in SYPRO Ruby protein blot stain for 15 min, and then washed at least six times for 1 min each in deionized water to remove excess dye. Stained membranes were monitored using UV epi-illumination periodically to determine if the background fluorescence had been washed away. Fluorescence image acquisition was carried out using the laser-scanning instrument Molecular Imager FX (Bio-Rad). The blots were excited with a 532 nm laser. These images were acquired as digital TIF files using the Quality One program and analyzed with MELANIE 3.0 computer software (Bio-Rad; see Section 2.5.2 for details).

10.2.7
MALDI-TOF and MALDI-TOF MS analyses of spots

The gel spots of interest were manually excised from micropreparative gels by biopsy punches, placed in an Eppendorf tube, and washed twice with double-distilled water. Proteins for analysis were in-gel reduced, alkylated and digested with trypsin (Roche Molecular Biochemicals, Indianapolis, IN, USA) according to the procedure published by Sechi and Chait [42]. After digestion, the supernatant was collected and 1 µL was spotted to a plate (96 × 2 spot teflon-coated plates; PerSeptive Biosystems, Framingham, MA, USA) and allowed to air-dry for 10 min at room temperature. Subsequently, 0.4 µL of matrix (3 mg/mL CHCA (Sigma) diluted in 0.1% TFA-ACN/H$_2$O (1:1, v/v)) was added to the dried peptide digest spots and allowed to air-dry for another 10 min at room temperature. The samples were analyzed with a MALDI-TOF MS model Voyager DE™ STR instrument fitted with a 337 nm nitrogen laser shots under threshold irradiance (PerSeptive Biosystems), with an accelerating voltage of 20 000 V. This is a highly sensitive system which can give M_r information at the attomole level (10^{-14} g for a 10 kDa molecule) [43]. All MALDI spectra were externally calibrated using a standard peptide mixture: angiotensin I (1296.7), adenocorticotropic hormone fragment 18–39 (2465.2) and 1–17 (2093.1) (Sigma). Peptides from the autodigestion of trypsin were used for the internal calibration. The analysis by MALDI-TOF MS produced peptide mass fingerprints. The peptides observed can be collated and represented as a list of monoisotopic M_r. For MS analyses, a monoisotopic peak is selected and fragmented. MS/MS sequencing analyses were carried out using a MALDI-tandem TOF mass spectrometer 4700 Proteomics Analyzer (Applied Biosystems, Framingham, MA, USA).

10.2.8
Database search

The monoisotopic PMF data obtained from MALDI-TOF MS were used to search for protein candidates in two sequence databases: the Swiss-Prot/TrEMBL non-redundant protein database (www.expasy.ch/sprot) and a nearly complete *C. albicans* genomic database, *Candida* DB (genolist.Pasteur.fr/CandidaDB) using the following software programs: MS-Fit (www.prospector.ucsf.edu), Profound (www.prowl.rockefeller.edu) and MASCOT (www.matrixscience.com). The parameters used for the search were as follows: modifications were considered (Cys as S-carboamidomethyl derivative and Met as oxidized methionine), allowing for one missed cleavage site; a restriction was placed on pI (3–10) and a protein M_r range from 10 to 100 kDa was accepted. Positive identifications were accepted when at least five peptide masses matched, and at least 20% of the peptide coverage of the theoretical sequences matched within a mass accuracy of 50 ppm or 25 ppm with internal calibration. Peptides were excluded if their masses corresponded to those for trypsin, human keratins or other irrelevant proteins.

10.3
Results

10.3.1
Vaccination assays with different mutant strains and generation of immune sera

Different *C. albicans* mutant strains were used in an attempt to protect mice from a subsequent infection with wild-type SC5314 (Tab. 1). Strains were chosen because of their lower virulence with respect to their parental strains. The doses employed for the vaccination procedures were chosen keeping in mind that the immunized animals would render a 90 to 100% rate of survival, according to previous studies by our group [38, 39, 44]. At day +15, mice were infected intravenously with a lethal dose of wild-type *C. albicans* (1×10^6 cells). In Tab. 1, the protection capacity of the different strains is represented by the percentage of survival and the MST reached by the mice. The different mutant strains used, the doses and the protocols of vaccination were as follows: (i) Two doses (5×10^4 cells and 1×10^5 cells) of the mutant strain CM1613 were used. When infected with SC5314 (wild-type), all the mice died during the same week independently of the dose used. (ii) Two different doses (1×10^6 cells and 1×10^7 cells) of the mutant strain *C. albicans* 92' were used. After 15 days, mice were infected with a lethal dose (1×10^6 cells) of *C. albicans* 1001 (wild-type). The virulence of *C. albicans* 1001 is similar to that of SC5314 [38, 39]. Again, all the mice died during a one week time period. (iii) Two different doses (1×10^5 cells and 1×10^6 cells) of the CNC13 mutant strain were employed. Mice were infected with *C. albicans* SC5314 (wild-type). Depending on the vaccination dose, the level of protection reached was different: 40% of survival was reached when using 1×10^5 cells and 60–70% of survival was reached with 1×10^6 cells.

Tab. 1 Summary of vaccination conditions with C. albicans mutant strains and generation of the different immune sera

C. albicans strain	Dose	Serum extraction (+15 d)	Infection	MST	% Survival	Serum extraction (+45 d)
92'	1×10^7	Serum 92'	1×10^6 1001	4	0	–
	1×10^6	–	1×10^6 1001	4	0	–
CM1613	1×10^5	Serum CM1613	1×10^6 SC5314	4	0	–
	5×10^4	–	1×10^6 SC5314	4	0	–
CNC13	1×10^6	Serum CNC13[a]	1×10^6 SC5314	> 30	60–70	Serum CNC13/SC
	1×10^5	–	1×10^6 SC5314	> 30	40	–

The MST from mice infected with C. albicans wild-type strains 1001 and SC5314 is four days.
a) Fernández-Arenas et al. [32]

According to the level of protection (Tab. 1), the CNC13 strain is the only candidate that is able to vaccinate. In order to study the antibody profile induced by the different immunization protocols (see above), immune sera were generated from mice immunized with the highest dose of each C. albicans mutant strain and extracted at day +15. Sera were named according to the strain used: (i) Serum CM1613 from mice immunized with 1×10^5 CM1613. (ii) Serum 92': from mice immunized with 1×10^7 92'. (iii) Serum CNC13: from mice immunized with 1×10^6 CNC13 [32]. (iv) Serum from mice who survived the lethal dose of SC5314 and were sacrificed on day +30 after the wild-type infection (identified as serum CNC13/SC).

10.3.2
Importance of cellular immunity in vaccination with C. albicans CNC13

10.3.2.1 Role of Th1/Th2 cytokines in CNC13 immune response
To ensure the effect of Th1 versus Th2 responses in the development of CNC13 protection in mice, cellular immune response was evaluated in a mouse model of systemic infection using KO mice. To measure the influence of certain cytokines important for the development of Th1 or Th2 responses, two strains of mice were employed: IFN-γ $R^{0/0}$ KO mice for Th1 responses and IL-4 KO mice for Th2 responses. The survival of the IFN-γ $R^{0/0}$ KO and IL-4 KO strains of mice (compared to their parental strain of mice) was followed over 30 days after intravenous challenge with 1×10^7 blastospores from the CNC13 strain (Fig. 1A). The survival of mice was different depending on the strain of mice: IFN-γ receptor depleted mice showed an important increase in sensitivity to infection with respect to the parental strain of mice: no survival vs. 20% for the parental strain, and differences in MST: 8 vs. 19 days. Less significant differences were observed when infecting IL-4 KO mice (14 vs. 19 days).

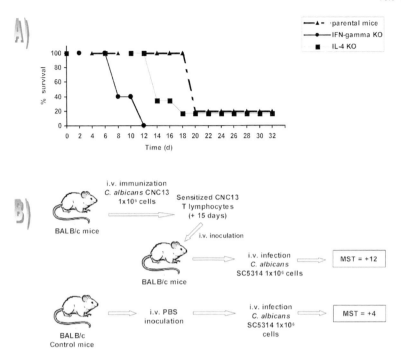

Fig. 1 (A) Survival of IFN-γ $R^{0/0}$ and IL-4 KO mice with respect to the parental immunocompetent strain after infection with 1×10^6 cells from the CNC13 *C. albicans* strain. (B) Effect of passive immunization with sensitized T lymphocytes on the MST of mice infected with a lethal dose of *C. albicans* SC5314. The results are representative of three different experiments.

10.3.2.2 Protection induced by passive transfer of sensitized CNC13 lymphocytes

Taking into account the relevance of the cellular immune contribution observed using KO mice and that the protection induced by whole cells from the CNC13 strain (Tab. 1) was much more significant (MST > 30 and 60–70% survival) than the protection induced by passive immunization with serum CNC13 (MST = 12 and no survival) [32], it was determined that the role of cellular immunity in the protection induced by the CNC13 strain needed to be explored. For this reason, sensitized T lymphocytes from CNC13 immunized mice were isolated and administered intravenously to BALB/c mice in order to evaluate their protective role (Fig. 1B) after challenge with the wild-type SC5314 *C. albicans* strain. The MST from control mice was +4 days. Two separate experiments were carried out and, in both cases, protection due to the passive transfer of sensitized lymphocytes was observed as an increase in the MST(from 4 to +16 days), similar to the protection conferred with CNC13 passive serum transfer, already described [32].

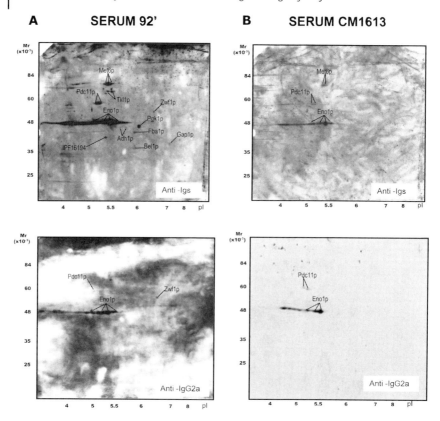

Fig. 2 Immunoblot analysis of *C. albicans* SC5314 extracts separated by 2-DE (250 μg of proteins) using different sera from BALB/c mice. Immunoreactive spots were detected by incubation with peroxidase-labeled anti-mouse Igs and IgG2a subtypes. (A) Serum 92'. (B) Serum CM1613.

10.3.3
Profile of *C. albicans* immunoreactive proteins in the different mutant strains

10.3.3.1 Detection and identification of the immunoreactive proteins

C. albicans cytoplasmic extracts were separated by 2-DE [12]. Two 2-D gels were prepared and run in parallel. One of the gels was silver-stained and used as a reference map (see below). The other gel was blotted onto NC membrane and used for the analyses. To ensure the reproducibility of the technique, sera from independent experiments were assayed at least twice. The described sera were revealed with antibodies of the anti-Igs and IgG2a subtype. A more exhaustive study was done using serum CNC13/SC, where anti-IgGs antibodies were also used. Representative Western blots corresponding to serum 92' and serum CM1613, which rendered nonvaccinated animals, are presented in Fig. 2. As can be observed, there was a considerable similarity in the pattern of antigens recognized by anti-IgG2a and the intensity of the spots between the sera. The reactivity of serum CNC13/SC was the highest, as shown in Fig. 3, when compared with 92' and CM1613 sera.

Immunogenic spots were identified in the reference map whenever possible. In some cases, it was difficult to find the correspondence between the spot in the immunoblot and the protein in the polyacrylamide gel. For this reason, the proteins electroblotted onto NC membranes were stained with SYPRO Ruby [45]. Protein spots were excised from Coomassie and silver-stained 2-D gels [41] and digested as described in Section 2.7. Identification was performed by MALDI-TOF and MALDI-TOF-TOF analyses. Fortunately, the existence of a nearly complete *C. albicans* genomic database, Candida DB (genolist. Pasteur.fr/CandidaDB), allowed us to classify previous nonconclusive protein identifications, and to verify the conclusive ones. Twelve new antigenic *C. albicans* proteins were identified in this research. Fig. 4 shows the silver-stained 2-DE reference map from a *C. albicans* SC5314 extract where all the immunogenic proteins are indicated by name. Using a broad range IPG strip of pH 3–10, the number of silver-stained spots in the gels ranged from between 1200 and 1300. This gel was chosen as the reference gel because of its high resolution and the large number of protein spots detected.

The new antigenic proteins identified in this work are the following: acetyl-CoA synthetase (Acs2p) [46], dihydrolipoamide dehydrogenase (Lpd1p) [47, 48], glucose-6-phosphate dehydrogenase (Zwf1p) [49, 50], ketol-acid reductoisomerase (Ilv5p) [47, 51], a probable alcohol dehydrogenase (Adh5p) [52], mitochondrial malate dehydrogenase precursor (Mdh1p) [53, 54], IPF14662, a D-xylose reductase by homology [55], IPF16194, a protein with unknown function that has homology to a member of the aldose 1-epimerase family [56], a protein transport protein, Sec13p [47], a ribosomal protein, Rpl13p [57], a putative phosphoglycerate mutase (Gpm1p) [11] and phosphomannomutase (Pmm1p) [58]. Some of these proteins have also been detected as antigens using human sera: Acs2p, Ilv5p, Mdh1p, Gpm1p and Pmm1p. The rest of the detected immunoreactive proteins have previously been identified as antigenic markers. They include glycolytic enzymes, like enolase (Eno1p), fructose-bisphosphate aldolase (Fba1p), triosephosphate isomerase (Tpi1p), glyceraldehyde-3-phosphate dehydrogenase (Gap1p), 3-phosphoglycerate kinase (Pgk1p) [12], pyruvate kinase (Cdc19p) [11] and DL-glycerol phosphatase (Rhr2p) [32]; metabolic enzymes: putative pyruvate decarboxylase (Pdc11p) [59, 60], IMP dehydrogenase (Imh3p), 5-methyltetrahydropteroyltriglutamate-homocysteine methyltransferase (Met6p), alcohol dehydrogenase (Adh1p) [12], and transketolase1 (Tkl1p) [32]. Finally, Bel1p, a component from the 40S ribosomal subunit was also detected [32].

The proteins recognized by the different sera generated in this work are summarized in Tab. 2. As can be observed, the sera obtained from the mice inoculated with the mutant strains that did not render protection (CM1613 and 92') recognized a different pattern of proteins than the CNC13 strain, the only one with protection capacity [32]. The reactivity of serum 92' with anti-Igs was highest (Fig. 2A, Tab. 2), and it strongly recognized the following proteins: four protein species of Eno1p, three protein species of Met6p, two protein species of Pdc11p, two protein species of Tkl1p and one protein specie of Pgk1p. It showed a moderate reaction

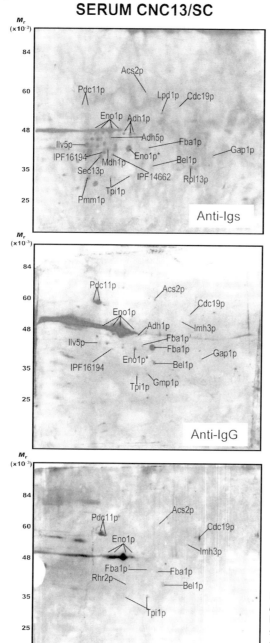

Fig. 3 Immunoblot analyses of C. albicans SC5314 extracts separated by 2-DE (250 μg protein) using serum CNC13/SC from BALB/c mice. Immunoreactive spots were detected by incubation with peroxidase-labeled anti-Igs, anti-IgG and anti-IgG2a antibodies.

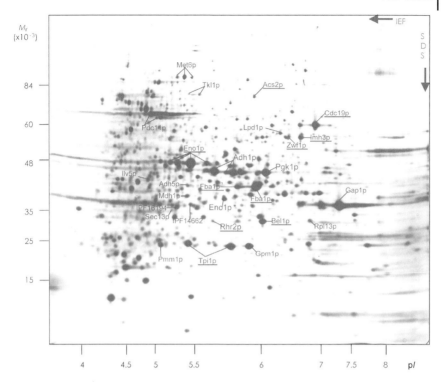

Fig. 4 2-D protein silver-stained map of C. albicans SC5314 extract (250 µg of protein). Underlined proteins are recognized by protective IgG2a antibodies. Twelve proteins were identified in this work and the rest have previously been identified and confirmed in this study. The indicated proteins were detected with different sera from BALB/c mice. More information can be found at the COMPLUYEAST web page.

against Zwf1p (exclusive to this serum), Fba1p and IPF16194, the protein with unknown function, and had a lower reaction against Bel1p, Gap1p and Adh1p. There was a lower reactivity with serum CM1613 when anti-Igs were used: four protein species of Eno1p and a very faint reactivity against two protein species of Met6p and Pdc11p (Fig. 2B, Tab. 2) were detected. When developing serum CNC13/SC with anti-Igs antibodies (Fig. 3, Tab. 2), striking differences were detected. A remarkable reaction against the following proteins was observed: Eno1p, Ilv5p, Sec13p, Rpl13p, Lpd1p, Pmm1p and Adh1p, and two new proteins with unknown function IPF16194 and IPF14662; and a lower reaction against Acs2p, Pdc11p, Cdc19p, Adh5, Fba1p, Mdh1p, Bel1p, Gap1p and Tpi1p (Fig. 3, Tab. 2).

Due to the importance of the isotype of the antibodies in the resolution of systemic candidiasis [32], all the sera were revealed with anti-IgG2a antibody. The pattern of reactivity was quite similar between nonvaccinated animals (Fig. 2, Tab. 2). The proteins detected in these sera were the following: in serum 92' four Eno1p protein species, Zwf1p and one Pdc11p protein specie, while in serum CM1613 only three protein species of Eno1p and one protein specie of Pdc11p (Fig. 2, Tab. 2) were detected. The reactivity of serum CNC13/SC was again higher (Fig. 3, Tab. 2) and displayed an important reaction against four protein species of Eno1p, two protein species of Pdc11p and Cdc19p, and less intense reaction against Bel1p, two protein species of Tpi1p and Fba1p, Rhr2p, Acs2p and Imh3p (Fig. 3, Tab. 2). Because of the immunoreactivity of this sera, we also used anti-IgGs isotype antibodies to detect new antigenic markers. The immunoreactive proteins detected were the following: four protein species of Eno1p, two protein species of Pdc11p and one protein specie of Fba1p with high intensity, while Ilv5p, IPF16194, a fragment of Eno1p*, Bel1p, protein species of Tpi1p and Fba1p', Gmp1p, Acs2p, Cdc19p, Adh1p, Gap1p and Imh3p were detected with lower intensity (Fig. 3, Tab. 2).

10.4
Discussion

10.4.1
Low virulent *C. albicans* strains as a tool to study the host immune response

In previous works, several mutants generated by our group have been employed to asses the involvement of certain mitogen activated protein (MAP) kinases in the pathogenicity of *C. albicans* and the differential immune response that distinct mutants induce in the host [61] (Molero et al., unpublished results). These mutants were the CM1613 strain, with a deletion in MAP kinase *MKC1* [62]; and the CNC13 strain, with a deletion in MAP kinase *HOG1* [34]. Mkc1p is an essential element of the cell integrity signalling pathway, and Hog1p is essential for the oxidative stress and hyperosmolarity responses [63, 64]. We also confirmed the importance of the morphological transition in the pathogenicity of *C. albicans* using the morphological mutant *C. albicans* 92', which is unable to form hyphae under all filament-inducing conditions. When the virulence of these strains was assayed in a murine model of systemic infection, all strains showed reduced virulence. Strain 92' displayed the lowest virulence, rapid clearance from organs and a lack of tissue damage [39]. The CM1613 mutant showed only diminished virulence and relatively high organ colonization [38], while the CNC13 mutant displayed very low virulence and a low but long lasting organ colonization [44], characteristics that could be responsible for the already described efficacy of the CNC13 strain in the generation of protective antibodies [32].

T cells and nonspecific cellular immunity are generally believed to provide the main defences against fungal infections [65]. In the protection induced by CNC13, the significant contribution of the cellular response was also confirmed by an important increase in the MST of the mice passively transferred with T lymphocytes primed with the CNC13 mutant (Fig. 1B). KO mice strains [37] have been used to delineate the role of certain specific cytokines in Th1/Th2 development (IFN-γ receptor deletion to study Th1 and IL-4 deletion to study Th2) [66–68]. These mutant strains of mice were infected with *C. albicans* wild-type and CNC13 strains. The results in Fig. 1A indicate that IFN-γ is necessary for better resolution of

Tab. 2 *C. albicans* immunogenic proteins detected by different sera from BALB/c mice

Antigenic proteins				Sera						
Name[a]	Mass[b] kDa	pI[b]	Accesion Number[a]	92'		CM1613		CNC13/SC		
				Igs	IgG2a	Igs	IgG2a	Igs	IgG2a	IgGs
Met6p	84	5.70	CA0653	+++	−	−	−	−	−	−
	84	5.49		+++	−	++	−	−	−	−
	84	5.38		+++	−	++	−	−	−	−
Tkl1p	75	5.73	CA3924	++	−	−	−	−	−	−
	75	5.61		++	−	−	−	−	−	−
Acs2p	74	5.92	CA2858	−	−	−	−	+	+	+
Pdc11p	66	5.16	CA2474	+++	+	+	+	+	+++	++
	65	5.23		+++	−	+	+	+	+++	+++
Cdc19p	58	7.00	CA3483	−	−	−	−	+	+++	+
Lpd1p	57	6.08	CA2998	−	−	−	−	++	−	−
Zwf1p	56	6.45	CA2634	++	++	−	−	−	−	−
Imh3p	56	6.70	CA1245	−	−	−	−	−	+	+
Eno1p	48	5.76	CA3874	++++	++	+	+	++	+++	++++
	48	5.68		++++	+++	++	++	+++	+++	++++
	48	5.50		++++	+++	++	++	+++	++++	++++
	48	5.24		++++	++	+	−	++	+++	+++
Adh1p	46	5.90	CA4765	+	−	−	−	++	−	−
	46	5.89		+	−	−	−	++	−	+
	46	5.85		−	−	−	−	++	−	−
Pgk1p	46	5.93	CA1691	+++	−	−	−	−	−	−
Ilv5p	44	5.05	CA1983	−	−	−	−	+++	−	+
Adh5p	44	5.68	CA2391	−	−	−	−	+	−	−
Fba1p	42	6.00	CA5180	++	−	−	−	+	+	++++
Fba1p[c]	42	5.80	CA5180	−	−	−	−	−	+	+
Eno1p[d]	39	5.84	CA3874	−	−	−	−	+++	−	++
Mdh1p	37	5.59	CA5164	−	−	−	−	+	−	−
Rhr2p	36	5.72	CA5788	−	−	−	−	−	+	−
Gap1p	35	6.68	CA5892	+	−	−	−	−	−	+
IPF14662	34	5.68	CA1592	−	−	−	−	++	−	−
Bel1p	33	6.09	CA4589	+	−	−	−	+	++	++
IPF16194	33	5.37	CA3352	++	−	−	−	+++	−	+
Sec13p	33	5.33	CA3392	−	−	−	−	+++	−	−
Rpl13p	32	6.88	CA2818	−	−	−	−	++	−	−
Gpm1p	30	5.89	CA4671	−	−	−	−	−	−	+
Pmm1p	29	5.20	CA2198	−	−	−	−	++	−	−

Tab. 2 Continued

Antigenic proteins				Sera						
Name[a]	Mass[b] kDa	p*I*[b]	Accession Number[a]	92'		CM1613		CNC13/SC		
				Igs	IgG2a	Igs	IgG2a	Igs	IgG2a	IgGs
Tpi1p	29	5.68	CA5950	–	–	–	–	++	+	–
	29	5.90		–	–	–	–	+	++	+

a) Protein name and accession number according to CandidaDB (genolist.pasteur.fr/CandidaDB)
b) Experimental M_r and p*I* (Melanie 3.0)
c) Protein species
d) Protein fragment

CNC13 systemic infection, whereas the deletion of IL-4 is also relevant, showing the importance of Th2 responses in the resolution of CNC13 systemic infection. In the case of the SC5314 strain, none of the cytokine deletions used altered the course of the infection (data not shown).

10.4.2
C. albicans hog1 mutant induces protection in a vaccination assay

One reason for the use of live attenuated vaccines is to mirror as much as possible the natural infection, because the immunity elicited by non-alive vaccines usually differs from that elicited by natural infection. A vaccine trial employing the low virulent mutants tested up to now in our laboratory was carried out. We investigated if this capability was strain-specific or if it was shared by all reduced-virulence strains. As can be observed in Tab. 1, in spite of the low virulence of the mutant strains used in this work, only the CNC13 strain was able to protect the mice from a subsequent infection with the wild-type virulent strain of *C. albicans*. The MST of vaccinated animals was > 60 days and the survival of the mice was around the 60–70% (Tab. 1). This experiment demonstrated that, in this case, only the collaboration between cellular and humoral responses induced by the CNC13 mutant was able to elicit a long lasting and effective protection against a subsequent infection with the wild-type strain. Previous works have described that the protection induced by low virulent strains was largely dependent on the mutant strain employed: in some cases, only a Th1 response was detected [69–71], but in others, a collaboration between a Th1 response and a high IgG2a antibody reaction was described [72]. The protection due to a mixed Th1/Th2 response has also been demonstrated when the antibody isotype were compatible with a Th1 response [31, 73–75]. Although the CNC13 strain can develop protection, as this mutant is still slightly virulent, its employment as an attenuated vaccine may be controversial. Nevertheless, it can be useful to select immunoprotective antigens, alone or in combination with other components, that can be used for a future vaccine formulation, similar to what is described for many of the current bacterial vaccines. For these reasons we believe, as do Cutler and coworkers [76], that a future fungal vaccine formulation should consist of a limited number of fungal antigens, rather than a whole cell preparation. The usefulness of this approach has been validated by several groups, both for systemic [77–80] and vaginal [21] experimental candidiasis.

10.4.3
C. albicans new antigenic proteins

We employed a proteomic approach in order to identify the antigens present in the sera obtained from mice inoculated with the two nonvaccinating strains (92' and CM1613) and from vaccinated animals (Tab. 1). Using anti-Igs antibodies, 25 C. albicans immunogenic proteins were detected and identified by MS (Tab. 2, Fig. 4). Twelve new antigenic proteins were also detected and some of them have also been detected as antigens with human sera (Pitarch et al., submitted). Five of these new identified proteins belong to the main pathways of carbohydrate utilization: Mdh1p [53], Adh5p [81], Pmm1p [58], Acs2p [82], and IPF14662 [83]. Two proteins, Lpd1p and Ilv5p, are involved in amino acid biosynthesis [47, 84], and Imh3p is involved in nucleotide metabolism [85]. In addition, some of the proteins belong to other pathways of energy production: Gmp1 which is involved in glycolysis [86]; Lpd1p and Mdh1p which are components of the tricarboxylic-acid pathway [87]; Adh5p which is involved in fermentation [52] and Acs2p which is involved in acetate metabolism [46]. Rpl13p is involved in protein synthesis [57, 88], Sec13p in protein transport [47] and Pmm1p in protein modification [58]. Significantly, the S. cerevisiae homologue of IPF14662, Gre3p, is an aldose reductase stress induced protein and is regulated by the *HOG* pathway [89]. Lpd1p is also involved in the stress response, and C. albicans *LPD1* gene encodes a protein homologous to a bacterial virulence determinant [48]. Proteins Lpd1p and IPF14662, involved in the stress response, could be differentially expressed in the CNC13 mutant inducing a different antibody response. An implication in pathogenicity and a possible role as a target for antifungal drugs has been described for Pmm1p in *Cryptococcus neoformans* [90]. The last protein identified, IPF16194, still has no known biological function in C. albicans, but has high similarity to an uncharacterized S. cerevisiae Ymr099p, related to resistance to fluconazole [91].

10.4.4
Antibody profile linked to successful vaccination against systemic candidiasis

We have detected striking differences between the serum of vaccinated animals treated with CNC13/SC and the sera that did not render protection (Figs. 2 and 3, Tab. 2). Serum CM1613 has low reactivity. It only recognizes Met6p, Pdc11p and Eno1p when revealed both with anti-Igs and anti-IgG2a. This antibody response is similar to the already described profile for the nonprotective sera [32]. On the contrary, the CNC13/SC serum, when revealed with anti-Igs, preserves the antigenic profile already described for the protective serum CNC13+15 [32], but detects more proteins: Fba1p, Gap1p, Lpd1p, Adh5p, IPF14662, IPF16194, Sec13p, Rpl13p, Acs2p, Ilv5p, Mdh1p, Gpm1p and Pmm1p. Due to the already described relationship of IgG2a antibody isotype with protection [22, 32, 92], the antigens detected when revealing with this second antibody could be postulated as being protective. In this case, serum CNC13/SC detected all the proteins previously described for the protective serum CNC13 +15d [32]: Eno1p, Cdc19p, Pdc11p, Bel1p, Tpi1p, Rhr2p and Fba1p and two new protective antigens: Imh3p and Acs2p (see above). The vast majority of the proteins detected with the serum from vaccinated animals are glycolytic enzymes and have already been described as key immunogens during candidiasis (Pitarch *et al.*, submitted) [12, 93].

Unexpectedly, serum 92' shares antibodies with protective sera: Pdc11p, Eno1p, Adh1p, Bel1p, Gap1p, Fba1p and with nonprotective sera: Met6p Tkl1p and Pgk1p [12, 32] when revealed with anti-Igs. When using anti-IgG2a, a new antigen exclusive from this strain, Zwf1p, was detected. Its possible role in protection needs to be evaluated. It belongs to the pentose phosphate pathway and its important protective role during oxidative stress [94–96] and in detoxification have been described [97]. Glycolytic and fermentative enzymes which are also present in the cell surface [61, 98, 99] are major antigenic determinants [11, 93] and play a protective role in the resolution of systemic candidiasis [32, 100]. In order to confirm cell surface recognition of the protective serum from vaccinated animals immunofluorescence detection has been performed (data not shown). This capability has also been detected in other immunoprotective sera demonstrating its role in the resolution of systemic candidiasis. The possible role of this protective serum in fixing complement or increasing the phagocytosis of the yeast is underway at the moment.

10.5
Concluding remarks

This work demonstrates that the reduced virulence of a certain mutant *C. albicans* strain does not confer vaccination capacity. This capability might be linked to a different antigenic expression resulting from a specific genetic background which determines the stimulation of the protective/nonprotective immune response. Once more, the utility of the proteomic approach has allowed us to identify new antigenic markers and to define the immunomes induced by these low virulent *C. albicans* mutant strains. We have described the vaccination capability of the CNC13 mutant. The relationship between protection and the presence of protective antibodies has again been established. The antigenic proteins described in this work that promote protective antibodies are excellent candidates for a fungal vaccine; their heterologous expression and vaccine design is currently underway.

We thank M. L. Hernáez and M. D. Gutierrez from the Centro de Genómica y Proteómica, Universidad Complutense for excellent technical support and Dr. J. Pla for providing the C. albicans CNC13 and CM1613 mutant strains. We also thank Dr. Manuel

Fresno for providing the KO mice for virulence studies. This work was supported by Grants SAF 2000-0108 and BIO 2003-00030 from Comisión Interministerial de Ciencia y Tecnología (CYCIT, Spain); CPGE 1010/2000 for Strategic Groups from Comunidad Autónoma de Madrid (Spain) and a Grant from Fundación Ramón Areces of Spain. E. Fernández-Arenas is the recipient of a fellowship from Ministerio de Educación Cultura y Deporte. C. Nombela is the director of the Merck Sharp and Dohme special chair in genomics and proteomics.

10.6
References

[1] Pfaller, M. A., Jones, R. N., Messer, S. A., Edmond, M. B. et al., *Diagn. Microbiol. Infect. Dis.* 1998, *31*, 327–332.

[2] Polonelli, L., Casadevall, A., Han, Y., Bernardis, F. et al., *Med. Mycol.* 2000, *38*, 281–292.

[3] Grandi, G., *Trends Biotechnol.* 2001, *19*, 181–188.

[4] Casadevall, A., Pirofski, L. A., *J. Exp. Med.* 2003, *197*, 1401–1404.

[5] Marshall, T., Williams, K. M., *Br. J. Biomed. Sci.* 2002, *59*, 47–64.

[6] Lemos, J. A., Giambiagi-Demarval, M., Castro, A. C., *J. Med. Microbiol.* 1998, *47*, 711–715.

[7] Teixeira-Gomes, A. P., Cloeckaert, A., Bezard, G., Bowden, R. A. et al., *Electrophoresis* 1997, *18*, 1491–1497.

[8] Jungblut, P. R., Grabher, G., Stoffler, G., *Electrophoresis* 1999, *20*, 3611–3622.

[9] Vytvytska, O., Nagy, E., Bluggel, M., Meyer, H. E. et al., *Proteomics* 2002, *2*, 580–590.

[10] Pitarch, A., Pardo, M., Jiménez, A., Pla, J. et al., *Electrophoresis* 1999, *20*, 1001–1010.

[11] Pardo, M., Ward, M., Pitarch, A., Sánchez, M. et al., *Electrophoresis* 2000, *21*, 2651–2659.

[12] Pitarch, A., Diez-Orejas, R., Molero, G., Pardo, M. et al., *Proteomics*. 2001, *1*, 550–559.

[13] Cassone, A., Boccanera, M., Adriani, D., Santoni, G. et al., *Infect. Immun.* 1995, *63*, 2619–2624.

[14] Giger, D. K., Domer, J. E., Moser, S. A., McQuitty, J. T. Jr., *Infect. Immun.* 1978, *21*, 729–737.

[15] Mourad, S., Friedman, L., *J. Bacteriol.* 1961, *81*, 550–556.

[16] Banerjee, U., Mohapatra, L. N., Kumar, R., *Indian J. Med. Res.* 1985, *81*, 454–458.

[17] Eckstein, M., Barenholz, Y., Bar, L. K., Segal, E., *Vaccine* 1997, *15*, 220–224.

[18] Levy, R., Segal, E., Barr-Nea, L., *Mycopathologia* 1985, *91*, 17–22.

[19] Mencacci, A., Torosantucci, A., Spaccapelo, R., Romani, L. et al., *Infect. Immun.* 1994, *62*, 5353–5360.

[20] Tavares, D., Ferreira, P., Arala-Chaves, M., *Microbiology* 2003, *149*, 333–339.

[21] De Bernardis, F., Boccanera, M., Adriani, D., Girolamo, A. et al., *Infect. Immun.* 2002, *70*, 2725–2729.

[22] Cardenas-Freytag, L., Cheng, E., Mayeux, P., Domer, J. E. et al., *Infect. Immun.* 1999, *67*, 826–833.

[23] Romani, L., Mencacci, A., Cenci, E., Spaccapelo, R. et al., *J. Immunol.* 1993, *150*, 925–931.

[24] Bistoni, F., Vecchiarelli, A., Cenci, E., Puccetti, P. et al., *Infect. Immun.* 1986, *51*, 668–674.

[25] Fidel, P. L. Jr., *AIDS Patient Care STDS*. 1998, *12*, 359–366.

[26] Cenci, E., Mencacci, A., Del Sero, G., Bacci, A. et al., *J. Infect. Dis.* 1999, *180*, 1957–1968.

[27] Romani, L., *Curr. Opin. Microbiol.* 1999, *2*, 363–367.

[28] Bromuro, C., Torosantucci, A., Chiani, P., Conti, S. et al., *Infect. Immun.* 2002, *70*, 5462–5470.

[29] Montagnoli, C., Bacci, A., Bozza, S., Gaziano, R. et al., *J. Immunol.* 2002, *169*, 6298–6308.

[30] Rao, K. V., He, Y. X., Kalyanasundaram, R., *Clin. Diagn. Lab Immunol.* 2003, *10*, 536–541.

[31] Velikovsky, C. A., Goldbaum, F. A., Cassataro, J., Estein, S. et al., *Infect. Immun.* 2003, *71*, 5750–5755.

[32] Fernández-Arenas, E., Molero, G., Nombela, C., Diez-Orejas, R. et al., *Proteomics* 2004, *4*, 1204–1215.

[33] Gillum, A. M., Tsay, E. Y. H., Kirsch, D. R., *Mol. Gen. Genet.* 1984, *198*, 179–182.

[34] San José, C., Monge, R. A., Pérez-Díaz, R. M., Pla, J. et al., *J. Bacteriol.* 1996, *178*, 5850–5852.

[35] Navarro-García, F., Pérez-Díaz, R. M., Magee, B. B., Pla, J. et al., *J. Med. Vet. Mycol.* 1995, *33*, 361–366.

[36] Nombela, C., Pomés, R., Gil, C., *CRC Crit. Rev. Microbiol.* 1987, *15*, 79–85.

[37] Huang, S., Hendriks, W., Althage, A., Hemmi, S. et al., *Science* 1993, *259*, 1742–1745.

[38] Diez-Orejas, R., Molero, G., Navarro-García, F., Pla, J. et al., *Infect. Immun.* 1997, *65*, 833–837.

[39] Diez-Orejas, R., Molero, G., Ríos-Serrano, I., Vázquez, A. et al., *FEMS Microbiol. Lett.* 1999, *176*, 311–319.

[40] Bjellqvist, B., Sanchez, J. C., Pasquali, C., Ravier, F. et al., *Electrophoresis* 1993, *14*, 1375–1378.

[41] Shevchenko, A., Wilm, M., Vorm, O., Mann, M., *Anal. Chem.* 1996, *68*, 850–858.

[42] Sechi, S., Chait, B. T., *Anal. Chem.* 1998, *70*, 5150–5158.

[43] McLafferty, F. W., Fridriksson, E. K., Horn, D. M., Lewis, M. A. et al., *Science* 1999, *284*, 1289–1290.

[44] Alonso-Monge, R., Navarro-Garcia, F., Molero, G., Diez-Orejas, R. et al., *J. Bacteriol.* 1999, *181*, 3058–3068.

[45] Kemper, C., Berggren, K., Diwu, Z., Patton, W. F., *Electrophoresis* 2001, *22*, 881–889.

[46] Sheridan, R., Ratledge, C., Chalk, P. A., *FEMS Microbiol. Lett.* 1990, *57*, 165–169.

[47] Enjalbert, B., Nantel, A., Whiteway, M., *Mol. Biol. Cell* 2003, *14*, 1460–1467.

[48] Cheng, S., Clancy, C. J., Checkley, M. A., Handfield, M. et al., *Mol. Microbiol.* 2003, *48*, 1275–1288.

[49] Nogae, I., Johnston, M., *Gene* 1990, *96*, 161–169.

[50] Tavanti, A., Gow, N. A., Senesi, S., Maiden, M. C. et al., *J. Clin. Microbiol.* 2003, *41*, 3765–3776.

[51] Petersen, J. G., Holmberg, S., *Nucleic Acids Res.* 1986, *14*, 9631–9651.

[52] Lan, C. Y., Newport, G., Murillo, L. A., Jones, T. et al., *Proc. Natl. Acad. Sci. USA* 2002, *99*, 14907–14912.

[53] Thompson, L. M., Sutherland, P., Steffan, J. S., McAlister-Henn, L., *Biochemistry* 1988, *27*, 8393–8400.

[54] Steffan, J. S., Minard, K. I., McAlister-Henn, L., *Arch. Biochem. Biophys.* 1992, *293*, 93–102.

[55] Nidetzky, B., Bruggler, K., Kratzer, R., Mayr, P., *J. Agric. Food Chem.* 2003, *51*, 7930–7935.

[56] Huh, W. K., Falvo, J. V., Gerke, L. C., Carroll, A. S. et al., *Nature* 2003, *425*, 686–691.

[57] Tait, E., Simon, M. C., King, S., Brown, A. J. et al., *Fungal Genet. Biol.* 1997, *21*, 308–314.

[58] Smith, D. J., Cooper, M., DeTiani, M., Losberger, C. et al., *Curr. Genet.* 1992, *22*, 501–503.

[59] Killenberg-Jabs, M., Konig, S., Hohmann, S., Hubner, G., *Biol. Chem. Hoppe-Seyler* 1996, *377*, 313–317.

[60] Pitarch, A., Sanchez, M., Nombela, C., Gil, C., *Mol. Cell Proteomics* 2002, *1*, 967–982.

[61] Diez-Orejas, R., Molero, G., Moro, M. A., Gil, C. et al., *Med. Microbiol. Immunol.* 2001, *189*, 153–159.

[62] Navarro-Garcia, F., Alonso-Monge, R., Rico, H., Pla, J. et al., *Microbiology* 1998, *144*, 411–424.

[63] Navarro-Garcia, F., Eisman, B., Roman, E., Nombela, C. et al., *Med. Mycol.* 2001, *39* Suppl. 1:87–100, 87–100.

[64] Alonso-Monge, R., Navarro-Garcia, F., Roman, E., Negredo, A. I. et al., *Eukaryot. Cell* 2003, *2*, 351–361.

[65] Levitz, S. M., *Clin. Infect. Dis.* 1992, *14*, S37–S42.

[66] Brummer, E., Morrison, C. J., Stevens, D. A., *Infect. Immun.* 1985, *49*, 724–730.

[67] Nishikawa, Y., Inoue, N., Makala, L., Nagasawa, H., *Vet. Parasitol.* 2003, *20;116*, 175–184.

[68] Vecchiarelli, A., Puliti, M., Torosantucci, A., Cassone, A. et al., *Cell Immunol.* 1991, *134*, 65–76.

[69] Cenci, E., Romani, L., Vecchiarelli, A., Puccetti, P. et al., *Infect. Immun.* 1989, *57*, 3581–3587.
[70] Romani, L., Howard, D. H., *Curr. Opin. Immunol.* 1995, *7*, 517–523.
[71] Mencacci, A., Spaccapelo, R., Del Sero, G., Enssle, K. H. et al., *Infect. Immun.* 1996, *64*, 4907–4914.
[72] Romani, L., Mocci, S., Bietta, C., Lanfaloni, L. et al., *Infect. Immun.* 1991, *59*, 4647–4654.
[73] Seder, R. A., Paul, W. E., *Annu. Rev. Immunol.* 1994, *12*, 635–673.
[74] Deepe, G. S. Jr., *Clin. Microbiol. Rev.* 1997, *10*, 585–596.
[75] Cannas, A., Naguleswaran, A., Muller, N., Gottstein, B. et al., *J. Parasitol.* 2003, *89*, 44–50.
[76] Cutler, J. E., Granger, B. L., Han, Y., in: Calderone, R. A. (Ed.), *Candida and Candidosis*, ASM Press, Washington D.C., USA 2002, pp. 243–256.
[77] Han, Y., Morrison, R. P., Cutler, J. E., *Infect. Immun.* 1998, *66*, 5771–5776.
[78] Han, Y., Ulrich, M. A., Cutler, J. E., *J. Infect. Dis.* 1999, *179*, 1477–1484.
[79] Bystricky, S., Paulovicova, E., Machova, E., *Immunol. Lett.* 2003, *85*, 251–255.
[80] Han, Y., Cutler, J. E., *Infect. Immun.* 1995, *63*, 2714–2719.
[81] Yamamoto, H., Matsuyama, A., Kobayashi, Y., *Biosci. Biotechnol. Biochem.* 2002, *66*, 925–927.
[82] Van den Berg, M. A., Steensma, H. Y., *Eur. J. Biochem.* 1995, *231*, 704–713.
[83] Hernández, R., Nombela, C., Diez-Orejas, R., Gil, C., *Proteomics* 2004, *4*, 374–382.
[84] Bateman, J. M., Perlman, P. S., Butow, R. A., *Genetics* 2002, *161*, 1043–1052.
[85] Köhler, G. A., White, T. C., Agabian, N., *J. Bacteriol.* 1997, *179*, 2331–2338.
[86] White, M. F., Fothergill-Gilmore, L. A., *Eur. J. Biochem.* 1992, *207*, 709–714.
[87] McAlister-Henn, L., Thompson, L. M., *J. Bacteriol.* 1987, *169*, 5157–5166.
[88] Jacq, C., Alt-Morbe, J., Andre, B., Arnold, W. et al., *Nature* 1997, *387*, 75–78.
[89] Aguilera, J., Prieto, J. A., *Curr. Genet.* 2001, *39*, 273–283.
[90] Wills, E. A., Roberts, I. S., Del Poeta, M., Rivera, J. et al., *Mol. Microbiol.* 2001, *40*, 610–620.
[91] Anderson, J. B., Sirjusingh, C., Parsons, A. B., Boone, C. et al., *Genetics* 2003, *163*, 1287–1298.
[92] Londono, L. P., Jones, H. B., Vie, A. T., McPheat, W. L. et al., *FEMS Immunol. Med. Microbiol.* 2000, *27*, 117–125.
[93] Martínez, J. P., Gil, M. L., López-Ribot, J. L., Chaffin, W. L., *Clin. Microbiol. Rev.* 1998, *11*, 121–141.
[94] Minard, K. I., McAlister-Henn, L., *Free Radical Biol. Med.* 2001, *31*, 832–843.
[95] Slekar, K. H., Kosman, D. J., Culotta, V. C., *J. Biol. Chem.* 1996, *271*, 28831–28836.
[96] Juhnke, H., Krems, B., Kotter, P., Entian, K. D., *Mol. Gen. Genet.* 1996, *252*, 456–464.
[97] Salvemini, F., Franze, A., Iervolino, A., Filosa, S. et al., *J. Biol. Chem.* 1999, *274*, 2750–2757.
[98] Chaffin, W. L., Lopez-Ribot, J. L., Casanova, M., Gozalbo, D. et al., *Microbiol. Mol. Biol. Rev.* 1998, *62*, 130–180.
[99] Urban, C., Sohn, K., Lottspeich, F., Brunner, H. et al., *FEBS Lett.* 2003, *544*, 228–235.
[100] Van Deventer, A. J. M., Goessens, W. H., van Vliet, H. J., Verbrugh, H. A., *Clin. Microbiol. Infect.*, 1996, *2*, 36–43.
[101] Pitarch, A., Abian, J., Carrascal, M., Sánchez, M. et al., *Proteomics* 2004, *4*, in press. DOI 10.1002/pmic.200400903.

11
Proteomic analysis of the sarcosine-insoluble outer membrane fraction of the bacterial pathogen *Bartonella henselae*[*]

Thomas A. Rhomberg, Olof Karlberg, Thierry Mini, Ursula Zimny-Arndt, Ulrika Wickenberg, Marlene Röttgen, Peter R. Jungblut, Paul Jenö, Siv G. E. Andersson and Christoph Dehio

Bartonella henselae is an emerging zoonotic pathogen causing a wide range of disease manifestations in humans. In this study, we report on the analysis of the sarcosine-insoluble outer membrane fraction of *B. henselae* ATCC 49882 Houston-1 by one-dimensional sodium dodecyl sulfate-polyacrylamide gel electrophoresis (1-D SDS-PAGE) and two-dimensional nonequilibrium pH gradient polyacrylamide gel electrophoresis (2-D NEPHGE). Protein species were identified by matrix-assisted laser desorption/ionization-time of flight-mass spectrometry (MALDI-TOF-MS) and subsequent database query against the *B. henselae* genome sequence. Subcellular fractionation, application of the ionic detergent lauryl sarcosine, assessment of trypsin sensitivity, and heat modifiability of surface-exposed proteins represented valuable tools for the analysis of the outer membrane subproteome of *B. henselae*. 2-D NEPHGE was applied to display and catalogue a substantial number of proteins associated with the *B. henselae* sarcosine-insoluble outer membrane fraction, resulting in the establishment of a first 2-D reference map of this compartment. Thus, 53 distinct protein species associated with the outer membrane subproteome fraction were identified. This study provides novel insights into the membrane biology and the associated putative virulence factors of this pathogen of increasing medical importance.

11.1
Introduction

Bartonellae are Gram-negative, facultative intracellular bacteria belonging to the α-subdivision of proteobacteria. To date, seven of the nineteen *Bartonella* species described are recognized as human pathogens causing a broad spectrum of clinical

---

[*] Originally published in Proteomics 2004, 10, 3021–3033.

Proteomics of Microbial Pathogens. Edited by Peter R. Jungblut and Michael Hecker
Copyright © 2007 WILEY-VCH Verlag GmbH & Co. KGaA, Weinheim
ISBN: 978-3-527-31759-2

manifestations [1]. The zoonotic species *B. henselae* is the causative agent of cat-scratch disease, a persistent, but usually self-limiting inflammation of the lymph nodes. In immunocompromised patients, *B. henselae* causes bacillary angiomatosis peliosis (BAP), a pathological disorder of the human vasculature characterized by bacteria-induced proliferation of endothelial cells. Stimulation of neovascularization in response to bacterial infection is a phenomenon uniquely associated with the genus *Bartonella* [2].* Present address: Federal Research Centre for Virus Diseases of Animals, Institute for Vaccines, Tübingen, Germany The molecular mechanism of this pathological angiogenesis process is unknown. However, *B. henselae* is able to stimulate the proliferation of primary human umbilical vein endothelial cells (HUVEC) *in vitro* upon cocultivation, even without direct contact [3]. A particulate, noncytosolic fraction of disrupted *B. henselae* is able to mimic this stimulation of HUVEC proliferation and its stimulatory activity can be inactivated by trypsin [4]. Taken together, these data suggest the existence of a proteinaceous angiogenic factor localized to the outer membrane (OM) of *B. henselae*. Based on 1-DE analysis, the OM of *B. henselae* contains at least nine protein species, with M_r of 28, 30, 35, 43, 58, 61, 79, 92, and 171 kDa [5]. Among those, a 43 kDa outer membrane protein (OMP) represents an adhesin for HUVEC [5]. Furthermore, a 31 kDa OMP, originally described as a phage-associated membrane protein in *B. henselae* (named Pap31) [6], was later demonstrated to be an OM hemin-binding protein [7].

Likely, additional pathogenicity factors are targeted to the OM of *B. henselae*, which mediate important steps in the infection of the host such as adhesion, invasion, intracellular survival, and replication as seen in other Gram-negative human-pathogenic bacteria. In particular, *B. henselae* possesses two type IV secretion systems (T4SS), VirB/VirD4 and Trw [8, 9]. These versatile transporters are evolutionarily derived from bacterial conjugation machines. They span both inner membrane (IM) and OM forming a macromolecular secreton for the delivery of yet unknown effector molecules into human endothelial cells [10]. Both systems are essential for *Bartonella* spp. pathogenicity [9, 11]. However, *B. henselae*-triggered human endothelial cell proliferation *in vitro* was recently found to be independent from a functional VirB/VirD4 T4SS [12].

The aim of this study was to characterize the OM subproteome of *B. henselae* by proteomic means to assess the protein composition of this subcellular compartment, which appears fundamental for the pathogenesis of this bacterium. The method of choice for separation of these protein species, inherently difficult to handle [13], was two-dimensional nonequilibrium pH gradient polyacrylamide gel electrophoresis (2-D NEPHGE) [14–16]. Subsequent identification of protein species was achieved by peptide mass fingerprinting (PMF) using matrix-assisted laser desorption/ionization-time of flight-mass spectrometry (MALDI-TOF-MS). The concomitant completion of the *B. henselae* genome sequence [17] allowed for the first time the broad-scale assignment of *B. henselae* proteins in general and proteins associated with the OM in particular. Thus, 53 distinct protein species were unambiguously identified. Out of these, seven protein species proved to represent prototypical OMPs.

11.2
Materials and methods

11.2.1
Strains and culture conditions

B. henselae typing strain ATCC 49882 Houston-1 [18] was routinely grown for 3 days on Columbia blood agar containing 5% defibrinated sheep blood (CSB-agar) in a humidified atmosphere with 5% CO_2 at 35°C.

11.2.2
Enrichment of *B. henselae* OMPs

Subcellular fractionation was performed by sonication, differential lysis of total bacterial membranes with the ionic detergent lauryl sarcosine, and ultracentrifugation. In brief, bacteria were harvested from 16 CSB-agar plates, washed in phosphate-buffered saline, and pelleted twice for 15 min in a Heraeus Biofuge Stratos with a #3046 rotor at 4000 rpm at 4°C. The pellet was resuspended in 4 mL hyperosmolar buffer (0.2 M Tris pH 8.0, 0.5 M sucrose, 250 µg/mL lysozyme, 1 mM EDTA) and incubated on ice for 1 h. Subsequently, 500 µL protease inhibitor cocktail (Roche Complete) were added and bacteria were broken by repeated sonication using a Branson Sonifier (single bursts at maximum level for no longer than 5 s with cooling on ice-cold water to avoid heat denaturation), until the solution turned opaque. Cell debris was removed by centrifugation for 30 min at 4000 rpm. The supernatant containing membrane vesicles was then cleared by centrifugation in a Centrikon T1075 ultracentrifuge with a TFT 80.13 FA rotor for 90 min at 40 000 rpm at 4°C. The resulting total membrane pellet was resuspended in lysis buffer (10 mM HEPES pH 7.4, 1% w/v lauryl sarcosine), incubated at room temperature for 20 min, and pelleted by ultracentrifugation. The lauryl sarcosine-insoluble pellet – that contains the outer membrane-peptidoglycan complex – was washed twice in 10 mM HEPES, pH 7.4, to remove residual detergent and pelleted again in a Heraeus Biofuge Stratos with a #3331 rotor at 20 000 rpm for 30 min at 4°C, collected, and stored at −70°C.

11.2.3
Protease exposure

Enzymatic surface treatment of intact bacteria was included into the subcellular fractionation protocol before sonication. Bacterial pellets were resuspended in 2.7 mL phosphate-buffered saline, 300 µL digestion buffer (0.25% w/v trypsin, 0.05% w/v EDTA, 5 mM glucose in phosphate-buffered saline) was added and bacteria were incubated for 90 min at 37°C. After digestion, 3 mL of trypsin neutralizing solution (PromoCell) were added and incubation at 37°C was continued for 30 min. Control experiments were performed by incubating bacteria at 37°C either in presence of both digestion buffer and trypsin neutralizing solution (mock treated) or simply in phosphate-buffered saline (untreated) for 2 h. Bacteria were subsequently pelleted for 20 min in a Heraeus Biofuge Stratos with a #3046 rotor at 4000 rpm at 4°C and processed for sonication as mentioned above.

11.2.4
1-D SDS-PAGE

1-D SDS-PAGE was performed following standard procedures [19] on small-scale (Mini Protean II system; BioRad, Hercules, CA, USA) or on large-scale format (Protean II xi system, BioRad) as recommended by the supplier.

11.2.5
Protein solubilization and protein quantitation

Protein concentration was determined by the modified method of Lowry [20] in a BioPhotometer (Eppendorf) with a commercial kit (Pierce, Rockford, IL, USA). For 1-D SDS-PAGE, protein samples were resuspended in 10 mM HEPES, pH 7.4, and quantitated. Up to 50 µg of protein were dissolved in SDS sample buffer and denatured by boiling at 100°C for 5 min, immediately loaded onto the gel or stored at −20°C. For heat modification experiments, samples were prepared accordingly at temperatures indicated. For 2-D NEPHGE, outer membranes were resuspended in solubilization buffer (9 M urea, 4% CHAPS, 70 mM DTT, 2% v/v carrier ampholytes pH 2–4) and incubated for 30 min. To avoid precipitation of urea, the suspension was then centrifuged in a Heraeus Biofuge Stratos with a #3331 rotor at 20 000 rpm for 30 min at 21°C. The supernatant containing solubilized OMPs was collected, protein content quantified, and immediately used or stored at −70°C.

11.2.6
2-D NEPHGE

2-D NEPHGE was performed with two different 2-DE systems (WITA and Millipore/2-D Investigator). 2-D NEPHGE on the WITA system was performed as described [15, 16]. This protocol was adapted for 2-D NEPHGE on the Millipore/2-D Investigator system as follows. First-dimensional IEF gels (9 M urea, 4% w/v acrylamide, 0.3% w/v bisacrylamide, 2% v/v ampholytes pH 2–11, 5% w/v glycerol, 0.06% v/v TEMED) were cast without a capping gel, allowed to polymerize, and submerged in 800 mL IEF anode buffer (100 mM *ortho*-phosphoric acid) and 4 L cathode buffer (100 mM sodium hydroxide). Solubilized protein samples (200 µg) were applied at the anode with a Hamilton syringe and covered with overlay solution (5 M urea, 5% w/v glycerol, 2% v/v ampholytes pH 2–4). First-dimensional IEF gels were run for a total of 8820 Vh at a current of 110 µA *per* gel. Thereafter, IEF gels were extruded from the glass tubes, incubated for 10 min in equilibration buffer (125 mM Tris/phosphate, pH 6.8, 40% w/v glycerol, 65 mM DTT, 3% w/v SDS), and stored at −20°C, if not used directly. IEF gels were then embedded on the second-dimensional slab gels and immobilized (1% w/v agarose, 125 mM Tris/phosphate, pH 6.8, 0.1% w/v SDS). For the second-dimensional run, voltage was set at 500 V and power at 16 000 mW per gel for no longer than 6 h. Staining with silver nitrate was performed according to a modified procedure of Heukeshoven and Dernick [21]. Dye staining with Coomassie Brilliant Blue (CBB) R-250 or G-250 was performed as described [22].

11.2.7
MALDI-TOF-MS

Protein processing and measurements were carried out as described [23, 24]. In brief, protein spots excised from CBB R-250 or CBB G-250-stained polyacrylamide gels were each washed five times for 1 min with 30 µL 40% n-propanol followed by five washes for 1 min with 30 µL each of 0.2 M NH_4HCO_3 containing 50% v/v acetonitrile. Gel pieces were then dried completely in a SpeedVac concentrator. To the dried gel pieces, 0.5 µg trypsin (specific activity 18 700 U/mg, Promega) in 10 µL 100 mM NH_4HCO_3 was added. After reswelling, an appropriate volume of digestion buffer (100 mM NH_4HCO_3) was added to completely immerse the gel pieces in the liquid which typically resulted in a digestion volume of 15 µL. Digestion was performed at 37°C for 2 h. Supernatants were collected and gel pieces extracted twice with first 15 µL 0.1% formic acid for 5 min followed by 15 µL acetonitrile for 1 min. To remove supernatants from the gel pieces, two 500 µL Eppendorf tubes arranged concentrically were used. The tube containing the gel pieces was pierced with a needle to generate a hole large enough to allow the liquid to be centrifuged into the lower tube, but retaining the gel pieces. To avoid possible cross contamination, all buffers and wash solutions were pipetted with clean Hamilton syringes, ensuring that the needle never touched the gel pieces. Finally, all supernatants were pooled and dried in a SpeedVac concentrator. For mass spectral analysis, peptides were redissolved in 10 µL 0.1% TFA. Peptides for MALDI analysis were desalted on micro C18 ZipTips (P10 tip size; Millipore, Bedford, MA, USA). Elution was done with 1.5 µL 80% acetonitrile, 0.1% TFA, containing 1 µg/mL α-cyano-4-hydroxycinnamic acid (CHCA; Aldrich, Milwaukee, WI, USA). 300 nL of the elute was deposited onto anchor spots of a Scout 400 µm/36 sample support (Bruker Daltonik, Bremen, Germany) and the droplet was left to dry at room temperature. Mass spectra were recorded on a Bruker Scout 26 Reflex III instrument (Bruker Daltonik). The instrument was calibrated with angiotensin II, substance P, bombesin, and adrenocorticotropic hormone ($ACTH_{18-39}$). Peptides were analyzed in reflector mode using delayed ion extraction with a total acceleration voltage of 23 kV. 50–100 single-shot spectra were acquired to improve signal-to-noise ratio. Spectrum calibration and peak assignment was carried out with the XMASS 5.0 software package provided by the manufacturer. All mass spectra were inspected for trypsin and keratin fragments and, if present, removed from the mass list used for data bank searching.

11.2.8
Database query

Peptide mass fingerprints were queried for identity using a local copy of MS-Fit (Protein Prospector, http://prospector.ucsf.edu/) as an interface [25] for the *B. henselae* ATCC 49882 genome sequence [17]. Searches were conducted using a p*I* range of pH 3–10 and a M_r range of 1 kDa to 100 kDa. The mass tolerance was set to 0.2 Da. Neither cysteine- nor N-/C-terminal modifications were considered in the searches which were conducted with monoisotopic masses. For a match to be considered successful, a minimal set of four matching peptides was set as scale of rating. In cases of matching more than one spot in a spot series to a single protein, a threshold of three peptides necessary was considered sufficient.

11.2.9
***In silico* analysis**

Identification of Pfam domains was conducted using the hmmpfam program in the HMMER package (http://www.hmmer.wustl.edu/) with the Pfam HMM library (http://www.sanger.ac.uk/Software/Pfam/) [26]. The program SignalP (http://www.cbs.dtu.dk/services/SignalP-2.0) was used to predict the presence of a membrane anchor or a secretory signal peptide at the N-terminus of the identified proteins [27, 28]. SignalP has been shown to be the most successful program in correctly assigning signal peptides, signal anchors, and nonsecretory proteins [29]. The program PSORT-B (http://www.psort.org/) was used to predict the subcellular localization of OMP candidates [30]. Grand average of hydropathy (GRAVY) values of OMP candidates were calculated as the sum of the hydropathy values of all amino acids, divided by the length of the amino acid sequence [31].

11.3
Results

11.3.1
Enrichment of *B. henselae* OMPs

Assessment of the quality of subcellular fractionation applied in this study was an indispensable prerequisite for the subsequent proteomic analysis of the OM subproteome of *B. henselae*. The use of the ionic detergent lauryl sarcosine was found to be the most critical step. Lauryl sarcosine was used in numerous studies because of its ability to efficiently and selectively solubilizing the IM of ruptured Gram-negative bacteria [32]. Inner membrane proteins (IMPs) are released in a soluble form, but neither the OM peptidoglycan complex nor the OMPs are affected, which enables their recovery by ultracentrifugation after detergent exposure. To determine the purity of subcellular fractions, proteins were separated by small-format SDS-PAGE and visualized by CBB R-250 staining (Fig. 1). A subtractive pattern for

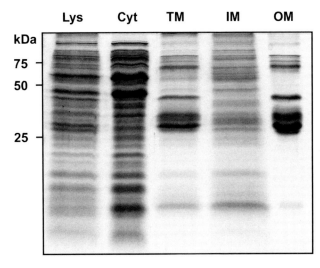

Fig. 1 CBB R-250-stained 1-D SDS-PAGE displaying the different subcellular fractions of B. henselae. Lys, total cell lysate; Cyt, cytoplasmic fraction; TM, total membrane fraction; IM, sarcosine-soluble inner membrane fraction; OM, sarcosine-insoluble outer membrane fraction.

total membrane *versus* IM and OM can be observed. The sarcosine-insoluble OM fraction (OM) reveals prominent protein bands at 85 kDa, 70 kDa, 60 kDa, 43 kDa and a cluster of protein bands between 27 kDa and 35 kDa. These protein species can also be detected in the total membrane fraction (TM), but they are excluded from the sarcosine-soluble IM fraction (IM). In agreement with this finding, prominent IMPs cannot be detected in the OM fraction.

11.3.2
1-D SDS-PAGE of B. henselae OMPs and protein assignment by PMF

To further elucidate the composition of the OM subproteome, trypsin sensitivity and heat modifiability of the sarcosine-insoluble OMP fraction was tested. Trypsin exposure of intact bacteria before subcellular fractionation will exclusively cleave surface-exposed proteins – namely OMPs protruding from the outer leaflet of the OM. In the course of subcellular fractionation membrane inserted polypeptide remnants can be recovered and should be detectable after 1-D SDS-PAGE due to a decrease in the apparent M_r.

Electrophoretic mobility of prototypical OMPs on 1-D SDS-PAGE is dependent on the solubilization temperature applied during the denaturation process [33, 34]. Therefore, heat modifiability of the sarcosine-insoluble OMP fraction was tested at two temperatures, 50°C and 100°C. Micropreparative amounts of the sarcosine-insoluble OM fraction (50 μg), subjected to trypsin exposure and/or heat modification, were separated by large-format 15% SDS-PAGE and visualized by CBB R-250 staining (Fig. 2A). Trypsin exposure led to the detection of at least nine protein species which differed in electrophoretic mobility upon trypsin exposure (Fig. 2A, lane 2) when compared to mock (Fig. 2A, lane 4) or untreated fractions (Fig. 2A, lane 6). Two proteins at approximately 60 kDa are missing from the protein pattern after trypsin exposure while the same proteins (marked by arrowheads) are clearly

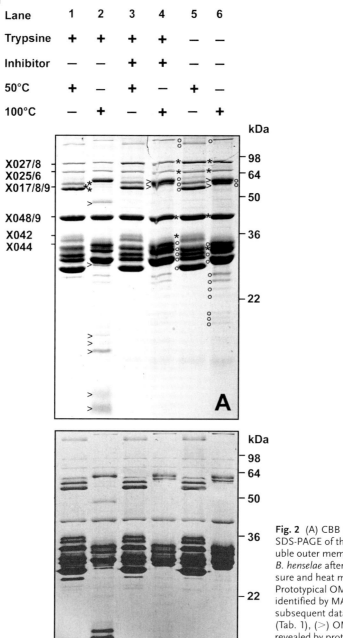

Fig. 2 (A) CBB R-250-stained 1-D SDS-PAGE of the sarcosine-insoluble outer membrane fraction of B. henselae after protease exposure and heat modification. (*) Prototypical OMPs positively identified by MALDI-TOF-MS and subsequent database query (Tab. 1), (>) OMP candidates revealed by protease exposure, (°) OMP candidates revealed by heat modification. (B) Silver-stained 1-D SDS-PAGE of the sarcosine-insoluble outer membrane fraction of B. henselae treated accordingly to Fig. 1A.

detectable in mock and untreated fractions (Fig. 2A, lanes 4 and 6). On the contrary, seven protein species (marked by arrowheads) ranging from 10 kDa to 48 kDa seem not to arise from a detectable loss in complexity of the pattern revealed in mock and untreated fractions (Fig. 2A, lanes 4 and 6), but instead emerged newly following trypsin exposure, leading to an increased complexity (Fig. 2A, lane 2).

Heat modification of sarcosine-insoluble OM fraction led to the generation of an even more complex pattern (Fig. 2A, lanes 5 and 6). Only three abundant protein species at approximately 90 kDa, 80 kDa, and 40 kDa appear not to be affected by differential heat conditioning and display identical electrophoretic mobility at 50°C and at 100°C (Fig. 2A, lanes 5 and 6). Ten distinct abundant protein species (marked by circles) detected at 50°C (Fig. 1B, lane 5) are not visible at 100°C, whereas 13 distinct protein species (marked by circles) detected at 100°C (Fig. 2A, lane 6) cannot be tracked at 50°C.

To corroborate the findings from both trypsin exposure and heat modification revealed by staining with CBB R-250, analytical amounts of the sarcosine-insoluble OM fraction (15 µg) were separated by large-format 15% SDS-PAGE and visualized by silver staining for more sensitive protein detection (Fig. 2B), revealing a consistent protein pattern. Forty-nine relevant protein bands were excised from the gel shown in Fig. 2A and submitted to PMF. Five components of the *B. henselae* OM subproteome (marked by asterisk) could be unambiguously identified (Tab. 1). Three of these, namely Omp43 (X048/9), HbpA (X044) and HbpD (X042), have previously been described in *Bartonella* spp. [7, 35, 36]. Furthermore, HutA (X025/6), and Omp89 (X027/8) have been identified successfully. Omp89 appears to be one of the trypsin-sensitive OMP candidates revealed by 1-D SDS-PAGE. Protein assignment led to the unambiguous identification of Omp89 as a distinct protein species (X017/8/9), migrating at approximately 60 kDa after trypsin exposure. The protein species, identified at approximately 90 kDa as Omp89 in the untreated fraction (X027/8, lanes 5 and 6), appears to be equally abundant in the mock treated fraction (lanes 3 and 4), but shows a light decrease in abundance after trypsin exposure (lanes 1 and 2). These data indicate that trypsin cleavage sites in Omp89 are exposed on the OM surface, but that cleavage by trypsin is far from completion.

All of these five proteins, HbpA, HbpD, HutA, Omp43, and Omp89, are predicted to fulfil the necessary criteria for prototypical OMPs (Tab. 1). However, it was not possible to unambiguously identify other trypsin-sensitive or heat-modifiable protein species of the OM subproteome of *B. henselae* by 1-D SDS-PAGE followed by MALDI-TOF-MS alone. In addition, clustering of several protein species in the M_r range of approximately 30 kDa to 36 kDa prevented identification. Single protein bands were hardly discriminated, clearly indicating the need for the separation of these protein species in two dimensions.

11.3.3
2-D NEPHGE of *B. henselae* OMPs and protein assignment by PMF

Numerous proteomic studies of OMPs from Gram-negative bacteria demonstrate the feasibility and limitations of combining 2-DE and PMF [16, 37–39]. Despite a wealth of experimental progress, OMPs remain inherently difficult to handle and

Tab. 1 Prototypical *B. henselae* OMPs identified in this study by PMF using MALDI-TOF-MS and subsequent database query including *in silico* analysis by assignment of Pfam domains (PFAM), predictions of N-terminal signal peptides (SignalP), predictions of subcellular localization (PSORT-B), and calculation of GRAVY

Gene	Matched masses	Coverage (%)	Spot ID	Locus	Protein description	Predicted mass	pI	Pfam model	Pfam E-Value	SignalP[a] signal peptide	SignalP[a] probability	PSORT-B[b] localization	PSORT-B[b] probability	GRAVY[c] value
hbpA	3	21	A005	BH02560	Hemin-binding protein A	29917	5.37	Porin	0.0019	Yes	1	OM	9.67	−0.205
	4	27	A007											
	3	20	B005											
	3	20	B007											
	3	15	B008											
	3	20	B009											
	3	15	X044											
hbpD	3	13	A025	BH04810	Hemin-binding protein D	30270	9.06	Membrane	0.056	Yes	1	OM	7.78	−0.158
	5	22	B043											
	4	18	X042											
hutA	8	19	A086	BH04970	Heme receptor precursor	82613	9.34	Receptor	3.70E-69	Yes	1	OM	9.67	−0.611
	12	24	B037											
	13	24	X025											
	11	21	X026											
omp89	20	30	A081	BH06280	OMP 89	88986	9.01	Surface Ag	2.30E-098	Yes	0.967	OM	9.60	−0.255
	10	12	A082											
	14	24	A083											
	8	15	B036											
	5	9	X017											

Tab. 1 Continued

Gene	Matched masses	Coverage (%)	Spot ID	Locus	Protein description	Predicted mass pI	Pfam model	Pfam E-Value	SignalP[a] signal peptide	SignalP[a] probability	PSORT-B[b] localization	PSORT-B[b] probability	GRAVY[c] value
	6	11	X018										
	5	9	X019										
	11	17	X027										
	9	16	X028										
omp43	12	32	A131	BH12500	OMP 43 precursor	44225 8.52	Porin	3.60E-057	Yes	1	OM	9.67	−0.254
	7	17	A132										
	6	18	B041										
	4	18	B042										
	4	11	B048										
	4	17	X048										
	3	10	X049										
plp	3	13	A112	BH02000	Peptidyl-prolyl cis-trans isomerase	35597 5.80	Rotamase	4.80E-041	Yes	1	OM	9.60	−0.442
	5	20	A113										
	3	9	B012										
–	5	8	A087	BH00450	Hypothetical protein	71580 9.46	Surface Ag	5.70E-058	Yes	0.801	OM	9.60	−0.133

a) Designations of signal peptide predictions based on SignalP. Numbers refer to probability assigned to the identification of a signal peptide (from 0 to 1).
b) Designations of subcellular localization based on PSORT-B. Numbers refer to localization probability (from 0 to 10). OM, outer membrane.
c) Grand average of hydropathy (Kyte/Doolittle).

have been found to be under-represented in standard 2-DE gels due to poor solubility and low abundance. We employed one particular separation technology, 2-D NEPHGE, to assess the *B. henselae* OM subproteome at a global level and at high resolution. 2-D NEPHGE resolves proteins in the first dimension in a time-dependent manner according to charge and not to their isoelectric point, thus enabling the separation of even highly alkaline proteins (pH > 8). OM fractions were subjected to 2-DE to identify and catalogue as many OMPs as possible. Thus, allowing for the establishment of a first 2-D reference map of the *B. henselae* OM subproteome.

Preparative amounts of sarcosine-insoluble OM fraction (200 µg) were subjected to 2-D NEPHGE [15] on two different commercial 2-DE systems (WITA and Millipore/2-D Investigator) and visualized by CBB G-250 (Figs. 3A and B, respectively). As demonstrated, the *B. henselae* OM subproteome is reproducibly resolved and conserved over a pI range of pH 4–9 and an M_r range of 10–100 kDa. However, due to technical limitations, basic OMP candidates appear to be lost at high pH due to cathodic drift in the gel, run on the Millipore 2-D Investigator system (Fig. 3B). There is an increased number of protein species dominating both 2-D gels in an M_r range of 30 kDa to 80 kDa. This finding is consistent with results obtained by 1-D SDS-PAGE (Fig. 2A).

To survey protein composition displayed by 2-D NEPHGE, 148 distinct protein spots from the gel depicted in Fig. 3A and 48 protein spots from the gel depicted in Fig. 3B were excised and submitted to protein assignment by PMF. This first global query allowed for the total identification of 53 different protein species, originating from 105 distinct protein spots (Tables 1 and 2). Thus, in a first round of assignment, based on a genome sequence in the final step of annotation, 54% of the protein spots analyzed (105 out of 196) were found to match a corresponding ORF. This analysis resulted in the assignment of six prototypical OMPs (HbpA, HbpD, HutA, Omp43, Omp89, and Plp; Tab. 1).

To further substantiate this analysis, *in silico* predictions by means of Pfam, PSORT-B, and SignalP were carried out. Thus, an additional OMP species, BH00450 (A087), was assigned. The gene product of BH00450 is predicted by SignalP to possess an N-terminal secretion signal, fits an appropriate Pfam model and PSORT-B predicts its subcellular localization to the OM. The corresponding locus has been assigned 'hypothetical' in the genome annotation of *B. henselae* [17] and the identification of its product as an OMP increases the number of OMPs identified by proteome analysis to seven (Tab. 1). Additionally, another set of four proteins (BH00190, BH05430, BH13580, and BH14010) were assigned as 'hypothetical' in the genome annotation [17]. Two of these, encoded by loci BH00190 and BH05430, are predicted to contain signal peptides with high probability (1.000 and 0.999, respectively) and thus are likely to represent periplasmic proteins or OMPs as well (data not shown). However, for the proteins encoded by loci BH13580 and BH14010, no signal peptides were predicted, leaving the question of their subcellular localization open. In addition to the identified OMPs, two paralogous periplasmic proteins (HtrA1, HtrA2) were recovered along with periplasmic SdhA.

Based on database entries of IMPs of other bacterial systems, six of the proteins identified may be considered to represent nonsecretory IMPs in *B. henselae* (AtpA, AtpD, BH08630, BH11380, BH15840, and PdhA; Tab. 2). In addition, a wealth of

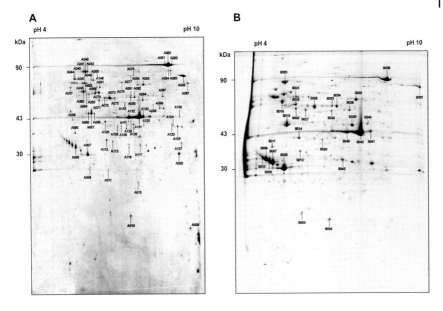

Fig. 3 (A) CBB G-250-stained 2-D NEPHGE master gel of the sarcosine-insoluble outer membrane subproteome of B. henselae (mock treated) displayed by a WITA system. (B) CBB G-250-stained 2-D NEPHGE master gel of the sarcosine-insoluble outer membrane subproteome of B. henselae (mock treated) displayed by a Millipore/ 2D Investigator system. Protein species positively identified by PMF using MALDI-TOF-MS and subsequent database query are indicated (Tables 1 and 2).

37 proteins of either cytoplasmic origin or unknown subcellular localization were identified (Tab. 2). Thus, miscellaneous protein species cofractionating with the OM subproteome represent up to 70% of the total protein species identified (37 out of 53) and fall into several subcategories: (i) synthesis of murein and lipidA (MurC, MurE, and LpxD), (ii) transcription (HrcA, NusA, NusG, RpoA), (iii) translation (FusA, RpsA, Tig, EF-TU, TypA), and (iv) heat shock (DnaJ, DnaK, GroEL). The 2-DE data presented here are available on-line on the proteome databases for microbial research (http://www.mpiib-berlin.mpg.de/2D-PAGE/) [40].

11.4 Discussion

Here, we present the establishment of a first 2-D reference map of the sarcosine-insoluble OM subproteome of B. henselae ATCC 49882. The aim of this study was to identify OMPs with a function in virulence and endothelial cell proliferation [2, 3]. These proteins could serve as novel drug targets, diagnostic probes or possible vaccine candidates in the future. 2-DE of enriched OMP fractions was followed by PMF using MALDI-TOF-MS and subsequent database query against the B. henselae

Tab. 2 Miscellaneous *B. henselae* proteins identified in this study by PMF using MALDI-TOF-MS and subsequent database query not predicted to be localized to the OM by PSORT-B.

Gene	Masses matched	Coverage (%)	Spot ID	Locus	Protein description	Predicted mass p*I*	PSORT-B[a] localization	PSORT-B[a] probability
accA	7	21	A116	BH16340	Acetyl-CoA carboxylase, carboxyl transferase α-subunit	34949 6.27	UNK	7.31
adk	4	13	A015	BH10300	Adenylate kinase	21481 6.63	CYT	9.78
asd	13	40	A136	BH12890	Aspartate-semialdehyde dehydrogenase	37070 6.20	UNK	7.31
atpA	9	26	A072	BH15340	ATP synthase α-chain	55512 5.91	UNK	2.00
	7	18	A073					
	4	9	A075					
	6	17	B028					
atpD	9	26	A069	BH15320	ATP synthase β-chain	56595 5.44	UNK	2.00
	9	26	A070					
	3	7	B024					
carA	5	12	A099	BH11590	Carbamoyl-phosphate synthase small chain	43758 7.31	UNK	2.00
	4	9	B040					
cgtA	7	30	A123	BH01560	GTP-binding protein	37297 8.69	UNK	7.31
	6	24	A124					
dnaJ1	4	11	A102	BH00660	Heat shock protein	42249 8.58	CYT	9.78
dnaK	5	12	A044	BH00650	Heat shock protein 70	68262 4.93	UNK	7.31
	7	16	A045					
fusA	5	8	A041	BH10540	Elongation factor EF-G	76334 5.12	CYT	9.78
	8	20	A042					
	11	20	A043					
	12	24	B020					
ftsZ	5	13	A049	BH11180	Cell division protein	62353 5.25	CYT	8.91

Tab. 2 Continued

Gene	Masses matched	Coverage (%)	Spot ID	Locus	Protein description	Predicted mass pI	PSORT-B[a] localization	PSORT-B[a] probability
fumC	10	25	A050	BH13550	Fumarate hydratase C	50152 6.83	CYT	9.78
gcvT	5	11	A098	BH12840	Glycine cleavage system protein	40555 6.21	UNK	7.31
glnA2	6	18	A139	BH10160	Glutamine synthetase	52674 5.47	CYT	9.78
	3	11	A071					
	7	17	B025					
glpX	6	20	A105	BH10020	Glycerol-inducible protein	35164 5.51	UNK	7.31
	5	22	B014					
guaB	8	24	A091	BH01800	Inosine-5'-monophosphate dehydrogenase	52817 6.30	UNK	7.31
	4	16	A092					
	8	22	B033					
	8	27	B034					
htrA1	6	18	A088	BH04770	Serine protease	54114 8.56	PP	9.21
	5	8	A089					
	6	10	B035					
htrA2	7	16	A097	BH10250	Serine protease	50563 9.02	PP	9.21
hrcA	3	12	A108	BH00550	Heat-inducible transcription repressor	39440 5.79	UNK	7.31
	3	14	A109					
ilvE	9	26	A125	BH10010	Branched-chain amino acid aminotransferase	41341 8.63	UNK	2.00
lpxD	10	40	A120	BH06290	UDP-3-O-[3-hydroxymyristoyl] glucosamine N-acyltransferase	36837 6.66	UNK	2.00
mopA	8	18	A062	BH13530	Chaperonin protein GroEL	57626 5.10	CYT	9.78
	5	13	A063					
	8	18	B016					

Tab. 2 Continued

Gene	Masses matched	Cover-age (%)	Spot ID	Locus	Protein description	Predicted mass p/	PSORT-B[a] localization	PSORT-B[a] probability
murC	5	11	A143	BH11230	UDP-N-acetylmuramate-L-alanine ligase	52079 5.90	CYT	9.78
murE	11	25	A093	BH11290	UDP-N-acetylmuramoylalanyl-D-glutamate-2,6-diaminopimelate ligase	52855 6.61	CYT	9.78
	13	28	A094					
	15	37	B038					
mutL	9	24	A077	BH02690	DNA mismatch repair protein	66926 6.21	UNK	7.31
	12	29	A078					
ntrX	8	21	B027	BH05850	Nitrogen regulation protein	50571 5.62	CYT	9.78
nusA	8	19	A057	BH02170	N utilization substance protein A	59078 4.78	UNK	7.31
nusG	4	35	B003	BH06050	Transcription antitermination protein	20165 5.61	UNK	7.31
parB	9	41	A127	BH16650	Chromosome partitioning protein	33835 8.81	UNK	7.31
pdhA	7	21	A111	BH05750	Pyruvate dehydrogenase E1 component α-subunit	38033 6.04	IM	9.02
	7	25	B030					
pdhD2	7	23	B032	BH16520	Dihydrolipoamide dehydrogenase	49419 6.05	CYT	9.78
rpiA	5	31	A008	BH06410	Ribose 5-phosphate isomerase	25161 4.97	UNK	7.31
rpoA	7	23	A059	BH10270	DNA-directed RNA polymerase α-chain	37672 4.85	UNK	7.31
rpsA	6	12	A046	BH00930	30S ribosomal protein S1	62819 5.17	UNK	7.31
	6	14	A047					
sdhA	5	13	A076	BH15780	Succinate dehydrogenase, flavoprotein subunit	66973 6.04	PP	8.19
tig	4	11	A064	BH05940	Trigger factor	53223 5.13	UNK	7.31
	11	26	A065					

Tab. 2 Continued

Gene	Masses matched	Cover-age (%)	Spot ID	Locus	Protein description	Predicted mass p*I*	PSORT-B[a] localization	PSORT-B[a] probability
tuf1	10	30	A066	BH06020	Elongation factor EF-TU	42866 5.28	CYT	9.78
tuf2	5	13	A067	BH10530			CYT	9.78
	9	31	A068					
	7	26	B015					
typA	6	13	A148	BH01700	GTP-binding protein	67223 5.30	CYT	9.78
–	4	14	A060	BH04920	Phage-related protein	36163 5.00	UNK	2.00
	7	30	B011					
–	5	42	A028	BH00190	Hypothetical protein	18715 9.76	UNK	2.35
–	5	7	A084	BH05430	Hypothetical protein	90181 9.08	UNK	2.35
	15	23	A085					
–	5	24	A011	BH08630	ATP-dependent transporter	27854 5.61	UNK	5.22
–	8	17	A051	BH11380	ABC transporter, ATP-binding protein	60988 5.36	UNK	7.31
	11	20	B023					
–	4	20	A019	BH13580	Hypothetical protein	19697 6.13	UNK	2.00
	5	26	B004					
–	9	32	A133	BH14010	Hypothetical protein	41032 6.98	UNK	2.00
	9	33	B039					
–	4	8	A117	BH15840	SapC-related protein	31209 6.51	UNK	2.00

a) Designations of subcellular localization based on PSORT-B. Numbers refer to localization probability (from 0 to 10). PP, periplasmic; CYT, cytosolic; UNK, unknown genome sequence [17]. Our analysis led to the assignment of 53 distinct protein species, including seven prototypical OMPs (HbpA, HbpD, HutA, Omp43, Omp89, Plp, and BH00450; Tab. 1). The first five OMP species are abundantly expressed and represent the major protein constituents of the B. henselae OM subproteome.

HbpA and HbpD are orthologous products of a five-member gene family in B. quintana, the hbpCAB/D/E locus [36], and HbpA (Pap31) has been shown to function as hemin binding protein in B. henselae [7]. Furthermore, 2-DE revealed that at least five HbpA isoforms comigrate in a diagnostic cluster (pH 4–5, 29–35 kDa). HutA is an iron-regulated TonB-dependent heme receptor. In vitro growth of B. henselae is dependent on media containing either hemin or blood suggesting that an efficient hemin uptake system in the absence of a hemin synthesis system is essential and represents a rate-limiting step for in vitro growth and successful propagation in the host. Thus, the expression of the three OMPs HbpA, HbpD, and HutA together with as yet unidentified gene products involved in iron-acquisition appears to be of fundamental importance for B. henselae survival.

Omp43 (380 amino acids in length after signal peptide removal), a porin with a predicted 16-stranded β-barrel membrane topology, binds endothelial cells and appears to be a homolog of Brucella spp. Omp2b porin (38% amino acid identity) [35]. Plp is a parvulin-like peptidyl-prolyl cis-trans isomerase precursor identified only by means of 2-DE. The homologous protein is found in two α-proteobacterial relatives of B. henselae, namely the obligate intracellular pathogens Rickettsia prowazekii [41] and Rickettsia conorii [42]. In addition, another bona fide OMP (BH00450) was assigned by in silico predictions.

Out of the six putative IMPs (AtpA, AtpD, BH08630, BH11380, BH15840, PdhA), only PdhA is predicted by PSORT-B to localize to the IM. AtpA and AtpD represent the α- and β-subunit of the F_1 ATP synthase [43]. Based on homology, the proteins encoded by the loci BH08630 and BH11380 represent members of the ATP-binding cassette (ABC) transporter superfamily [44]. Finally, the gene product of BH15840 putatively represents a SapC-related protein functioning as an IM peptide permease [45]. Taken together, these proteins give insight into diverse molecular functions such as electron transport across the inner membrane, oxidative phosphorylation, and finally nutrient import or export of lipids, toxic substances, proteins, and peptides across the IM (Tab. 2).

Finally, we have classified 37 proteins to be of cytoplasmic origin or of unknown subcellular localization. However, it is not excluded that also this subset contains proteins that may at least occasionally be associated with the IM. For example, ribosome-associated proteins (FusA, RpsA, Tig, EF-TU, TypA) may be transiently membrane-associated during the biosynthesis of proteins destined for the periplasm or the OM [46]. Likewise, heat shock proteins (DnaJ, DnaK, GroEL) may be membrane-associated while functioning as chaperones [47]. Out of these, the most interesting is GroEL (A062, A063, B016), one of the products from the groES-groEL operon. Only recently, GroEL from Bartonella bacilliformis, a human-specific species of Bartonella and the causative agent of Oroya fever and verruga peruana, has been reported to act as a mitogenic effector on HUVECs in vitro in a dose-dependent manner and may constitute an actively secreted protein [48]. Upon subcellular fractionation, B. bacilliformis GroEL copellets with both IM and OM fractions and is heat- as well as trypsin-sensitive. In agreement with this finding, *B. henselae* GroEL copellets with the OM fraction, but, in contradiction, has not been found to be a heat- and/or trypsin-sensitive component.

Phage-related and chromosome-partitioning proteins are additional candidates for proteins that may putatively be membrane-associated. However, present in this subset are also *bona fide* cytoplasmic proteins, such as Adk. Contaminations with cytoplasmic proteins are an inherent problem in subcellular fractionation, reported in many other studies on the identification of OMPs. For example, a recent global survey of the *H. influenzae* proteome reported on the identification of several nonmembrane proteins in the total membrane fraction [49]. Amongst these, the investigators assigned several proteins whose orthologs can be found in *B. henselae* as well (e.g., Adk, DnaJ, GroEL, HtrA, MurC). Already earlier, a study on the OM subproteome of *Escherichia coli* had similarly reported on the identification of distinct protein species co-isolated with the OMP fraction and separated by 2-D NEPHGE (e.g., AtpA, AtpD, DnaK, GroEL, RpsA) [50]. In contrast, two studies focussing on the OM subproteomes of *E. coli* [38] and *Caulobacter crescentus* [39] identified 80% (21 proteins identified of 26 predicted) and 76% (41 proteins of 54 analyzed) respectively, to represent prototypical OMPs. In these studies, an alternative approach for sample preparation, namely sodium carbonate extraction, was applied. This strategy may prove beneficial as well in further attempts to characterize the *B. henselae* OM subproteome.

It is well-known that OMPs tend to be depleted from standard 2-DE gels displaying total cell extracts [13]. The general under-representation of OMPs is not only due to the poor solubility but also to low expression levels, which renders detection and thus identification difficult. Therefore, it is crucial to design the appropriate isolation procedure to enrich for these protein species. Our strategy to assess the OM subproteome from *B. henselae* was to perform subcellular fractionation and subsequent enrichment of OMPs by the ionic detergent lauryl sarcosine. This method has already been adapted to *B. henselae* previously [5] and is well established in other bacterial systems as well [51].

Sonication and ultracentrifugation allowed for the isolation of total membranes, whereas detergent partitioning and ultracentrifugation allowed for the separation of the outer OM from the IM. In contrast, separation of the IM and the OM by ultracentrifugation over sucrose-step gradient was found to substantially increase the complexity of the protein pattern and reduced the purity of both the IM and OM fractions due to cross-contamination with the cytosolic fraction (data not shown). Assessment of trypsin sensitivity and heat modifiability proved to be instrumental in the analysis of the composition and surface accessibility of the sarcosine-insoluble OM fraction. Especially trypsin exposure of intact bacteria – applied in the course of the preparation of OMs – in the absence (exposed) or presence of an inhibitor (mock treated), represents a valuable tool for the analysis of (i) the quality of the preparation and (ii) for the acquisition of topological protein information when compared to mock-treated preparations.

The general under-representation of OMPs on 2-D gels is not solely attributed to the isolation procedure, but is also related to problems with the separation process. For example, prototypical, integral OMPs are inherently difficult to solubilize into aqueous-based conditions used for isoelectric focusing and they typically possess alkaline p*I*s [52]. In effect, they are not visible on standard 2-D gels resolving acidic to weakly basic proteins (pH 4–8) and are therefore not amenable to the separation procedure. These problems were the greatest challenge when trying to separate the *B. henselae* OMPs by 2-DE.

Although 2-D NEPHGE normally displays an excellent resolution over a p*I* range of approximately pH 4–9 including highly alkaline proteins and an M_r range of 10–100 kDa, we were unable to display the OMPs as clearly as achieved with the Millipore/2-D Investigator System (compare Figs. 3B to 3A). This may be due to the minor modifications we introduced when adapting the protocol from Klose and Kobalz [15], namely the omission of the capping gel. The use of immobilized pH gradients (IPGs) is another commonly used technology for protein separation in the first dimension [53–55]. In a complementary approach to 2-D NEPHGE, we employed 2-D IPG (pH 3–10) using denaturing solutions containing varying concentrations of urea and thiourea. However, we were not successful in sufficiently resolving alkaline OMPs such as Omp43 and Omp89 (data not shown).

The production of a high-resolution reference map of the *B. henselae* OM displayed in Fig. 3 was performed concomitantly with the genome sequence analysis [17]. Beyond the identification of the protein composition of the OM, there is another benefit resulting from complementing genomics with proteomics. One of the main problems in genomics is to distinguish ORFs, that occur by chance, from real genes that encode functional proteins. Five ORFs, annotated in the genome sequence as 'hypothetical' were successfully recovered as protein species, a fact that provides convincing evidence that these ORFs indeed correspond to protein-encoding genes, thereby expanding the genome analysis from a structural to a functional level.

A whole-genome membrane protein prediction analysis of *B. henselae* suggests that the relative fraction of genes encoding membrane proteins amounts to 25%, which corresponds to approximately 400 genes (data not shown). In comparison, the few OMPs identified in this study may appear disappointingly low. However, it should be recalled that *B. henselae* was cultivated on standard CSB-agar, which allowed simple propagation of sufficient quantities of bacteria and isolation of pure bacterial membranes devoid of contamination by host material. It is also well established that several OMPs or membrane–associated proteins of *Bartonella* are synthesized exclusively during interaction with host cells. These include the two essential T4SS, VirB [11, 12], and Trw [9], as well as a class of autotransporters termed Iba (<u>i</u>nducible *<u>B</u>artonella* <u>a</u>utotransporters [1, 56]. Consequently, none of these proteins were identified in the study, but may be recovered in future studies by performing 2-DE analysis with OMPs isolated from *B. henselae* during infection of endothelial cells. We believe that the 2-D reference map of the OM subproteome of *B. henselae* established in this study will show a wide utility in future proteome analyses of *B. henselae* in this context of host-pathogen interaction.

We thank R. Hiestand for excellent technical assistance, B. Grünenfelder, S. Tawfilis, and I. Christiansen for valuable suggestions, and H. L. Saenz for critical reading of the manuscript. This work was supported by grants from the Swiss Federal Office for Educa-

tion and Science (00.0422-1) and the 5th EU-framework program 'Quality of Life' (EBP European Network on Bacterial Proteomes contract QLK-2000-01536) to C. Dehio and S. G. E. Andersson. Financial support was also provided to C. Dehio from the Swiss National Science Foundation (SNF, 3100–0061777.00/1) and to S. G. E. Andersson from the Wallenberg Foundation (KAW), the Foundation for Strategic Research (SSF), and the Swedish Science Research Council (VR).

11.5 References

[1] Dehio, C., *Trends Microbiol.* 2001, *9*, 279–285.
[2] Dehio, C., *Curr. Opin. Microbiol.* 2003, *6*, 61–65.
[3] Maeno, N., Oda, H., Yoshiie, K., Wahid, M. R. et al., *Microb. Pathog.* 1999, *27*, 419–427.
[4] Conley, T., Slater, L., Hamilton, K., *J. Lab. Clin. Med.* 1994, *124*, 521–528.
[5] Burgess, A. W., Anderson, B. E., *Microb. Pathog.* 1998, *25*, 157–164.
[6] Bowers, T. J., Sweger, D., Jue, D., Anderson, B., *Gene* 1998, *206*, 49–52.
[7] Zimmermann, R., Kempf, V. A., Schiltz, E., Oberle, K. et al., *J. Bacteriol.* 2003, *185*, 1739–1744.
[8] Schmiederer, M., Anderson, B., *DNA Cell Biol.* 2000, *19*, 141–147.
[9] Seubert, A., Hiestand, R., de la Cruz, F., Dehio, C., *Mol. Microbiol.* 2003, *49*, 1253–1266.
[10] Cascales, E., Christie, P. J., *Nat. Rev. Microbiol.* 2003, *1*, 137–149.
[11] Schulein, R., Dehio, C., *Mol. Microbiol.* 2002, *46*, 1053–1067.
[12] Schmid, M. C., Schulein, R., Dehio, M., Denecker, G. et al., *Mol. Microbiol.* 2004, *52*, 81–92.
[13] Santoni, V., Molloy, M., Rabilloud, T., *Electrophoresis* 2000, *21*, 1054–1070.
[14] Garrels, J. I., *J. Biol. Chem.* 1979, *254*, 7961–7977.
[15] Klose, J., Kobalz, U., *Electrophoresis* 1995, *16*, 1034–1059.
[16] Jungblut, P. R., Bumann, D., Haas, G., Zimny-Arndt, U. et al., *Mol. Microbiol.* 2000, *36*, 710–725.
[17] Alsmark, C. M., Frank, A. C., Karlberg, E. O., Antoine-Legault, B. et al., *Proc. Natl. Acad. Sci. USA* 2004, *101*, 9716–9721.

[18] Regnery, R. L., Anderson, B. E., Clarridge III, J. E., Rodriguez-Barradas, M. C. et al., *J. Clin. Microbiol.* 1992, *30*, 265–274.
[19] Laemmli, U. K., *Nature* 1970, *227*, 680–685.
[20] Lowry, O. H., Rosbrough, N. J., Farr, A. L., Randall, R. J., *J. Biol. Chem.* 1951, *193*, 267–275.
[21] Jungblut, P. R., Seifert, R., *J. Biochem. Biophys. Methods* 1990, *21*, 47–58.
[22] Doherty, N. S., Littman, B. H., Reilly, K., Swindell, A. C. et al., *Electrophoresis* 1998, *19*, 355–363.
[23] Beavis, R. C., Chait, B. T., *Methods Enzymol.* 1996, *270*, 519–551.
[24] Bonenfant, D., Schmelzle, T., Jacinto, E., Crespo, J. L. et al., *Proc. Natl. Acad. Sci. USA* 2003, *100*, 880–885.
[25] Clauser, K. R., Baker, P., Burlingame, A. L., *Anal. Chem.* 1999, *71*, 2871–2882.
[26] Bateman, A., Birney, E., Cerruti, L., Durbin, R. et al., *Nucleic Acids Res.* 2002, *30*, 276–280.
[27] Nielsen, H., Engelbrecht, J., Brunak, S., von Heijne, G., *Int. J. Neural Syst.* 1997, *8*, 581–599.
[28] Nielsen, H., Brunak, S., von Heijne, G., *Prot. Eng.* 1999, *12*, 3–9.
[29] Menne, K. M., Hermjakob, H., Apweiler, R., *Bioinformatics* 2000, *16*, 741–742.
[30] Gardy, J. L., Spencer, C., Wang, K., Ester, M. et al., *Nucleic Acids Res.* 2003, *31*, 3613–3617.
[31] Kyte, J., Doolittle, R. F., *J. Mol. Biol.* 1982, *157*, 105–132.
[32] Filip, C., Fletcher, G., Wulff, J. L., Earhart, C. F., *J. Bacteriol.* 1973, *115*, 717–722.
[33] Hancock, R. E., Carey, A. M., *J. Bacteriol.* 1979, *140*, 902–910.

[34] Urakami, H., Ohashi, N., Tsuruhara, T., Tamura, A., *Infect. Immun.* 1986, *51*, 948–952.

[35] Burgess, A. W., Paquet, J. Y., Letesson, J. J., Anderson, B. E., *Microb. Pathog.* 2000, *29*, 73–80.

[36] Minnick, M. F., Sappington, K. N., Smitherman, L. S., Andersson, S. G. *et al.*, *Infect. Immun.* 2003, *71*, 814–821.

[37] Nouwens, A. S., Cordwell, S. J., Larsen, M. R., Molloy, M. P. *et al.*, *Electrophoresis* 2000, *21*, 3797–3809.

[38] Molloy, M. P., Herbert, B. R., Slade, M. B., Rabilloud, T. *et al.*, *Eur. J. Biochem.* 2000, *267*, 2871–2881.

[39] Phadke, N. D., Molloy, M. P., Steinhoff, S. A., Ulintz, P. J. *et al.*, *Proteomics* 2001, *1*, 705–720.

[40] Pleissner, K. P., Eifert, T., Buettner, S., Schmidt, F. *et al.*, *Proteomics* 2004, *4*, 1305–1313.

[41] Andersson, S. G., Zomorodipour, A., Andersson, J. O., Sicheritz-Ponten, T. *et al.*, *Nature* 1998, *396*, 133–140.

[42] Ogata, H., Audic, S., Renesto-Audiffren, P., Fournier, P. E. *et al.*, *Science* 2001, *293*, 2093–2098.

[43] Huang, F., Parmryd, I., Nilsson, F., Persson, A. L. *et al.*, *Mol. Cell. Proteomics* 2002, *1*, 956–966.

[44] Locher, K. P., Lee, A. T., Rees, D. C., *Science* 2002, *296*, 1091–1098.

[45] Parra-Lopez, C., Baer, M. T., Groisman, E. A., *EMBO J.* 1993, *12*, 4053–4062.

[46] Herskovits, A. A., Shimoni, E., Minsky, A., Bibi, E., *J. Cell Biol.* 2002, *159*, 403–410.

[47] Cordwell, S. J., Nouwens, A. S., Walsh, B. J., *Proteomics* 2001, *1*, 461–472.

[48] Minnick, M. F., Smitherman, L. S., Samuels, D. S., *Infect. Immun.* 2003, *71*, 6933–6942.

[49] Langen, H., Takacs, B., Evers, S., Berndt, P. *et al.*, *Electrophoresis* 2000, *21*, 411–429.

[50] Link, A. J., Robison, K., Church, G. M., *Electrophoresis* 1997, *18*, 1259–1313.

[51] Baik, S. C., Kim, K. M., Song, S. M., Kim, D. S. *et al.*, *J. Bacteriol.* 2004, *186*, 949–955.

[52] Molloy, M. P., Phadke, N. D., Chen, H., Tyldesley, R. *et al.*, *Proteomics* 2002, *2*, 899–910.

[53] Bjellqvist, B., Ek, K., Righetti, P. G., Gianazza, E. *et al.*, *J. Biochem. Biophys. Methods* 1982, *6*, 317–339.

[54] Görg, A., Obermaier, C., Boguth, G., Harder, A. *et al.*, *Electrophoresis* 2000, *21*, 1037–1053.

[55] Grunenfelder, B., Rummel, G., Vohradsky, J., Roder, D. *et al.*, *Proc. Natl. Acad. Sci. USA* 2001, *98*, 4681–4686.

[56] Seubert, A., Schulein, R., Dehio, C., *Int. J. Med. Microbiol.* 2002, *291*, 555–560.

12
The influence of *agr* and σ^B in growth phase dependent regulation of virulence factors in *Staphylococcus aureus*

Anne-Kathrin Ziebandt, Dörte Becher, Knut Ohlsen, Jörg Hacker, Michael Hecker and Susanne Engelmann

The expression of many virulence determinants in *Staphylococcus aureus* is tightly coordinated generally by global regulatory elements such as accessory gene regulator (*agr*), staphylococcal accessory regulator and the alternative sigma factor σ^B. We have compared the two-dimensional (2-D) protein pattern of extracellular protein extracts of wild-type cells with the 2-D patterns of the respective regulatory mutants in order to identify proteins whose amount is influenced by a mutation in *agr* or *sigB*. In order to quantify changes in the level of interesting proteins we used the Ettan-fluorescence difference gel electrophoresis technique (Amersham Biosciences). As in most bacteria, the amount of extracellular proteins was strongly regulated and increased mainly in the stationary phase of growth at high cell densities. By comparing the extracellular protein pattern of the RN6390 *rsbU* strain with that of an isogenic *agr* mutant RN6911 we show that the level of about 70 protein spots changed in the mutant. To analyze the role of σ^B in virulence gene expression an RsbU$^+$ (RN6390 RsbU$^+$) derivative was included in this study. The protein pattern of the RsbU$^+$ strain (RN6390 RsbU$^+$) was very similar to that of the $\Delta agr/\Delta rsbU$ mutant strain (RN6911) indicating an opposing effect of *agr* and *rsbU* on the expression of the same genes.

12.1
Introduction

Over the last few years genome sequences of many pathogenic bacteria have become available and these should provide an excellent basis for a more comprehensive understanding of the mechanisms of bacterial pathogenicity. In particular, the use of proteomics to analyze all members of those regulons known to contain virulence

---

* Originally published in Proteomics 2004, 10, 3034–3047

Proteomics of Microbial Pathogens. Edited by Peter R. Jungblut and Michael Hecker
Copyright © 2007 WILEY-VCH Verlag GmbH & Co. KGaA, Weinheim
ISBN: 978-3-527-31759-2

genes provides an opportunity to discover new virulence functions. Visualizing all members of a regulon in this way also opens a new dimension in the comprehensive understanding of regulatory networks important in the pathogenicity of *Staphylococcus aureus* [1]. *S. aureus* is a Gram-positive human pathogen and is the major cause of nosocomial infections. It may cause a wide spectrum of diseases ranging from wound infections to osteomyelitis, endocarditis, pneumonia, septicaemia or toxic shock syndrome. One cause of its pathogenic diversity is the fact that it produces a large number of different virulence factors many of which are extracellular proteins. The analysis of the extracellular proteome of *S. aureus* may thus provide a means of identifying new virulence factors and elucidating their regulation [2–6]. A large number of extracellular and surface associated proteins, *e.g.*, α-toxin, coagulase, lipases, hemolysins, enterotoxins, protein A, and fibronectin-binding protein are already known to be involved in the virulence of *S. aureus* and their expression is regulated in a coordinated fashion during the growth cycle.

So far the best characterized regulators of virulence gene expression are accessory gene regulator (*agr*) [7–9] and Staphylococcal accessory regulator (SarA) [10, 11]. The *sarA* locus encodes a DNA-binding protein that influences the amount of fibronectin- and fibrinogen–binding protein as well as immunodominant antigen IsaA, protein A, β-hemolysin, autolysin Aly, aureolysin, staphopain, V8 protease, and lipases Lip and Geh [12–15]. This may be caused by (i) SarA binding to the target gene promoters; (ii) downstream effects of SarA controlled gene products on other regulons; or (iii) degradation of proteins by *sarA*-dependent proteases. The *sar* locus is believed to be necessary for the activation of the *agr* locus [16–18].

The *agr* operon acts as a quorum sensing system and enhances the production of extracellular proteins and represses the synthesis of cell wall adhesins. RNAIII appears to be the major effector molecule of the *agr* system and is thought to regulate most target genes at the level of transcription, although it has also been shown to affect the translation of some genes [19–21]. The mechanism of interaction of RNAIII with its target genes, however, is not fully understood and needs to be further analyzed. Recent studies indicate that the alternative sigma factor σ^B may also contribute to virulence gene expression in Gram-positive bacteria by interfering with SarA and *agr* activity [15, 22, 23]. Up to now, our understanding of the interaction of these regulons within the pathogenicity network is incomplete since most of the data described so far were obtained using various strains and a range of different experimental conditions. For a complete characterization of proteins/genes involved in the pathogenicity of *S. aureus* the identification of all members of the virulence associated regulons and the analysis of their contribution to pathogenicity in defined *in vitro* models or in infection systems are required. In the present study we have focused on the *agr* and *sigB* regulons of *S. aureus*.

12.2
Material and methods

12.2.1
Bacterial strains and culture conditions

The bacterial strains used in this study were *S. aureus* RN6390 [8], which is a derivate of NCTC 8325, the isogenic *agr* mutant RN6911 [21] and the *rsbU*-complemented strain RN6390 RsbU$^+$ (kindly provided by M. Bischoff). The *sigB* operon of GP268 [24] was phage transduced into RN6390 resulting in a RsbU$^+$ strain and the construction was confirmed by Southern hybridization and Western blotting with polyclonal antibodies against RsbU. Cells were grown with agitation at 37°C in tryptic soya broth (TSB) medium.

12.2.2
Preparation of the extracellular protein fraction

For the preparation of extracellular protein extracts, bacteria were grown in TSB medium. At the optical density values given cells of 250 mL culture were centrifuged for 20 min at 4°C, 7000 × g to remove bacteria and the extracellular proteins in the supernatant were precipitated with 10% w/v TCA at 4°C overnight. The precipitates were harvested by centrifugation of 4°C, 7000 × g, 1 h, washed with 96% v/v ethanol several times and dried. The protein extracts were resolved in an appropriate volume of a solution containing 8 M urea and 2 M thiourea [15]. The protein concentration was determined using the Nanoquant assay kit (Bio-Rad, Hercules, CA, USA).

12.2.3
Analytical and preparative PAGE

Preparative 2-DE was performed using the IPG technique described by Bernhardt *et al.* [25]. The protein samples were separated using immobilized, linear IPG strips (Amersham Biosciences, Piscataway, NJ, USA) in the pH range of 3 to 10. For identification of proteins by MALDI-TOF MS, protein samples (500 µg) were separated by preparative 2-D gels and the proteins were stained with CBB G-250. The resulting peptide mass fingerprints were analyzed using MS-Fit software (http://prospector.ucsf.edu), GPMAW 4.10 (Lighthouse data) and the genome sequences of *S. aureus* N315, Mu50 and COL. Protein spots with low M_r were identified using sequence data obtained from MS/MS experiments. The measurement was carried out with a HPLC-ESI-Q-TOF mass spectrometer (LC Packings, Amsterdam, The Netherlands; Applied Biosystems, Foster City, CA, USA) as described in Eymann *et al.* [50].

12.2.4
Quantitation of protein spots

For quantitation prior to separation by 2-DE, protein extracts of the respective strains were labeled with CyDye DIGE Cy3 or Cy5 from Amersham Biosciences and a protein standard (a mixture of all protein extracts included in the analysis) labeled with Cy2. A mixture of 50 µg of two different protein extracts labeled with Cy3 or Cy5, and 50 µg of the Cy2 labeled protein standard were fractionated on one gel. The different 2-D gels were compared and spots were quantified using DeCyder software from Amersham Biosciences. Volume ratios in the range of 1 to ∞ indicate an increase of the volume of the respective protein spot and volume ratios in the range -1 to $-\infty$ indicate a decrease of the volume of the respective protein spot. Only volume ratios ≥ 2 or ≤ 2 were defined as significant changes between the different strains.

12.2.5
Transcriptional analyses

Total RNA from *S. aureus* cells was isolated using the acid-phenol method with some modifications [26]. Digoxygenin labeled RNA probes were prepared by *in vitro* transcription with T7 RNA polymerase by using PCR generated fragments as templates. The PCR fragments were generated by using chromosomal DNA of *S.aureus* COL isolated with the chromosomal DNA isolation kit (Promega, Madison, WI, USA) according to the manufacturer's recommendations and the respective oligonucleotides. Primers used for PCR are listed in Tab. 1. Reverse primers contain the T7 RNA polymerase recognition sequence at the 5' end. Northern blot analyses were carried out as previously described [27]. The digoxygenin labeled RNA M_r marker I (Roche, Indianapolis, IN, USA) was used as a standard to calculate the sizes of the transcripts. For the quantitation of specific mRNAs serial dilutions of total RNA were transferred onto a positively charged nylon membrane by slot blotting and hybridized with gene-specific, digoxygenin labeled RNA probes according to the manufacturer's instructions (Boehringer, Mannheim, Germany). The hybridization signals were detected using a the Lumi-Imager (Boehringer) and analyzed using the software package LumiAnalyst (Boehringer). The induction ratios were calculated by setting the value of the control to 1.

12.3
Results

12.3.1
Growth phase dependent regulation of extracellular proteins

For the extracellular proteome analyses presented here *S. aureus* RN6390 was grown in complex medium TSB. Extracellular proteins were prepared from the supernatants of this strain at different cell densities in order to analyze their growth

Tab. 1 Oligonucleotides

Name[a]	Sequence (5' → 3')
GLPQF	GCTGCTTCTGCTGTTTTTAC
GLPQR	<u>CTAATACGACTCACTATAGGGAGA</u>CAATCGCATAAGAGCGTATC
ISAAF	TCATTAGCAGTGGCATTAGG
ISAAR	<u>CTAATACGACTCACTATAGGGAGA</u>AGTTGGAGCTGAAACAGCTT
LIPF	CGGTGTTGACGAACAGCAAC
LIPR	<u>CTAATACGACTCACTATAGGGAGA</u>GTTCAACTGCGCGGTCATAG
PLCF	GGTTCACATGATAGTGGCTC
PLCR	<u>CTAATACGACTCACTATAGGGAGA</u>CTATTAAATGCGCTGCCTCC
RNAIIIF	AGGAAGGAGTGATTTCAATG
RNAIIIR	<u>CTAATACGACTCACTATAGGGAGA</u>ACTCATCCCTTCTTCATTAC
SARAF	TAGGGAGGTTTTAAACATGG
SARAR	<u>CTAATACGACTCACTATAGGGAGA</u>GTTGTTTGCTTCAGTGATTC
SSAAF	GCTCATGCTTCTGAGCAAGA
SSAAR	<u>CTAATACGACTCACTATAGGGAGA</u>CTGGGCCATAACCATAGTTC
SSPF	GATCGTCACCAAATCACAGA
SSPR	<u>CTAATACGACTCACTATAGGGAGA</u>AGCCATTGTCTGGATTATCAG
SA0570F	GTTCTGTTGTAATGGGAGTA
SA0570R	<u>CTAATACGACTCACTATAGGGAGA</u>CTGAGGCAAATTGATAAGTC
SA2097F	CTACGTTAACAGCAGGAATC
SA2097R	<u>CTAATACGACTCACTATAGGGAGA</u>GTAGTTACCTGCTTCACTTG

a) Oligonucleotides with R contain the recognition sequence for T7 at the 5' end

phase dependent regulation. The proteins were separated in a linear pH range of 3 to 10. Forty-seven proteins were identified and assigned to the ORF number defined in the *S. aureus* N315 genome sequencing project [28] (Fig. 1, Tab. 2). Only one protein could not be found in the *S. aureus* N315 protein database. The protein sequence obtained from it corresponded to a hypothetical protein in *S. aureus* MW2 which is similar to the Ear protein. Eight extracellular proteases including the metalloprotease aureolysin Aur, V8 protease SspA, cystein protease SspB and the serine proteases SplA, SplB, SplC, SplE, and SplF were identified. Furthermore, four different proteins Geh, GlpQ, Plc, and Lip with putative lipase activity were detected. Remarkably, at least 18 proteins were identified in more than one 2-D gel spot, as a result of p*I* variation or fragmentation. The most extensively modified *S. aureus* proteins included the autolysin Aly (6 spots), the lipases Geh (6 spots), GlpQ (4 spots) and Lip (9 spots), α-hemolysin Hla (4 spots), immunodominant antigen IsaA (4 spots), and V8 protease SspA (4 spots) (Fig. 1, Tab. 2). Whereas some of these spots differed only in their p*I* values, other protein spots showed additional changes in their M_r. Clear deviations of the p*I* values of the same proteins might reflect PTM. The nature of these modifications is currently under investigation. Deviations in the M_r of one protein indicate specific processing of these proteins. In particular, lipases and proteases are known to be secreted as proproteins, which may be subsequently processed into an active form

Tab. 2 Extracellular proteins identified on 2-D gels of S. aureus

Protein	Function	M_r/pI mature	ORFID S. aureus N315 NCBI (NC 002745)	Accession (NCBI)	Identification	Quantitation[a] RN6390/ RN6911	Quantitation[a] $rsbU^-/rsbU^+$
AhpC[b]	Alkyl Hydroperoxide Reductase Subunit C	21/4.9	SA0366 (ahpC)	BAB41593	MALDI-MS	Not changed	Not changed
AhpF[b]	Alkyl Hydroperoxide Reductase Subunit F	54.7/6.7	SA0365 (ahpF)	O05204	MALDI-MS	Not changed	Not changed
Aly	Hypothetical protein similar to Autolysin	66.3/5.8	SA2437	BAB43742	MALDI-MS	2.5	15.5
Aly^1	Hypothetical protein similar to Autolysin	n.d.	SA2437	BAB43742	MALDI-MS	2.5	14.2
Aly^2	Hypothetical protein similar to Autolysin	n.d.	SA2437	BAB43742	MALDI-MS	3	11.6
Aly^F	Hypothetical protein similar to Autolysin	n.d.	SA2437	BAB43742	MALDI-MS	−2.7	Not changed
Aly^{F1}	Hypothetical protein similar to Autolysin	n.d.	SA2437	BAB43742	MALDI-MS	−2.3	Not changed
Aly^{F1}	Hypothetical protein similar to Autolysin	n.d.	SA2437	BAB43742	MALDI-MS	Not changed	11.7
Aur	Aureolysin	33.4/4.8	SA2430 (aur)	P81177	MALDI-MS/ N-terminal	−5.3	−9.1
DeoD[b]	Purine nucleoside phosphorylase	25.9/4.7	SA1940 (deoD)	BAB43224	MALDI-MS	Not changed	Not changed
DnaK[b]	DnaK protein	66.4/4.7	SA1409	P45554	MALDI-MS	Not changed	Not changed
EarH	Hypothetical protein similar to Ear protein	16/5.2	–	BAB95622.1	LC-MS/MS	−6.8	−4.5
EF-G[b]	Translational elongationfactor G	79.3/4.7	SA0505 (fus)	P81683	MALDI-MS	Not changed	Not changed
EF-TU[b]	Translational elongationfactor TU	43.1/4.7	SA0506 (tufA)	BAB41737	MALDI-MS	Not changed	Not changed
Eno[b]	Enolase	47.1/4.4	SA0731	BAB41964	MALDI-MS	Not changed	Not changed
Fhs[b]	Formyltetrahydrofolate synthetase	59.9/5.7	SA1553 (fhs)	BAB42821	MALDI-MS/ N-terminal	−3.1	Not changed
Fhs^1	Formyltetrahydrofolate synthetase	59.9/5.7	SA1553 (fhs)	BAB42821	MALDI-MS/ N-terminal	−2.2	Not changed
Gap[b]	Glyceraldehyde-6 phosphate Dehydrogenase	36.3/4.8	SA0727	BAB41960	MALDI-MS	Not changed	Not changed
Geh	Glycerolesterhydrolase	72.4/8.9	SA0309 (geh)	BAB41533	MALDI-MS/ N-terminal	2.4	8.0
Geh^F	Glycerolesterhydrolase	n.d.	SA0309 (geh)	BAB41533	MALDI-MS	−113.9	−6.5

Tab. 2 Continued

Protein	Function	M_r/pI mature	ORFID S. aureus N315 NCBI (NC 002745)	Accession (NCBI)	Identification	Quantitation[a] RN6390/ RN6911	Quantitation[a] rsbU−/rsbU+
Geh[F1]	Glycerolesterhydrolase	n.d.	SA0309 (geh)	BAB41533	MALDI-MS	−58.3	−6.4
Geh[F2]	Glycerolesterhydrolase	n.d.	SA0309 (geh)	BAB41533	MALDI-MS	−139.9	−6.5
Geh[F3]	Glycerolesterhydrolase	n.d.	SA0309 (geh)	BAB41533	MALDI-MS	−70.8	−3.5
Geh[F4]	Glycerolesterhydrolase	n.d.	SA0309 (geh)	BAB41533	MALDI-MS	−88.3	−4.1
GlpQ	Glycerophosphoryl diester phosphodiesterase	32.2/8	SA0820 (glpQ)	BAB42059	MALDI-MS/ N-terminal	−3.1	−2.4
GlpQ[1]	Glycerophosphoryl diester phosphodiesterase	n.d.	SA0820 (glpQ)	BAB42059	MALDI-MS	−3.7	−2.4
GlpQ[2]	Glycerophosphoryl diester phosphodiesterase	n.d.	SA0820 (glpQ)	BAB42059	MALDI-MS	−7.1	Not changed
GlpQ[3]	Glycerophosphoryl diester phosphodiesterase	n.d.	SA0820 (glpQ)	BAB42059	MALDI-MS	−15.7	−7.9
GroEL[b]	GroEL protein chaperonin, 60 kDa	57.6/4.4	SA1836 (groEL)	BAB43116	MALDI-MS	Not changed	Not changed
GuaB[b]	Inositol-Monophosphat-Dehydrogenase	52.9/5.5	SA0375 (guaB)	BAB41602	MALDI-MS	Not changed	Not changed
Hla	Alpha hemolysin precursor	33/7.9	SA1007	P09616	MALDI-MS/ N-terminal	−25.4	Not changed
Hla[1]	Alpha hemolysin precursor	n.d.	SA1007	P09616	MALDI-MS	−18.4	Not changed
Hla[2]	Alpha hemolysin precursor	n.d.	SA1007	P09616	MALDI-MS	−13.6	Not changed
Hla[3]	Alpha hemolysin precursor	n.d.	SA1007	P09616	MALDI-MS	−11.2	Not changed
Hlb	Beta hemolysin precursor	33.7/7.3	SA1811 (hlb)	S15324	MALDI-MS/ N-terminal	−8.7	Not changed
Hlb[1]	Beta hemolysin precursor	n.d.	SA1811 (hlb)	S15324	MALDI-MS	−6.3	Not changed
Hlb[2]	Beta hemolysin precursor	n.d.	SA1811 (hlb)	S15324	MALDI-MS	−5	Not changed
HlgB	Gamma hemolysin Subunit B (LukF)	33.9/9.3	SA2209 (hlgB)	BAB43511	MALDI-MS	−7.9	Not changed
HlgC	Gamma hemolysin Subunit C (LukS)	32.5/9.1	SA2208 (hlgC)	BAB43510	MALDI-MS	−4.3	Not changed
IsaA	Immunodominant antigen A	21.5/5.3	SA2356 (isaA)	BAB43660	MALDI-MS/ N-terminal	30.7	12.1

Tab. 2 Continued

Protein	Function	M_r/pI mature	ORFID S. aureus N315 NCBI (NC 002745)	Accession (NCBI)	Identification	Quantitation[a] RN6390/ RN6911	Quantitation[a] $rsbU^-/rsbU^+$
IsaA[1]	Immunodominant antigen A	n.d.	SA2356 (isaA)	BAB43660	MALDI-MS	13.8	4.9
IsaA[2]	Immunodominant antigen A	n.d.	SA2356 (isaA)	BAB43660	MALDI-MS	20.6	13.6
IsaA[3]	Immunodominant antigen A	n.d.	SA2356 (isaA)	BAB43660	MALDI-MS	Not changed	Not changed
Lip	Triacylglycerol lipase precursor	72.9 /6.2	SA2463 (lip)	BAB43769	MALDI-MS/ N-terminal	Not changed	12.7
Lip[1]	Triacylglycerol lipase precursor	n.d.	SA2463 (lip)	BAB43769	MALDI-MS	Not changed	9.7
Lip[2]	Triacylglycerol lipase precursor	n.d.	SA2463 (lip)	BAB43769	MALDI-MS	Not changed	14.4
Lip[3]	Triacylglycerol lipase precursor	n.d.	SA2463 (lip)	BAB43769	MALDI-MS	−11.5	Not changed
Lip[4]	Triacylglycerol lipase precursor	n.d.	SA2463 (lip)	BAB43769	MALDI-MS	−2.8	Not changed
Lip[F]	Triacylglycerol lipase precursor	n.d.	SA2463 (lip)	BAB43769	MALDI-MS	−156.1	−6.2
Lip[F1]	Triacylglycerol lipase precursor	n.d.	SA2463 (lip)	BAB43769	MALDI-MS	−106.7	−5.6
Lip[F2]	Triacylglycerol lipase precursor	n.d.	SA2463 (lip)	BAB43769	MALDI-MS	−11.2	−4.7
Lip[F3]	Triacylglycerol lipase precursor	n.d.	SA2463 (lip)	BAB43769	MALDI-MS	−11.5	−2.4
LukD	Leucotoxin LukD	34.2/9	SA1637 (lukD)	BAB95632	MALDI-MS	−19.5	Not changed
LytM	Peptidoglycanhydrolase	31.7/6	SA0265 (lytM)	BAB41489	MALDI-MS	2.2	Not changed
PdhD	Dihydrolipoamide dehydrogenase component of pyruvate dehydrogenase E3	49,4/4,8	SA0946 (pdhD)	BAB42192	MALDI-MS	Not changed	Not changed
Plc	1-Phosphatidylinositol phosphodiesterase precursor	34.2 /6.4	SA0091 (plc)	BAB41310	MALDI-MS/ N-terminal	−16.9	−2
Sak	Staphylokinase	15.5/6.2	SA1758 (sak)	BAB43032	LC-MS/MS	−6.2	Not changed
SasD	Surface protein SasD	16.7/8.3	SA0129	BAB41349	LC-MS/MS	4	−2.2
SasD[1]	Surface protein SasD	n.d.	SA0129	BAB41349	LC-MS/MS	Not changed	−3.1
SasD[2]	Surface protein SasD	n.d.	SA0129	BAB41349	LC-MS/MS	Not changed	−5.2
SceD	Hypothetical protein similar to SceD precursor	21.5/4.9	SA1898	BAB43182	MALDI-MS	3.5	3.6

Tab. 2 Continued

Protein	Function	M_r/pI mature	ORFID S. aureus N315 NCBI (NC 002745)	Accession (NCBI)	Identification	Quantitation[a] RN6390/ RN6911	Quantitation[a] $rsbU^-$/$rsbU^+$
Spa	Immunoglobulin G binding protein A precursor	45.3/5.2	SA0107 (spa)	BAB41326	MALDI-MS	8.9	Not changed
SplA	Serinprotease SplA	21.9/8.6	SA1631 (splA)	BAB42899	MALDI-MS	−8.6	−2.9
SplA[1]	Serinprotease SplA	n.d.	SA1631 (splA)	BAB42899	MALDI-MS	−9.5	−3.2
SplB	Serinprotease SplB	22.4/9.1	SA1630 (splB)	BAB42898	MALDI-MS	−6	−2.7
SplB[1]	Serinprotease SplB	22.4/9.1	SA1630 (splB)	BAB42898	MALDI-MS	−22.1	−2.9
SplC	Serinprotease SplC	22.4/6.4	SA1629 (splC)	BAB42897	MALDI-MS	−6.6	−3.8
SplC[1]	Serinprotease SplC	22.4/6.4	SA1629 (splC)	BAB42897	MALDI-MS	−3.6	−2.4
SplE	Serinprotease SplE	22/9.2	−	AAF97929	MALDI-MS	−4.3	−2.4
SplF	Serinprotease SplF	21.9/8.9	SA1627 (splF)	BAB42895	MALDI-MS	−3.6	−2.7
SplF[1]	Serinprotease SplF	n.d.	SA1627 (splF)	BAB42895	MALDI-MS	−4.3	−2.2
SplF[2]	Serinprotease SplF	n.d.	SA1627 (splF)	BAB42895	MALDI-MS	−3.2	−2
SsaA	Secretory antigen precursor SsaA homologue	26.7/8.7	SA2093 (ssaA)	BAB43391	MALDI-MS	6.1	2.4
SspA	Serine protease SspA, V8 protease, glutamyl endopeptidase	33.4/4.8	SA0901 (sspA)	BAB42146	MALDI-MS/ N-terminal	−14.1	−16
SspA[1]	Serine protease SspA, V8 protease, glutamyl endopeptidase	n.d.	SA0901 (sspA)	BAB42146	MALDI-MS	−17	−8.6
SspA[2]	Serine protease SspA, V8 protease, glutamyl endopeptidase	n.d.	SA0901 (sspA)	BAB42146	MALDI-MS	−17.7	−6.4
SspA[3]	Serine protease SspA, V8 protease, glutamyl endopeptidase	n.d.	SA0901 (sspA)	BAB42146	MALDI-MS	−14.4	−7.2
SspA[4]	Serine protease SspA, V8 protease, glutamyl endopeptidase	n.d.	SA0901 (sspA)	BAB42146	MALDI-MS	−7.6	−7.8
SspB	Cysteine protease SspB	40.7/5.3	SA0900 (sspB)	BAB42145	MALDI-MS/ N-terminal	−46.5	−10.7
SspB[1]	Cysteine protease SspB	n.d.	SA0900 (sspB)	BAB42145	MALDI-MS	−28.2	−4.2

Tab. 2 Continued

Protein	Function	M_r/pI mature	ORFID S. aureus N315 NCBI (NC 002745)	Accession (NCBI)	Identification	Quantitation[a] RN6390/ RN6911	Quantitation[a] $rsbU^-/rsbU^+$
SspB	Cysteine protease SspB	n.d.	SA0900 (sspB)	BAB42145	MALDI-MS	−14.6	−4.5
Tkt[b]	Transketolase	72.3/5	SA1177 (tkt)	BAB42435	MALDI-MS	Not changed	Not changed
TrxB[b]	Thioredoxin reductase	33.6/5.2	SA0719 (trxB)	BAB41952	MALDI-MS	Not changed	Not changed
Yfnl	Hypothetical protein similar to anion-binding protein	32.2 /6.1	SA0674	BAB41907	MALDI-MS/ N-terminal	Not changed	Not changed
Yfnl[1]	Hypothetical protein similar to anion-binding protein	32.2 /6.1	SA0674	BAB41907	MALDI-MS	Not changed	Not changed
Yfnl[2]	Hypothetical protein similar to anion-binding protein	32.2/6.1	SA0674	BAB41907	MALDI-MS	Not changed	Not changed
SA0570	Hypothetical protein	15.9/9.2	SA0570	BAB41802	MALDI-MS	Not changed	5.8
SA0620	Secretory antigen SsaA homologue	25.6/5.6	SA0620	BAB41853	MALDI-MS	7.3	2.4
SA1737	Conserved hypothetical protein	38.5/4.9	SA1737	BAB43007	MALDI-MS	−8.3	−2.4
SA1812[b]	Hypothetical protein similar to synergohymenotropic toxin precursor	38.7/8.6	SA1812	BAB43092	MALDI-MS	−2.9	8.2
SA1812[1]	Hypothetical protein similar to synergohymenotropic toxin precursor	38.7/8.6	SA1812	BAB43092	MALDI-MS	−11.2	3.3
SA2097	Hypothetical protein similar to secretory antigen precursor SsaA	14.7/5.1	SA2097	BAB43396	MALDI-MS	5.2	7
SA2097[F]	Hypothetical protein similar to secretory antigen precursor SsaA	n.d.	SA2097	BAB43396	MALDI-MS	−24.9	−19.4

a) Volume ratios in the range of 1 to ∞ indicate an increase of the volume of the respective protein spot and volume ratios in the range −1 to −∞ indicate a decrease of the volume of the respective protein spot. Only volume ratios ≥ 2 or ≤ −2 were defined as significant changes between the different strains

b) Proteins without a typical signal sequence

n.d., pI and M_r could not be determined because these spots represent fragments of the respective protein

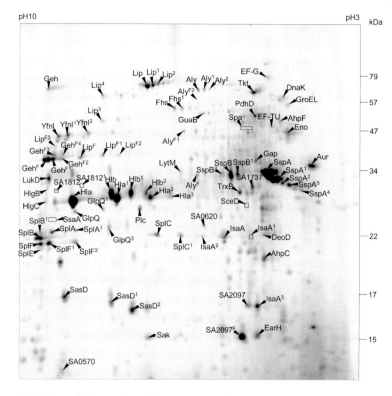

Fig. 1 Extracellular proteins of *S. aureus* RN6390. Proteins (50 µg) isolated from the supernatant of cells grown in TSB medium to an optical density of 5 at 540 nm were separated by 2-DE. Prior to separation proteins were labeled with Cy2 (Amersham Biosciences). The identified proteins are indicated by arrows and listed in Tab. 2.

outside the bacteria [5, 29, 30]. In addition to these genuine extracellular proteins some cytoplasmic proteins, probably released by autolysed cells, were also identified in the supernatants (Fig. 1, Tab. 2).

The expression of most of the known cell wall adhesins and extracellular virulence factors in *S. aureus* is coordinately regulated during the growth phase. Striking differences are seen when we compare the extracellular protein pattern of exponentially growing cells (OD_{540} = 1) with that of stationary phase cells (OD_{540} = 5; Fig. 2). As shown for other bacteria, the amount of most of the extracellular proteins increased at high cell densities (Figs. 2 and 3A and C). In general, extracellular proteins can be divided into at least two groups based on their growth phase dependent expression: (i) proteins present in exponentially and post-exponentially growing cell cultures, but which are not secreted at high cell densities (Fig. 3B) and (ii) proteins only present in post-exponential and stationary phase cell cultures (Fig. 3C). SsaA, SceD, LytM, IsaA, SA0620, Aly, Spa, SA2097, SA0570, SasD, and Aur belong to the first group (Fig. 3B). Hla, Hlb, Plc, Sak, Geh, GlpQ, Lip, SasD, SplA, SplB, SplC, SplE, SplF, SspA, SspB, YfnI, HlgB, HlgC, and LukD are listed in the second group (Fig. 3C).

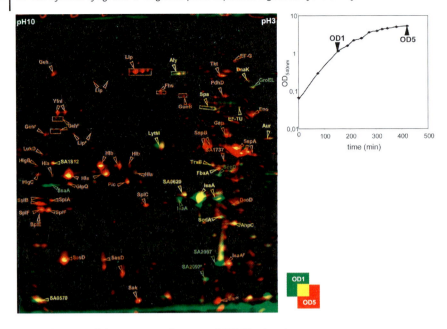

Fig. 2 The extracellular proteome of S. aureus RN6390 at low (green image) and high (red image) cell densities. Proteins (250 µg) of the supernatant of cells grown in TSB medium at $OD_{540} = 1$ or 5, were separated by 2-DE and stained with Sypro Ruby. Extracellular proteins whose amount is increased at high cell densities are labeled red and those proteins only present at low cell densities are labeled green.

12.3.2
The influence of *agr* and σ^B on the extracellular proteome

Since RNAIII is believed to be a global regulator of the synthesis of late virulence genes we analyzed the influence of a mutation in the quorum sensing system Agr on the extracellular protein pattern at high cell densities. By comparing the extracellular protein pattern of the wild-type RN6390 with that of an isogenic *agr* mutant (RN6911) we show that the level of about 70 protein spots changed in the mutant (Fig. 4A). Quantitation of these protein spots revealed that the amount of more than 50 protein spots was positively influenced by *agr* and an amount of about 15 protein spots was decreased in the presence of *agr* (Fig. 4A, Tab. 3). Besides Aur, Hla, SspA, SspB, Hlb, the mature lipase Lip^F and the serine proteases SplA, SplB, SplC and SplF, Geh, Plc, LytM, and Sak already shown to be affected by a mutation in *agr* [12, 14, 21, 31–36], we identified several other proteins including Aly, EarH, Fhs, GlpQ, HlgB, HlgC, LukD, SasD, and SceD which are putatively *agr* dependent (Fig. 4A, Tab. 3). Among the proteins whose level was decreased by *agr* we identified one spot of Geh, autolysin LytM, Spa, a protein with a high degree of similarity to autolysin Aly, SceD, the secretory antigen SsaA and the immunodominant antigen IsaA. As expected, the

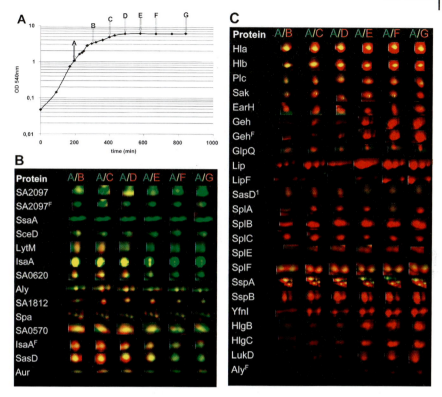

Fig. 3 (A) Growth of *S. aureus* RN6390 in TSB medium. The sampling is indicated by an arrow and a letter in the respective growth curve. (B, C) Patterns of the amount of extracellular proteins of *S. aureus* whose amount depends on the growth phase. The amount of the respective protein at an optical densitiy $OD_{540} = 1$ (green) of cells grown in TSB medium was compared with the amount of the protein at the higher optical densities (red). (B) Proteins only present at low cell densities. (C) Proteins only present at high cell densities. Proteins were stained with Sypro Ruby.

amount of those proteins under negative control of *agr* was decreased in the wild-type at high cell densities and conversely all proteins up-regulated by *agr* are present in higher amounts at high cell densities.

S. aureus RN6390 carries an 11 bp deletion in *rsbU*, which encodes a positive regulator of σ^B activity. Therefore, the results obtained here might be modified by the lack of σ^B. To overcome this problem, RN6390 complemented with *rsbU* from the *S. aureus* COL strain was included in the study (Fig. 4B, Tab. 3). Interestingly, the protein pattern of the RsbU⁺ strain was very similar to that of the RN6911 (Δ*agr*, Δ*rsbU*) strain indicating that *agr* and *rsbU* have opposing effects on the regulation of the same proteins (Fig. 4A and B). For instance Geh, GlpQ, Lip, Plc, SplA, SplB, SplC, SplE, SplF, SspA, and SspB were positively regulated by *agr* and negatively regulated by RsbU (Tab. 3). Furthermore, some

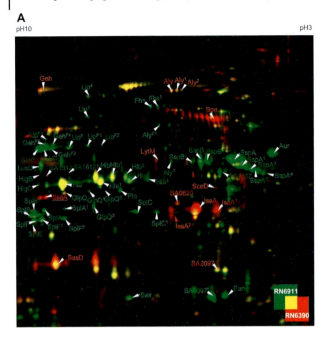

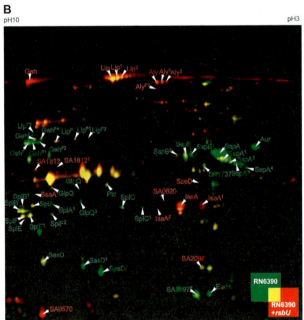

Fig. 4 Extracellular protein pattern of S. aureus RN6390 compared with that of (A) an isogenic agr mutant (RN6911) and (B) the rsbU complemented strain. Proteins (50 µg) from the supernatant of each strain grown to an optical density of OD_{540} = 5 in TSB medium were labeled with Cy2, Cy3 and Cy5, respectively, and separated by 2-DE.

proteins such as Aly, Geh, IsaA, SsaA and SA2097 (SsaA-homologue) were negatively regulated by *agr* but seemed to be induced in the RsbU$^+$ background (Tab. 3).

12.3.3
The influence of *agr* and σ^B on the transcription of virulence genes

Transcriptional studies were carried out to determine if the growth phase dependent effect on the amount of extracellular proteins in the different strains were due to changes in transcriptional initiation or to protein stability. As expected, Northern blot analyses of SA2097, SsaA, IsaA and SA0570 revealed that the transcription of these genes was regulated in a growth phase dependent manner showing high expression in growing cells and low expression in stationary phase cells (Fig. 5A). However, with the exception of SceD already shown to be transcribed by σ^B [15] neither *agr* nor *rsbU* showed any effect on the transcription of these genes (Fig. 5A). In contrast, for *sspAB*, *plC*, *lip*, and *glpQ* whose transcripts increased at high cell densities, the analysis indicated that regulation by *agr* and *rsbU* is at the transcriptional level (Fig. 5B). A mutation in *agr* prevents the stationary phase induced transcription of these genes. Conversely, lack of RsbU increases the amount of the respective transcripts (Fig. 5B).

12.3.4
Effect of σ^B on the transcription of *agr* and *sarA*

Northern blot analyses with a *sarA* gene probe showed strong *sarA* transcription, originating from the σ^A-dependent P_{A2} promoter, during exponential growth in TSB medium (OD_{540} = 1) which declined at high cell densities (Fig. 6A). A much weaker transcription, however, with a very similar time course was observed for the *sarA* transcript, originating from the σ^A-dependent P_{A1} promoter. In contrast, the σ^B-dependent transcript was detectable only in the *rsbU* complemented RN6390 strain. Transcription from the σ^B-dependent promoter increased as cells progressed from late exponential growth phase up to the stationary phase. Therefore, the amount of *sarA* specific transcripts at high optical densities was diminished in the RN6390 compared to the RN6390 RsbU$^+$ strain (Fig. 6A). A mutation in *agr* has no detectable influence on *SarA* transcription (Fig. 6A). Quantification of the growth phase dependent transcription of RNAIII by RNA Slot Blot experiments revealed low transcription levels in exponentially growing cells and an increased amount of transcripts at an optical density of 2 and higher. However, the presence of active RsbU in RN6390 did not change the amount of RNAIII strikingly (Fig. 6B).

12.4
Discussion

In the present study we could show that approximately 70 extracellular protein spots were affected by a mutation in the quorum sensing system *agr* (Fig. 4A, Tab. 3). As expected, the *agr* system positively influences the level of proteins which are only present at higher cell densities. In accordance with previous reports, we found many virulence factors among these proteins including the proteases Aur,

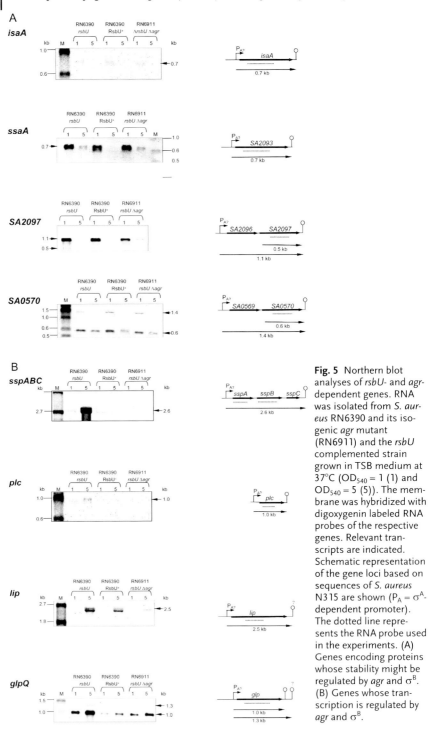

Fig. 5 Northern blot analyses of *rsbU*- and *agr*-dependent genes. RNA was isolated from *S. aureus* RN6390 and its isogenic *agr* mutant (RN6911) and the *rsbU* complemented strain grown in TSB medium at 37°C (OD_{540} = 1 (1) and OD_{540} = 5 (5)). The membrane was hybridized with digoxygenin labeled RNA probes of the respective genes. Relevant transcripts are indicated. Schematic representation of the gene loci based on sequences of *S. aureus* N315 are shown (P_A = σ^A-dependent promoter). The dotted line represents the RNA probe used in the experiments. (A) Genes encoding proteins whose stability might be regulated by *agr* and σ^B. (B) Genes whose transcription is regulated by *agr* and σ^B.

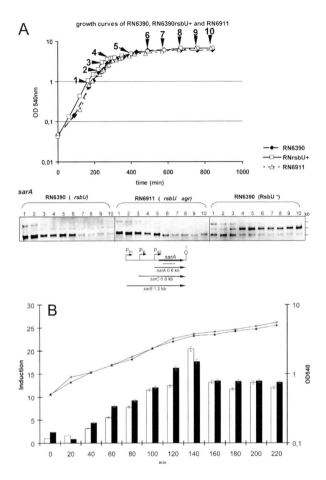

Fig. 6 Transcriptional analyses of agr and SarA. (A) Northern blot analysis of SarA. RNA was isolated from S. aureus RN6390 and its isogenic agr mutant (RN6911) and the rsbU complemented strain grown in TSB medium at 37°C. The sampling points are indicated in the growth curve by an arrow and the respective number. The membrane was hybridized with digoxygenin labeled RNA probes specific for SarA. Relevant transcripts are indicated. The dotted line represents the RNA probe used in the experiments. ($P_A = \sigma^A$-dependent promoter; $P_B = \sigma^B$-dependent promoter). (B) Schematic representation of changes in the RNA level of agr during growth in TSB medium (RN6390 ▲, RN6390 complemented with rsbU x). Total RNA was prepared from S. aureus RN6390 and the rsbU complemented strain from exponentially growing cells (OD_{540} = 0.5), post-exponentially growing cells, and stationary phase cells. Serial dilutions of total RNA were blotted onto a positively charged nylone membrane and hybridized with a digoxygenin labeled RNAIII probe. The hybridization signals were detected with a Lumi-Imager and quantitated using the Lumi-Analyst software from Boehringer. Induction ratios were calculated by setting the value of RN6390 at OD_{540} 0.5 to 1. The induction ratios of agr in RN6390 are shown in white bars and that of agr in the rsbU complemented RN6390 strain are shown in black bars.

Tab. 3 The influence of *agr* and σ^B on the amount of selected protein spots

Protein spot	*agr*	σ^B
Aur	+	−
EarH	+	−
Geh^F	+	−
Geh^{F1}	+	−
Geh^{F2}	+	−
Geh^{F3}	+	−
Geh^{F4}	+	−
GlpQ^1	+	−
GlpQ^2	+	−
GlpQ^3	+	−
Lip^F	+	−
Lip^{F1}	+	−
Lip^{F2}	+	−
Lip^{F3}	+	−
Plc	+	−
SplA	+	−
SplA^1	+	−
SplB	+	−
SplB^1	+	−
SplC	+	−
SplC^1	+	−
SplE	+	−
SplF	+	−
SplF^1	+	−
SplF^2	+	−
SspA	+	−
SspA^1	+	−
SspA^2	+	−
SspA^3	+	−
SspA^4	+	−
SspB	+	−
SspB^1	+	−
SspB^2	+	−
SA1737	+	−
SA2097^F	+	−
Aly^{F1}	+	+
Lip^3	+	+
Lip^4	+	+
Aly^F	+	0
Fhs	+	0
Fhs^1	+	0
GlpQ	+	0
Hla	+	0
Hla^1	+	0
Hla^2	+	0
Hla^3	+	0

Tab. 3 Continued

Protein spot	agr	σ^B
Hlb	+	0
Hlb^1	+	0
Hlb^2	+	0
HlgB	+	0
HlgC	+	0
LukD	+	0
Sak	+	0
Aly	−	+
Aly^1	−	+
Aly^2	−	+
Geh	−	+
IsaA	−	+
IsaA^1	−	+
IsaA^2	−	+
SceD	−	+
SsaA	−	+
SA2097	−	+
SasD	−	−
LytM	−	0
Spa	−	0
Lip	0	+
SA0570	0	+
SasD^1	0	−
SasD^2	0	−

+ positively influenced by the respective regulator
− negatively influenced by the respective regulator
0 no influence

SspA, SspB, SplA, SplB, SplC, and SplF, toxins Hla and Hlb, putative lipases Lip, Plc, and Geh, and staphylokinase Sak [12, 14, 21, 31–35]. Additionally, a set of new *agr*-dependent proteins were identified, which include a putative lipase (GlpQ), the leucocidines HlgB, HlgC, and LukD, a surface protein SasD, formyltetrahydrofolate synthase Fhs and EarH, a protein whose function is so far unknown. The level of these proteins was increased by the *agr* system at high cell densities. *Agr* seems also to be involved in repressing the amount of some proteins whose presence is largely restricted to exponentially growing cultures. Among these are proteins like IsaA, SsaA, SA2097, and SceD with no sequence similarities to proteins of known functions. Both SsaA and IsaA are solely described to be highly antigenic and elevated IgG antibody levels against them have been reported in patients with endocarditis or sepsis [37, 38]. Since the hypothetical protein SA2097 shows considerable similarities to SsaA it seems very likely that it may have analogous functions. Considering their growth phase dependent regulation, one might speculate that all these proteins are needed during the first stages of an infection for evasion of host defense mechanisms. In addition to these, autolysins like Aly and LytM and protein A belong to the group of proteins required at low cell densities. These autolysins are peptidoglycan hydrolases with autolytic activity -[39]. The *lytM* gene coding for a glycylglycine endopeptidase was shown to be highly expressed in early log phase cells and regulated by *agr*, SarA, the autolysis regulator RAT and the two component system YycG/YycF [36, 40–42].

In order to distinguish between effects by transcriptional regulation and those due to changes in protein stability we analyzed the transcription of selected genes. Transcriptional analyses for *sspABC, plC, lip* and *glpQ* revealed that the growth phase dependent regulation of gene expression occurs at the transcriptional level. As expected, the transcription of these genes was enhanced in stationary phase cells and was positively influenced by *agr* and negatively by active σ^B. In contrast, the level of gene-specific transcripts whose proteins were down-regulated by *agr* like *isaA, ssaA, SA2097*, and *SA0570* was diminished at high cell densities and a mutation in *agr* did not influence the amount of the gene-specific transcripts or the growth phase dependent kinetics. Therefore, it is very likely that the increased amount of these proteins in the *agr* mutant strain at high cell densities was due to protein stabilization possibly affected by the decreased level of proteases in the *agr* mutant. Proteases like staphopain, V8 protease, aureolysin, and various serine proteases which are up-regulated at high cell densities might contribute to the degradation of these proteins in the wild-type. On the other hand proteases produced by *S. aureus* may in some cases lead to activation of enzymes as has been suggested for lipases [30]. It has already been shown that proteases play a role in the transition of *S. aureus* cells from an adhesive to an invasive phenotype by degrading bacterial cell surface proteins [43, 44]. The serine protease SspA contributes to the degradation of fibronectin-binding protein and protein A [43]. Aureolysin has been shown to inactivate clumping factor ClfB by cleavage [45]. Our data indicate that extracellular proteins only present at low cell densities may also be targets of these proteases.

Clearly, active σ^B seems to prevent the expression of many extracellular proteins which are *agr* positively regulated like proteases, lipases, and toxins. This finding is supported by a recent study of Horsburgh *et al.* [23] demonstrating that the activities of SspA, a second unidentified protease and Hla were greatly reduced in the presence of RsbU which activates σ^B or by a lack of *agr* compared to the levels in the wild-type 8325-4. Previous results by Bischoff *et al.* [22] showed that the RNAIII level was lower in the presence of active σ^B, however, we could detect only very slight changes in the RNAIII amount in the RsbU$^+$ strain compared to RN6390 ($\Delta rsbU$). In contrast, the amount of specific *sarA* transcripts increased significantly if σ^B was active and it was noticeable that at high optical densities the transcription was almost solely initiated at the *sigB*-dependent promoter. In accordance with these data, we found that the SarA level is higher in a wild-type *S. aureus* COL compared to a *sigB* mutant [46] but the published data are still contradictory [23, 40, 47]. Contrary to our results Horsburgh *et al.* [23] demonstrated that the overall levels of transcription of *SarA* in strains 8325-4 and SH1000 (8325-4 *rsbU$^+$*) and *sigB* mutant derivatives of these strains were very similar. Cheung *et al.* [47] even published that a *sigB* mutant showed an elevated SarA level. Interestingly, most of the proteins whose amounts were up-regulated by *agr* and repressed by σ^B are also repressed by SarA ([15], unpublished observations). More recently, we found that the expression of *aur, stp, plc, nuc, sspA* and *sspB* by *agr* and SarA occurs at the transcriptional level [15]. Interestingly, in a *SarA* and *agr* deficient background these genes are transcribed as efficiently as in wild-type cells [15]. From these results, we conclude that for the transcriptional initiation of these genes *agr* is only needed in the presence of SarA. Furthermore, SarA seems to be a positive effector of some proteins like Aly, IsaA, and SA2097 which are negatively influenced by *agr* and positively influenced by σ^B ([15], unpublished observations). SarA is a DNA-binding protein that activates or represses the transcription of several target genes. The regions of SarA binding on the *agr, hla, spa, fnb,* and *sec* promoters have been mapped and a conserved SarA recognition motif (29 bp) has been postulated [13, 48, 49]. We therefore looked for this motif in front of all SarA dependent genes. However, the significance of this consensus search is not entirely clear as somewhat similar sequences are found in front of many other *S. aureus* genes.

12.5 Concluding remarks

These data strongly suggest that pathogenicity is established through a network of interacting regulons and is characterized by tightly controlled temporal gene expression patterns, ensuring the tight temporal control of the activity of the regulons and their proteins. At an early stage of infection SarA is mainly transcribed at the σ^A-dependent promoter P_{A2} and represses directly or indirectly the synthesis of late virulence factors ensuring adhesion to the host and possible biofilm formation. After reaching a critical cell density *agr* is activated and overcomes SarA dependent repression. The SarA dependent adhesion phase might be replaced by the induction of late virulence proteins initiating a more invasive phenotype. Simultaneously, the amount of early virulence factors is diminished due to a reduced transcriptional level by a mechanism independent of *agr*. Additionally, the level of some SarA dependent proteins might be decreased by the action of proteases synthesized in an *agr*-dependent manner. At this stage the transcription of *SarA* solely occurs from the σ^B-dependent promoter. Thus σ^B is involved in the tightly controlled temporal expression of virulence factors, probably, by enhancing the amount of SarA.

We thank Renate Gloger for excellent technical assistance. We are very grateful to Robert S. Jack and Jan Pané-Farré for critical comments on the manuscript and to Markus Bischoff for providing the rsbU complemented RN6390 strain. This work was supported by grants of the BMBF (031U107A), the Bayerische Forschungsstiftung, the Bildungsministerium des Landes Mecklenburg-Vorpommern (0100420), and the Fond der Chemischen Industrie to M.H.

12.6 References

[1] Hecker, M., Engelmann, S., Cordwell, S. J., *J. Chromatogr. B Analyt. Technol. Biomed. Life Sci.* 2003, *787*, 179–195.

[2] Bernardo, K., Fleer, S., Pakulat, N., Krut, O. et al., *Proteomics* 2002, *2*, 740–746.

[3] Herbert, S., Barry, P., Novick, R. P., *Infect. Immun.* 2001, *69*, 2996–3003.

[4] Kawano, Y. Ho, Y., Yamakawa, Y., Yamashino, T. et al., *FEMS Microbiol. Lett.* 2000, *189*, 103–108.

[5] Kawano, Y. Kawagishi, M., Nakano, M., Nase, K. et al., *Microbiol. Immunol.* 2001, *45*, 285–290.

[6] Vytvytska, O., Nagy, E., Bluggel, M., Meyer, H. E. et al., *Proteomics* 2002, *2*, 580–590.

[7] Morfeldt, E., Janzon, L., Arvidson, S., Lofdahl, S., *Mol. Gen. Genet.* 1988, *211*, 435–440.

[8] Peng, H. L., Novick, R. P., Kreiswirth, B., Kornblum, J., Schlievert, P., *J. Bacteriol.* 1988, *170*, 4365–4372.

[9] Recsei, P., Krewiswirth, B., O'Reilly, M., Schlievert, P. et al., *Mol. Gen. Genet.* 1986, *202*, 58–61.

[10] Cheung, A. L., Koomey, J. M., Butler, C. A., Projan, S. J., Fischetti, V. A., *Proc. Natl. Acad. Sci. USA* 1992, *89*, 6462–6466.

[11] Cheung, A. L., Projan, S. J., *J. Bacteriol.* 1994, *176*, 4168–4172.

[12] Chan, P. F., Foster, S. J., *J. Bacteriol.* 1998, *180*, 6232–6241.

[13] Chien, Y., Cheung, A. L., *J. Biol. Chem.* 1998, *273*, 2645–2652.

[14] Dunman, P. M., Murphy, E., Haney, S., Palacios, D. et al., *J. Bacteriol.* 2001, *183*, 7341–7353.

[15] Ziebandt, A. K., Weber, H., Rudolph, J., Schmid, R. et al., *Proteomics* 2001, *1*, 480–493.

[16] Chakrabarti, S. K., Misra, T. K., *J. Bacteriol.* 2000, *182*, 5893–5897.

[17] Manna, A. C., Bayer, M. G., Cheung, A. L., *J. Bacteriol.* 1998, *180*, 3828–3836.

[18] Morfeldt, E., Tegmark, K., Arvidson, S., *Mol. Microbiol.* 1996, *21*, 1227–1237.

[19] Janzon, L., Arvidson, S., *EMBO J.* 1990, *9*, 1391–1399.

[20] Morfeldt, E., Taylor, D., von Gabain, A., Arvidson, S., *EMBO J.* 1995, *14*, 4569–4577.

[21] Novick, R. P., Ross, H. F., Projan, S. J., Kornblum, J. et al., *EMBO J.* 1993, *12*, 3967–3975.

[22] Bischoff, M., Entenza, J. M., Giachino, P., *J. Bacteriol.* 2001, *183*, 5171–5179.

[23] Horsburgh, M. J., Aish, J. L., White, I. J., Shaw, L. et al., *J. Bacteriol.* 2002, *184*, 5457–5467.

[24] Giachino, P., Engelmann, S., Bischoff, M., *J. Bacteriol.* 2001, *183*, 1843–1852.

[25] Bernhardt, J., Büttner, K., Scharf, C., Hecker, M., *Electrophoresis* 1999, *20*, 2225–2240.

[26] Gertz, S., Engelmann, S., Schmid, R., Ohlsen, K. et al., *Mol. Gen. Genet.* 1999, *261*, 558–566.

[27] Wetzstein, M., Volker, U., Dedio, J., Lobau, S. et al., *J. Bacteriol.* 1992, *174*, 3300–3310.

[28] Kuroda, M., Ohba, T., Uchiyama, I., Baba, T. et al., *Lancet* 2001, *357*, 1225–1240.

[29] Massimi, I., Park, E., Rice, K., Muller-Esterl, W. et al., *J. Biol. Chem.* 2002, *277*, 41770–41777.

[30] Rollof, J., Normark, S., *J. Bacteriol.* 1992, *174*, 1844–1847.

[31] Vandenesch, F., Kornblum, J., Novick, R. P., *J. Bacteriol.* 1991, *173*, 6313–6320.

[32] Rice, K., Peralta, R., Bast, D., de Azavedo, J., McGavin, M. J., *Infect. Immun.* 2000, *69*, 159–169.

[33] Regassa, L. B., Novick, R. P., Betley, M. J., *Infect. Immun.* 1992, *60*, 3381–3388.

[34] Reed, S. B., Wesson, C. A., Liou, L. E., Trumble, W. R. et al., *Infect. Immun.* 2001, *69*, 1521–1527.

[35] Arvidson, S., Tegmark, K., *Int. J. Med. Microbiol.* 2001, *291*, 159–170.

[36] Ramadurai, L., Lockwood, K. J., Nadakavukaren, M. J., Jayaswal, R. K., *Microbiology* 1999, *145*, 801–808.

[37] Lang, S., Livesley, M. A., Lambert, P. A., Littler, W. A., Elliott, T. S., *FEMS Immunol. Med. Microbiol.* 2000, *29*, 213–220.

[38] Lorenz, U., Ohlsen, K., Karch, H., Thiede, A., Hacker, J., *Adv. Exp. Med. Biol.* 2000, *485*, 273–278.

[39] Shockman, G. D., Daneo-Moore, L., Kariyama, R., Massidda, O., *Microb. Drug Resist.* 1996, *2*, 95–98.

[40] Dubrac, S., Msadek, T., *J. Bacteriol.* 2004, *186*, 1175–1181.

[41] Ramadurai, L., Jayaswal, R. K., *J. Bacteriol.* 1997, *179*, 3625–3631.

[42] Ingavale, S. S., Van Wamel, W., Cheung, A. L., *Mol. Microbiol.* 2003, *48*, 1451–1466.

[43] Karlsson, A., Saravia-Otten, P., Tegmark, K., Morfeldt, E., Arvidson, S., *Infect. Immun.* 2001, *69*, 4742–4748.

[44] McGavin, M. J., Zahradka, C., Rice, K., Scott, J. E., *Infect. Immun.* 1997, *65*, 2621–2628.

[45] McAleese, F. M., Walsh, E. J., Sieprawska, M., Potempa, J., Foster, T. J., *J. Biol. Chem.* 2001, *276*, 29969–29978.

[46] Gertz, S., Engelmann, S., Schmid, R., Ziebandt, A. K. *et al.*, *J. Bacteriol.* 2000, *182*, 6983–6991.

[47] Cheung, A. L., Chien, Y. T., Bayer, A. S., *Infect. Immun.* 1999, *67*, 1331–1337.

[48] Chien, Y., Manna, A. C., Proja, S. J., Cheung, A. L., *J. Biol. Chem.* 1999, *274*, 37169–37176.

[49] Rechtin, T. M., Gillaspy, A. F., Schumacher, M. A., Breman, R. G. *et al.*, *Mol. Microbiol.* 1999, *33*, 307–316.

[50] Eymann, C., Dreisbach, A., Albrecht, D., Bernhardt, J. *et al.*, *Proteomics* 2004, DOI 10.1002/pmic.200400907.

13
Comparative proteome analysis of cellular proteins extracted from highly virulent *Francisella tularensis* ssp. *tularensis* and less virulent *F. tularensis* ssp. *holarctica* and *F. tularensis* ssp. *mediaasiatica**

Martin Hubálek, Lenka Hernychová, Martin Brychta, Juraj Lenčo, Jana Zechovská and Jiří Stulík

Francisella tularensis is the causative agent of the zoonotic disease tularemia. Four subspecies of this pathogen, namely ssp. *tularensis*, *mediaasiatica*, *holarctica*, and *novicida* are spread throughout the northern hemisphere. Although there are marked variations in their virulence to mammals, the subspecies are difficult to identify as they are closely genetically related. We carried out the comparative proteome analysis of cellular extracts from isolates representing the highly virulent subspecies *tularensis*, and the less virulent subspecies *mediaasiatica* and *holarctica* in order to identify new diagnostic markers and putative factors of virulence. We identified 27 protein spots that were either specifically present or at significantly higher abundance in ssp. *tularensis* strains, 22 proteins in ssp. *mediaasiatica* strains, and 26 proteins in ssp. *holarctica* strains. Subspecies *tularensis*-specific proteins might represent putative virulence factors. Of 27 identified *tularensis*-specific spots 17 represented charge and mass variants of proteins occurring in other subspecies, 7 spots were found to be present at higher abundance, and 3 spots were specifically present in *tularensis* strains. Amongst them, PilP protein, as a component neccessary for the biogenesis of the type IV pilus, virulence and adhesion factor for many human pathogen, was identified. Furthermore, the identification of additional 27 proteins common for ssp. *tularensis* and *mediaasiatica*, and 19 proteins shared by ssp. *mediaasiatica* and *holarctica* documented apparent closer genetic similarity between ssp. *tularensis* and *mediaasiatica*.

* Originally published in Proteomics 2004, 10, 3048–3060.

Proteomics of Microbial Pathogens. Edited by Peter R. Jungblut and Michael Hecker
Copyright © 2007 WILEY-VCH Verlag GmbH & Co. KGaA, Weinheim
ISBN: 978-3-527-31759-2

13.1
Introduction

Tularemia is a zoonotic bacterial disease caused by the small Gram-negative intracellular bacterium *Francisella tularensis*. *F. tularensis* is endemic in the nothern hemisphere with the highest frequency of occurrence in Scandinavia, northern America, Japan, and Russia [1]. Nevertheless, recent outbreaks of tularemia in Spain, Switzerland, Turkey, and Kosovo document wider distribution of the bacterium than originally assumed [2–4]. Several different subspecies of *F. tularensis* have been proposed. The division of *F. tularensis* into *F. tularensis* ssp. *tularensis* (type A) and *F. tularensis* ssp. *palaearctica* (type B) has been officially accepted in 1959 [5]. Currently, the ssp. *palearctica* is commonly referred to as *holarctica*. The division of *F. tularensis* subspecies is based on their virulence, citrulline ureidase activity, and glycerol fermentation activity [1]. Based on this, more virulent subspecies (with a mortality of 5–30% to humans, if not properly treated), able to ferment glycerol is designated type A, *F. tularensis* ssp. *tularensis* [6]. The distribution of *F. tularensis* ssp. *tularensis* was thought to be restricted to North America, however, it was recently isolated in Europe [7]. Comparing to type A, type B strains are less virulent and found in Eurasia and North America. Both subspecies *tularensis* and *holarctica* exhibit highly species-specific regions of 16S rRNA, and oligonucleotides complementary to such sequences are now used as reliable tool for their discrimination [8]. Two other subspecies *F. tularensis* ssp. *mediaasiatica* and *novicida* were described. The former was isolated in central Asia and seems to be more closely related to *F. tularensis* ssp. *tularensis* because it hybridizes to the same sequence of 16S rRNA as type A strains and is capable of glycerol fermentation [6]. However, in contrast to ssp. *tularensis* it exhibits only mild virulence for humans and rabbits. *F. tularensis* ssp. *novicida* was isolated from a water sample in Utah and it is capable to cause a tularemia-like illness in humans. *F. novicida* and *F. tularensis* are genetically indistinguishable [9]. Taken together, biochemical and genetic results do not correlate with differential virulence of individual subspecies. Additionally, the application of DNA microarray analysis based on clones from a shotgun library used for the sequencing of highly virulent type A strain Schu S4 further confirmed high genomic similarity between strains of ssp. *tularensis* and *mediaasiatica* [10].

In the current study, we employed comparative proteome analysis of the three *F. tularensis* ssp. *tularensis*, *mediaasiatica*, and *holarctica* in order to identify proteins that distinguish these three individual subspecies to be used for subspecies discrimination and, possibly, that can lead to progress in understanding of the virulence mechanism of this pathogen.

Tab. 1 *Francisella tularensis* strains included in the comparative proteome analysis

FSC No.[a]	Subspecies	Strain information	Alternative strain desigantion
13	tularensis	Isolated by Eigelsbach (no additional information available)	FAM standard
33	tularensis	Squirrel, CDC-standard, Georgia, USA	SnMF
41	tularensis	Tick, British Columbia, Canada,1935	Vavenby
199	tularensis	Mite, Slovakia, 1988	SE-221–38
237	tularensis	Human ulcer, Ohio, 1941	Schu-S4
35	holarctica	Beaver, Hamilton, Montana, 1976	B423 A
74	holarctica	Hare, Sweden, 1974	SVA T7
200	holarctica	Human, Ljusdal, Sweden, 1998	Rem nr 3001
247	holarctica	Human, Vosges, France, 1993	SVA T20
257	holarctica	Tick, Moscow area, Russia, 1949	503/840
147	mediaasiatica	Miday gerbil, Kazakhstan, 1965	543
148	mediaasiatica	Ticks, Central Asia, 1982	240
149	mediaasiatica	Hare, Central Asia, 1965	120

a) *Francisella* strain collection number at the Swedish defense research agency

13.2 Material and methods

13.2.1 Bacterial cultures and sample preparation for 2-DE

Thirteen different isolates of *F. tularensis* ssp. *tularensis* ($n = 5$), *holarctica* ($n = 5$), and *mediaasiatica* ($n = 3$) (Tab. 1) acquired from *Francisella* strain collection (FSC) were kindly provided by M. Forsman, Swedish Defense Research Agency, Umea and by A. Sjostedt, Umea University, Sweden. The strains were cultured on McLeod agar supplemented with bovine hemoglobin and IsoVitaleXTM (Becton/Dickinson, San José, CA, USA) for 20 h at 36.6°C. For the preparation of whole-cell lysate, 1×10^{10} microbes were suspended in cold PBS, centrifuged, and the pellets were resuspended in 1 mL of lysis buffer composed of 137 mM NaCl, 10% glycerol, 1% p-octyl-β-D-glucopyranoside, 50 mM NaF, 1 mM Na_3VO_4, and proteinase inhibitors Complete mini (Roche Diagnostics, Mannheim, Germany). Bacterial suspension was then subjected to ten cycles of freeze-thawing in liquid nitrogen and undisrupted microbes were removed by centrifugation at $12\,578 \times g$ for 15 min at 4°C. Bacterial proteins were precipitated overnight in 20% TCA in acetone (-18°C) containing 0.2% DTT [11], then solubilized in IEF buffer containing 9 M urea, 4% w/v CHAPS, 70 mM DTT, and 5% v/v carrier ampholytes pH 9–11 (Sigma, St. Louis, MO, USA). Protein concentration in IEF buffer was determined by a modified bicinchoninic acid (BCA) assay [12].

13.2.2
Two-dimensional gel electrophoresis

Two-dimensional gel electrophoresis (2-DE) was performed as previously described [13]. The proteins extracted from whole-cell lysates were separated using either non-linear pH 3–10 gradient or linear pH 6–11 gradient Immobiline DryStrips, 18 cm (Amersham Biosciences, Uppsala, Sweden), in the first dimension. 100 µg protein was applied for analytical purposes. In the case of wide-range pH gradients, the samples were first diluted to a total volume of 350 µL with rehydration buffer (2 M thiourea, 6 M urea, 4% w/v CHAPS, 40 mM Tris-base, 2 mM tributylphosphine (TBP), and 0.003% w/v bromophenol blue, 1% v/v Pharmalyte pH 3–10, and 0.5% v/v Pharmalyte pH 8–10.5) and then in-gel rehydration was performed overnight. The basic IPG strips were swollen in rehydration buffer containing 2% v/v IPG buffer pH 6–11 overnight and the samples were cup-loaded at the anodic side. In the second dimension, gradient 9–16% SDS-PAGE was poured out, the resolved IPG strip placed on top, and electrophoresis performed. The final dimension of the resolved 2-DE gel was 18×20 cm. The gels were then silver-stained. The isoelectric points and molecular weights of proteins were approximated using polypeptide 2-D SDS-PAGE standards (Bio-Rad, Hercules, CA, USA). The standards were separated together with sample and pI and M_r of several bacterial protein spots in sample (spots known to show low level of variation throughout all isolates in all samples) were determined by values of standard protein spots. These values were then used for calibration.

13.2.3
Statistical analysis

Thirteen wild isolates of *Francisella tularensis* (Tab. 1) were cultivated to prepare two independent samples for each strain. Each sample was used for preparation of one gel of wide pH gradient (3–10) and one basic pH gradient (6–11). Thus, the final set comprised 26 images for wide pH gradient and 26 gels for basic pH gradient. The gels were scanned by a CCD camera (Image Station 2000R; Eastman Kodak, Rochester, NY, USA) and the data were analyzed by Melanie 3 package. All gels included in the experiment were matched against each other both within the subspecies and among different subspecies. The matching quality was manually corrected to avoid mismatches due to the gel variation. The gels were then divided into three classes corresponding to the subspecies. Relative spot volumes (% vol), *i.e.*, digitized staining intensity integrated over the area of the individual spot divided by the sum of integrating staining intensities of all spots and multiplied by 100 were used for spot quantitation. Normalized data for the matched spots were analyzed by Student's t-test for each combination of subspecies (*tularensis vs. holarctica, tularensis vs. mediaasiatica, holarctica vs. mediaasiatica,* and reverse). Spots with a p-value < 0.01 were accepted as significantly different. All identified spots were again manually corrected for possible mismatches and reanalyzed. In the case the spot was present in all subspecies, only those groups showing relative spot volume differences more than 2-fold were accepted.

13.2.4
In-gel digestion

For the micropreparative 2-DE gels, 0.5 mg protein was loaded on a IPG strip. Selected spots stained with Colloidal Blue Stain Kit (Invitrogen, San Diego, CA, USA) were excised and covered with 200 µL 100 mM Tris-HCl, pH 8.5, in 50% acetonitrile for 20 min at 30°C. Then, 200 µL equilibration buffer (50 mM ammonium bicarbonate, pH 7.8, in 5% acetonitrile) was added to the gel pieces. The gel pieces were then vacuum-dried, covered with 0.1 µg of sequencing-grade trypsin (Promega, Madison, WI, USA) in 30 µL 50 mM ammonium bicarbonate, pH 7.8, and 5% acetonitrile and slowly shaken overnight at 37°C. Protein digests were removed from gel pieces by sequential extraction with 50 µL 2% TFA, 60% acetonitrile mixed in ratios of 3:2 and 2:3, and finally with 60% acetonitrile itself. The extract volume was reduced to approximately 10 µL and frozen at $-20°C$.

13.2.5
Mass spectrometric identification

The mass spectra were recorded in reflector mode on a MALDI mass spectrometer Voyager-DE STR (PerSeptive Biosystems, Framingham, MA, USA) equipped with delayed extraction. One µL peptide mixture was spotted onto the target plate, air-dried, and covered with 1 µL matrix solution (2,5-dihydroxybenzoic acid, 50 mg/mL in 33% acetonitrile, 0.3% TFA). The instrument was calibrated externally with a five-point calibration using peptide standards. Additionally, spectra containing autolytic tryptic peptide masses were also calibrated internally. Proteins were identified by peptide mass fingerprinting by searching the obtained spectral data against the translated *Francisella* genome, that was kindly provided by the *Francisella tularensis* strain Schu4 genome sequence consortium [14]. MS-FIT algorithm of ProteinProspector program (Version 3.4.1.; University of California, San Francisco Mass Spectrometry Facility) was applied to match experimentally measured peptide masses against the *Francisella* genome. The following parameters were used for protein identification: 100 ppm error, fixed carbamidomethylation of cystein residues, optional oxidation of methionine, optional formation of pyroglutamate at *N*-terminal Gln, optional protein *N*-terminus acetylation, and 1 possible miscleavage site. Results were evaluated according to the MOWSE score, MALDI coverage, number of matching peptides, and intensity of matching peaks. If MALDI coverage was higher than 25%, at least 4 peaks matched, and the MOWSE score reached at least 100, the ORF was considered successfully identified. If the MALDI coverage was not over 25%, the intensity of the peaks was evaluated. If three high intensity peaks in MALDI spectrum matched within the 50 ppm error tolerance, then the ORF was also considered as identified. If the identification was not succesfull, the procedure was repeated with a new gel spot from a different gel. However, in some cases, the spot did not yield any identified ORF. These spots are not reported in the current work and further attempt to identify them is ongoing using more powerful techniques. The BLAST program was used to find proteins with sequence similarity in nonredundant NCBI protein database or Uniprot database. *E*-value cutoff was set to 0.01 and proteins with lower similarity were assigned as No homolog. Simultaneously, the Conserved Domain Database (CDD) search was performed and proteins' family relationship recorded [15]. Similarity search led also to assigning the proteins to one of the functional categories included in cluster of orthologous groups (COGs) [16] and simple sorting of proteins into the categories was done.

13.3
Results and discussion

The whole-cell extracts of 13 different isolates of *Francisella tularensis* representing each of the three *F. tularensis* ssp. *tularensis*, *mediaasiatica*, and *holarctica* were used for comparative proteome analysis. The silver-stained patterns of proteins resolved along the nonlinear pH 3–10 gradients yielded about 1800 distinct spots. The basic protein patterns then encompassed approximately 500 spots that partially overlapped with the wide pH range protein spectrum. Representative 2-DE maps of individual subspecies are depicted in Fig. 1. We observed evident differences in protein spectra of individual *F. tularensis* subspecies that comprised both qualitative changes mostly due to the presence of charge variants and quantitative changes. On the other hand, there was almost no variation observed within the subspecies (unpublished results). This was reported also by Broekhuijsen *et al.* [10] on the genomic level. The detected alterations in protein expression among subspecies were classified as groups of spots specifically detected in individual *F. tularensis* subspecies and groups of spots common for ssp. *tularensis* and *mediaasiatica* (TM), ssp. *tularensis* and *holarctica* (TH), and ssp. *mediaasiatica* and *holarctica* (MH). In total, we identified 27 protein spots either specifically present or at significantly higher abundance in ssp. *tularensis* strains, 22 spots in ssp. *mediaasiatica* strains, and 26 spots in ssp. *holarctica* strains. The presence of these groups of spots on 2-DE maps of *F. tularensis* lysates enables unambigous discrimination of individual *F. tularensis* subspecies. Furthermore, 27 identified proteins occurred specifically or were at higher abundance in group TM, 19 proteins characterized group MH, and 9 proteins fell into group TH encompassing spots missing or diminished in *mediaasiatica* strains. All identified protein spots together with their expression profiles in individual *F. tularensis* subspecies are listed in Tab. 2 (see Addendum). Few spots did not identify any ORF by used technique. These spots are not reported either in Tab. 2 or Figs. 1–3.

Since no annotation of the *Francisella* genome was available at the time of the current study, we sought for protein homologs in other genomes to functionally assign identified ORFs by BLAST against nr database at NCBI. Most of the proteins encoded by the *Francisella* genome can be aligned with similar proteins and group into one of the COGs of proteins [16]. We found that beside the group involving proteins with unknown function (11 members), the largest number of proteins (10) was classified as group of post-translational modification, protein turnover, and

13.3 Results and discussion | 255

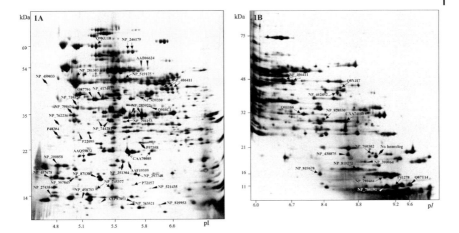

Fig. 1 Silver-stained 2-DE spot patterns of proteins from *F. tularensis* ssp. *tularensis*. (A) separated on wide pH gradient 3–10 and (B) basic pH gradient 6–11. All pictures are images of single gels representing each subspecies in either wide or basic pH range. The proteins are labeled by their accession numbers that correspond to accession number in the first column in Tab. 2. Spots indicated by a simple arrow designate proteins specific for each individual subspecies. Spots characteristic for subspecies *tularensis* and *mediaasiatica* are indicated by triangle, spots common to subspecies *holarctica* and *mediaasiatica* are indicated by block arrow and spots not detected in subspecies *mediaasiatica* and occurring in both subspecies *tularensis* and *holarctica* are denoted by diamond.

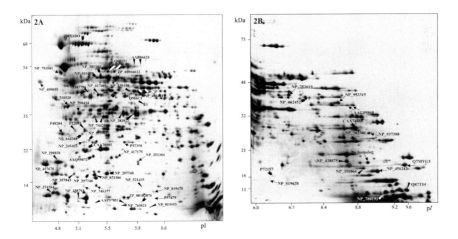

Fig. 2 Silver-stained 2-DE spot patterns of proteins from *F. tularensis* ssp. *mediaasiatica* separated on (A) wide pH gradient 3–10 and (B) basic pH gradient 6–11. Designations as in Fig. 1.

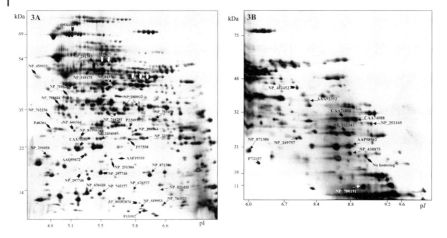

Fig. 3 Silver-stained 2-DE spot patterns of proteins from *F. tularensis* ssp. *holartica* separated on (A) wide pH gradient 3–10 and (B) basic pH gradient 6–11. Designations as in Fig. 1.

chaperones. This group was then followed by groups of proteins involved in translation, ribosomal structure, and biogenesis (9 proteins), energy production and conversion (7 proteins), and amino acid transport and metabolism (7 proteins). The groups containing proteins with general function or proteins engaged in carbohydrate, lipid, coenzyme, and nucleotide metabolism ranged from 4 to 7 proteins. Finally, the groups classified as transcription, cell envelope biogenesis, cell motility and secretion, predicted phosphatase/phosphohexomutase, inorganic ion transport and metabolism, defense mechanism, secondary metabolite biosynthesis and signal transduction mechanism contained 1–3 proteins. A possible explanation may be that this protein distribution into COG categories reflects the general distribution of proteins in all bacterial genomes. On the other hand, differences among genomes in bacteria, archae, and eukaryotes influenced by lifestyle, habitat, physiology, energy sources, and other factors concern mainly variation of transcription/translation factors, chaperone proteins, and genes encoding pathways of energy metabolism [17].

It was already mentioned that *F. tularensis* subspecies exhibit marked differences in their virulence but on the other hand they display a very close phylogenetic relationship and genomic similarity [6]. Broekhuijsen *et al.* [10] exploited the accessibility of the nearly completed genome sequence database of *F. tularensis* Schu4 strain for construction of DNA microarrays that were used for detection of regions of difference (RDs) between this strain and 27 strains of all 4 subspecies of *F. tularensis* differing in their virulence. The authors analyzed 8 RDs and identified 21 ORFs absent in ssp. *holarctica* as opposed to ssp. *tularensis*. None of these 21 ORFs was identified as different in the current proteomic study. Broekhuijsen's study also confirmed restricted genetic variation within *F. tularensis* species with maximally 3.7% of differentially hybridized probes. Nevertheless, the close genetic

Tab. 2 Protein spot variability among F. tularensis subspecies

Accession No.[a]	Protein name[b]	M_r/pI[c]	pH range of IPC strip[d]	MALDI coverage[e]	Score (bits)[f]	E-value[g]	Profile[h]
T – Proteins specific for ssp. tularensis (arrow designation in Figs. 1A and B)							
Q9KU18	ClpB protein	69/5.3	3–10	29	1021	0	DP
NP_246179	Transketolase 1	65/5.6; 64/5.7	3–10	20; 23	800	0	DP, DP
NP_281303	Putative periplasmic protein	54/5.1; 54/5.2	3–10	51; 43	117	1E-26	CV, CV
NP_519175	Probable 3-oxoacyl-(acylcarrier-protein) synthase II	52/5.6	3–10	42	165	1E-125	CV
NP_406411	Serine hydroxymethyltransferase	48/6.2; 48/6.5; 48/6.7	3–10; 6–11	26;30;22	236	1E-161	CV, CV, CV
Q8YJI7	4-Hydroxy-3-methylbut2-en-1-yl diphosphate synthase	43/8.6	6–11	39	508	1E-143	SP
NP_462452	γ-Glutamyltranspeptidase	40/8.6	6–11	30	101	1E-136	CV
NP_798153	Putative carbon-nitrogen hydrolase	32/5.6	3–10	38	187	3E-94	SP
Q01166	β-Lactamase precursor	32/7.4	6–11	65	107	3E-44	DP
NP_820530	Thymidylate synthase	31/8.4	6–11	47	193	1E-100	DP
CAA74088	Succinate dehydrogenase putative iron sulfur subunit	30/8.6	6–11	44	311	1E-93	CV
CAA70085	Hypothetical 23 kDa protein	22/5.7; 22/5.9	3–10	37; 55	405	1E-102	CV, CV
AAQ59872	Ribosome recycling factor	21/5.1	3–10	56	209	1E-47	CV
NP_760302	Tfp pilus assembly protein PilP	20/9.0	6–11	53	52	3E-04	SP
NP_297748	Nucleoside diphosphate kinase	17/5.7	3–10	52	210	2E-54	CV
NP_819272	Ribosomal protein L10	17/8.8	6–11	71	185	9E-40	DP
	Phosphoribosylaminoimidazole carboxylase catalytic subunit	16/5.7	3–10	50	218	2.9E-53	CV
NP_819678	Riboflavin synthase, β-subunit	16/8.0	6–11	70	153	5E-36	CV
NP_745377	Heat shock protein, HSP20 family	15/5.5; 14/5.3	3–10	29;44	85	1E-15	CV, CV

Tab. 2 Continued

Accession No.[a]	Protein name[b]	M_r/pI[c]	pH range of IPG strip[d]	MALDI coverage[e]	Score (bits)[f]	E-value[g]	Profile[h]
NP_790484	Ribosomal protein L24	12/9.1	6–11	47	119	5E-21	CV
P41278	50S ribosomal protein L23	11/9.2	6–11	50	98.6	3E-20	DP
M – Proteins specific for ssp. mediaasiatica (arrow designation in Figs. 2A and B)							
NP_793501	Trigger factor	56/5.0	3–10	46	115	1E-77	CV
NP_459033	Putative thiol-disulfide isomerase	51/4.4	3–10	59	105	2E-21	CV
ZP_00066612	Phosphomannomutase	50/5.5	3–10	27	352	1E-95	SP
NP_752619	Hypothetical aminotransferase ybdL	44/6.7	6–11	33	270	6E-74	DP
NP_953315	UDP-3-O-(3-hydroxymyristoyl) glucosamine N-acyltransferase	40/8.6	6–11	23	203	2E-48	SP
NP_710520	Glycerophosphoryl diester phosphodiesterase	39/5.0	3–10	60	71	9E-19	CV
AAG59860	Major acid phosphatase	39/8.7; 39/8.8	6–11	42;41	123	9E-28	SP, SP
NP_798434	Malonyl CoA-acyl carrier protein transacylase	35/5.1	3–10	42	145	3E-75	CV
NP_937380	Conserved hypothetical protein	28/9.0	6–11	35	49	2E-11	DP
P48204	GRPE protein	27/4.9	3–10	35	345	2E-80	CV
P57358	Orotidine 5'-phosphate decarboxylase	25/5.6	3–10	53	58	8E-19	CV
NP_245455	Grx2	25/5.2	3–10	39	117	1E-24	CV
NP_417175	Putative phosphatase	20/5.6	3–10	61	86	5E-20	SP
NP_251304	Periplasmic chaperone LolA	20/5.8	3–10	46	107	7E-23	CV
No homolog	No significant similarity to any protein in databases[i]	19/9.0	6–11	49	–	–	CV
Q7MYG3	50S ribosomal protein L5	18/9.6	6–11	50	265	5E-70	SP
NP_456242	Putative chorismate mutase	17/9.1	6–11	61	55	3E-07	DP

Tab. 2 Continued

Accession No.[a]	Protein name[b]	M_r/pI[c]	pH range of IPG strip[d]	MALDI coverage[e]	Score (bits)[f]	E-value[g]	Profile[h]
NP_745377	Heat shock protein, HSP20 family	13/5.2; 13/5.3	3–10	51;35	85	1E-15	CV, CV
NP_819678	Riboflavin synthase, β-subunit	13/6.2	3–10; 6–11	84	153	5E-36	CV
P57479	3-Dehydroquinate dehydratase	12/6.2	3–10	50	129	4E-28	CV
H – Proteins specific for ssp. holarctica (arrow designation in Figs. 3A and B)							
NP_841567	Thiolase	48/5.6	3–10	17	179	1E-129	DP
AAA92352	Peptidyl propyl cis/trans isomerase	40/7.9	6–11	23	46	9E-18	SP
NP_283922	Putative succinyl-CoA synthetase α-subunit	35/5.7; 35/5.9	3–10	47;41	244	1E-97	CV, CV
NP_744281	Universal stress protein family	32/5.5; 32/5.6	3–10	12;18	89	5E-16	CV, CV
NP_793495	Enoyl-(acyl-carrier-protein) reductase	31/6.2	3–10	15	210	8E-86	CV
CAA74088	Succinate dehydrogenase putative iron sulfur subunit	30/8.9	6–11	54	311	1E-93	CV
NP_252165	Cyclohexadienyl dehydratase precursor	29/9.0	6–11	72	85	5E-20	SP
P22095	Tryptophan synthase α-chain	28/5.7	3–10	38	322	1E-88	CV
NP_298764	ABC transporter ATP-binding protein	27/5.8	3–10	44	198	3E-83	CV
NP_245455	Grx2	25/5.5	3–10	46	117	1E-24	DP
NP_233439	Oxidoreductase, short-chain dehydrogenase/reductase family	25/5.8	3–10	58	304	2E-84	DP
NP_249757	Probable short-chain dehydrogenase	22/7.8	6–11	63	185	1E-45	SP
NP_290858	Inorganic pyrophosphatase	21/4.7	3–10	29	231	5E-58	CV
AAP58962	IglA	21/8.8	6–11	41	349	1E-79	CV
NP_438875	Transcription antitermination protein (NusG)	20/8.8	6–11	38	128	2E-27	CV
NP_871386	HlpA	18/6.2	3–10; 6–11	23	85	2E-08	DP

Tab. 2 Continued

Accession No.[a]	Protein name[b]	M_r/pI[c]	pH range of IPG strip[d]	MALDI coverage[e]	Score (bits)[f]	E-value[g]	Profile[h]
NP_670377	Hypothetical protein y3078	14/5.7	3–10	60	139	4E-34	SP
NP_521435	Probable ATP synthase ε-chain	14/6.5	3–10	42	82	3E-14	CV
NP_636420	Bacterioferritin	14/5.3	3–10	49	48	0.0008	SP
NP_745377	Heat shock protein, HSP20 family	14/5.5	3–10	54	85	1E-15	CV
NP_763521	Unknown	12/6.6	3–10	42	52	3E-06	CV
NP_819953	Conserved hypothetical protein	11/5.9	3–10	48	107	1E-20	CV
P15592	Probable sigma (54) modulation protein	11/6.0	3–10	66	107	3E-23	CV
NP_780191	50S ribosomal protein L25	11 / 9.0	6–11	39	103	3E-21	CV

TM – Proteins common to ssp. tularensis and mediaasiatica (triangle designation in Figs. 1A, 1B, 2A, 2B)

Accession No.[a]	Protein name[b]	M_r/pI[c]	pH range of IPG strip[d]	MALDI coverage[e]	Score (bits)[f]	E-value[g]	Profile[h]
AAB06624	Acid phosphatase	57/5.8; 57/5.9	3–10	38;34	345	0	SP, SP
NP_439330	S-Adenosylmethionine synthetase	46/5.8	3–10	42	248	1E-151	SP
O87796	Fructose-bisphosphate aldolase	43/5.3	3–10	56	288	1E-170	DP
NP_417401	Phosphoglycerate kinase	44/5.4	3–10	33	186	1E-122	DP
Q9I6C1	Signal recognition particle receptor FtsY	40/6.0	3–10	45	123	9E-99	DP
NP_283922	Putative succinyl-CoA synthetase α-subunit	35/5.6; 35/5.8	3–10	41;51	244	1E-97	CV, CV
NP_744281	Universal stress protein family	32/5.3; 32/5.4	3–10	27;30	89	5E-16	CV, CV
P22095	Tryptophan synthase α-chain	28/5.0; 28/5.1	3–10	37; 44	322	1E-88	CV, CV
NP_438875	Transcription antitermination protein (NusG)	20/8.7	6–11	46	128	2E-27	CV
NP_290858	Inorganic pyrophosphatase	19/4.8	3–10	33	231	5E-58	CV
NP_390064	Dihydrofolate reductase	18/8.9	6–11	54	149	3E-34	SP

Tab. 2 Continued

Accession No.[a]	Protein name[b]	M_r/pI[c]	pH range of IPG strip[d]	MALDI coverage[e]	Score (bits)[f]	E-value[g]	Profile[h]
NP_457678	Transcription elongation factor	17/4.7	3–10	56	182	8E-41	CV
NP_357843	30S Ribosomal protein S7	17/5.0	3–10	60	211	5E-49	CV
NP_871386	HlpA	17/5.3	3–10	30	85	2E-08	CV
NP_521435	Probable ATP synthase ε-chain	15/5.9	3–10	83	82	3E-14	CV
NP_274384	Thioredoxin	14.5/4.6	3–10	48	138	1E-32	CV
NP_458793	Conserved hypothetical protein	13/5.1	3–10	61	158	5E-40	SP
AAP97851	15.7 kDa putative exported protein	12/5.3; 12/5.7	3–10	40; 40	129	2E-29	SP, SP
NP_763521	Unknown	12/5.6	3–10	42	52	3E-06	CV
NP_819953	Conserved hypothetical protein	11/6.2	3–10	54	107	1E-20	CV
NP_780191	50S ribosomal protein L25	11/9.1	6–11	48	103	3E-21	CV
Q87T14	30S ribosomal protein S10	11/9.3	6–11	50	167	3E-41	SP
HM – Proteins common to ssp. holarctica and mediaasiatica (block arrow designation in Figs. 2A, 2B, 3A, 3B)							
Q9X6B0	Peroxidase/catalase	68/5.2	3–10	23	856	0	CV
Q60151	Glutathione reductase	53/5.6;53/5.7	3–10	22;27	121	1E-109	CV, CV
NP_281303	Putative periplasmic protein	53/5.4	3–10	37	117	1E-26	CV
NP_519175	Probable 3-oxoacyl-(acyl-carrier-protein) synthase II	49/5.3	3–10	30	165	1E-125	CV
NP_406411	Serine hydroxymethyltransferase	48/6.0; 48/6.2; 48/6.5	3–10	20; 21; 25	236	1E-161	CV, CV, CV
NP_462452	γ-Glutamyltranspeptidase	40/7.6	6–11	17	101	1E-136	CV
CAA74088	Succinate dehydrogenase putative iron sulfur subunit	30/8.8	6–11	44	311	1E-93	CV
NP_644364	Phenol hydroxylase	28/5.1	3–10	51	96	3E-21	CV
NP_819961	Hypothetical protein	25/5.2	3–10	46	56	6E-05	SP

Tab. 2 Continued

Accession No.[a]	Protein name[b]	M_r/pI[c]	pH range of IPC strip[d]	MALDI coverage[e]	Score (bits)[f]	E-value[g]	Profile[h]
NP_562341	Pyrrolidone-carboxylate peptidase	24/8.8	6–11	39	174	2E-45	CV
CAA70085	Hypothetical 23 kDa protein	22/5.2	3–10	30	405	1E-102	CV
AAQ59872	Ribosome recycling factor	20/5.3	3–10	54	209	2E-53	CV
NP_297748	Nucleoside diphosphate kinase	17/5.2; 17/5.4	3–10	46; 48	210	2E-54	CV, CV
P72157	Phosphoribosylaminoimidazole carboxylase catalytic subunit	17/6.2	6–11	50	218	3E-53	CV
ZP_00102874	Hypothetical protein	12/5.8	3–10	29	73	1E-13	SP
TH – Proteins common to ssp. tularensis and holarctica (diamond designation in Figs. 1A, 1B, 3A, 3B)							
NP_459033	Putative thiol-disulfide isomerase	50/4.6	3–10	48	105	2E-21	CV
NP_710520	Glycerophosphoryl diester phosphodiesterase	39/5.2	3–10	50	71	9E-19	CV
NP_798434	Malonyl CoA-acyl carrier protein transacylase	35/5.0	3–10	41	145	3E-75	CV
NP_762236	Isopenicillin N-synthase	30/4.8	3–10	36	151	7E-73	SP
P48204	GRPE protein	27/4.8	3–10	34	345	2E-80	CV
P57358	Orotidine 5'-phosphate decarboxylase	25/5.8	3–10	59	58	8E-19	CV
NP_251304	Periplasmic chaperone LolA	20/5.6	3–10	73	107	1E-22	CV
AAF19359	ATP synthase δ-subunit	20/5.6	3–10	47	110	2E-23	CV
No homolog	No significant similarity to any protein in databases[i]	19/9.0	6–11	49	–	–	CV

Table. 2:
Proteins were identified by peptide mass fingerprinting. MS-Fit program matched mass spectra against the *Francisella* translated genome. BLAST program was then applied to search for known proteins with sequence similarity in NCBI nr or Uniprot database. If no *Francisella* proteins matched, the most similar proteins of different microbial origin were recorded. Proteins are sorted by decreasing M_r and increasing pI for each of the six sections (T, M, H, TM, HM, TH).
a) Accession numbers according to NCBI nr or Uniprot databases
b) Name of the protein
c) Experimental molecular mass (kDa) and pI as calculated based on calibrated gel. In the case several spots on one 2-DE map identified one protein, the M_r/pI is recorded for all these spots in one cell
d) pH range of IPG strip where protein was identified; if there are two strips enclosed in record, the protein was found as overlapping on both 2-DE maps
e) MALDI coverage as recorded by MS-Fit, ProteinProspector 3.4.1
f) Normalized bit score as calculated by NCBI Blast program
g) BLAST E-value means probability of the match occurring by chance related to NCBI nr database
h) Differential profile of variable protein spots; SP, proteins specifically present in subspecies or in combination of subspecies and not detected elsewhere; CV, charge variant of protein identified in different subspecies at different position on gel; DP, differentially present spots detected at higher abundance than in other subspecies or combination of subspecies (all spots belong to category of significantly different at the p-level < 0.01 and minimaly 2-fold different in value of normalized volume)
i) No similarity with E-value < 0.01 found in NCBI nr or Uniprot databases

similarity between ssp. *tularensis* and *mediaasiatica* was revealed using principal component analysis applied for exploration of co-variances among strains. This result is also in agreement with our data because comparative proteome analysis proved the existence of 27 proteins significant for both ssp. *tularensis* and *mediaasiatica* comparing to 19 proteins shared by *holarctica* and *mediaasiatica* isolates. Conversely, the lesser degree of genetic similarity between *holarctica* and *tularensis* subspecies is corroborated by our observation that only 9 proteins were found to be characteristically expressed in *holarctica* and *tularensis* strains comparing to *mediaasiatica* isolates.

The existence of a high degree of genomic similarity between ssp. *tularensis* and *mediaasiatica*, however, does not correspond with high discrepancy in their virulence. We identified 27 protein spots specifically expressed in isolates of highly virulent *tularensis* subspecies. 17 of these protein spots had their charge and mass variant counterparts in less virulent subspecies and therefore represented *tularensis*-specific protein species (labeled as CV). This charge and mass heterogeneity probably reflects subspecies-specific amino acid substitutions, as we identified for variants of the hypothetical 23 kDa protein (unpublished observation), or differential post-translational protein modification. The existence of proteins differing in electrophoretic mobility was also documented for virulent and attenuated strains of *Mycobacterium tuberculosis* [18]. Some proteins exhibiting charge variants specific for *tularensis* subspecies were found to be associated with virulence in other bacteria. These proteins include enzymes involved in amino acid metabolism serine hydroxymethyltransferase and γ-glutamyltranspeptidase. Both enzymes exhibited a shift from more acidic positions in less virulent strains to more basic locations in highly virulent strains. Serine hydroxymethyltransferase catalyzes the interconversion of L-Ser and Gly using tetrahydrofolate as a cofactor. This enzyme is upregulated during an integral response to signals eliciting curli formation. Curli are fimbrial structures expressed by *Escherichia coli* and glycine makes 11.5% of the amino acid residues in total *E. coli* protein that specifically interact with matrix proteins such as fibronectin and laminin. Similar structures are also expressed by *Salmonella enteritidis* and have been denoted thin aggregative fimbriae. A simultaneous binding of fibrinolytic proteins and matrix proteins to fimbriae of *E. coli* and *S. enteritidis* could provide these pathogens with both adhesive and invasive properties [19, 20]. Recently, γ-glutamyltranspeptidase was shown to play a significant role in *Helicobacter pylori*-mediated apoptosis. Furthermore, γ-glutamyltranspeptidase-induced cell damage was associated with increased expression of COX-2 and EGF-related peptides [21, 22]. Nucleoside diphosphate kinase that also exhibits basic shift in ssp. *tularensis* belongs to ATP-utilizing enzymes that convert external ATP, presumably effluxed from macrophages, to various adenine nucleotides, which then activate purinergic receptors, such as P2Z, leading to enhanced macrophage cell death [23]. Basic shift in *tularensis* subspecies exerts also variants of the hypothetical 23 kDa protein that is critical for the intramacrophage growth of *F. tularensis* ssp. *novicida in vitro* [24] and the β-subunit of riboflavin synthase whose mutation abrogates fatal pleuropneumonia in swine induced by *Actinobacillus pleuropneumoniae* [25]. Acidic shift is observed in peroxidase/catalase that together with superoxide dismutase, glutathione peroxidase, and peroxiredoxins protects cells against reactive oxygen species (ROS). Protection against oxygen radicals mediated by this enzyme is also required by the intracellular pathogen *Legionella pneumonia* for its multiplication inside pulmonary macrophages [26]. On the contrary, a heat shock protein from HSP 20 family exerts both charge as well as mass heterogeneity. This HSP protein is a member of α-crystallin family proteins that are rapidly induced at reduced oxygen tensions and enable bacteria to survive under oxygen-limited conditions [27].

Additional subspecies *tularensis* specifically detected protein spots represent either variants that have no analogs in less virulent strains found so far (labeled as SP in Tab. 2) or variants at significantly higher abundance in comparison to their counterparts detected in *mediaasiatica* and *holarctica* strains (labeled as DP in Tab. 2). The latter group forms basic proteins like the β-lactamase precursor that exhibits a broad spectrum of hydrolytic activity, recognizing cephalosporins, penicillins, monobactams, and carbapenems as substrates [28], thymidylate synthase that is necessary for intracellular growth and survival of *Salmonella typhimurium in vitro* in both professional phagocytes and epithelial cells [29], and two ribosomal proteins, ribosomal protein L10 and 50S ribosomal protein L23. As for the former it was recently described that it may have an extraribosomal function in *E. histolytica* and its overexpression led to a 60% reduction in host cellular growth probably by destabilization of the activating protein 1 (AP-1) complex [30]. Subspecies *tularensis*-specific proteins that were found to be at significantly higher abundance in the acidic part of 2-DE gel then involved the ClpB protein whose gene interruption was associated with poor intramacrophage growth of *F. tularensis novicida* [24] and enzyme transketolase whose role in microbial pathogenicity is not understood.

The group of ssp. *tularensis* specifically presents proteins whose counterparts in less virulent subspecies were not detected, including 3 proteins: putative carbon-nitrogen hydrolase, 4-hydroxy-3-methylbut-2-en-1-yl diphosphate synthase, and Tfp assembly protein PilP. 4-Hydroxy-3-methylbut-2-en-1-yl diphosphate synthase is involved in the nonmevalonate terpenoid biosynthesis pathway [31]. From this point of view it is interesting to note that *in vivo* infection with *F. tularensis* leads to significant increase in levels of Vgamma9 Vdelta2 cells in 7–18 days after the onset of disease. Powerful stimuli of these cells are nonpeptidic pyrophosphorylated molecules [32]. To date no adherence factors have been identified on the protein level in *F. tularensis* despite significant homology of a set of ORFs to proteins needed for biogenesis of the type IV pilus in type A strain SchuS4. Type IV pilus is an adhesion and virulence factor expressed, *e.g.*, by *Pseudomonas aeruginosa*, *Neisseria gonorrhoeae*, and *Ralstonia solanacearum* [33, 34]. In this study, the Tfp assembly protein PilP, that together with PilK, PilM, PilO, and PilT proteins forms components of the pilin transport apparatus and thin-pilus basal body, was identified as ssp. *tularensis* specifically present protein [35].

13.4
Concluding remarks

In conclusion, our results show that despite close genetic similarity of *F. tularensis* subspecies these can be easily distinguished on the basis of a differential profile of protein spots. The majority of subspecies protein variants is represented by protein species differing in charge and molecular masses. Studies are currently underway to determine whether single amino acid substitution or different degrees of post-translational modifications are responsible for such a heterogeneity. Furthermore, comparison of protein patterns of highly virulent subspecies *tularensis* strains with the less virulent *mediaasiatica* and *holarctica* strains revealed potential new proteins involved in *Francisella* virulence. The construction of mutants using the recently described strategy for genetic manipulation of *F. tularensis* will then definitively confirm or exclude the participation of these candidates in the virulence mechanism of this bacterial pathogen [36].

We would like to thank Jana Michaličková and Alena Firychová for their excellent technical assistance. We also wish to thank Karin Hjalmarsson, the Head of the Dept. of NBC-Analysis, the FOI NBC-Defence (Umea, Sweden), for access to the FOI NBC-Defence Intranet database of genome sequence of F. tularensis Schu4 strain. The study was supported by Grant No. LN00A033 obtained from the Ministry of Education, Czech Republic. Sequencing of F. tularensis strain Schu4 was accomplished with the support from the US Department of the Army, Swedish Ministry of Defence, US Defence Advanced Research Projects Agency.

13.5 References

[1] Elllis, J., Oyston, P. C. F., Green, M., Titball, R. W., *Clin. Microbiol. Rev.* 2002, *15*, 631–646.

[2] Bachiller Luque, P., Perez Castrillon, J. L., Martin Luquero, M., Mena Martin, F. J. et al., *Rev. Clin. Esp.* 1998, *198*, 789–793.

[3] Reintjes, R., Dedushaj, I., Gjini, A., Jorgensen, T. R. et al., *Emerg. Infect. Dis.* 2002, *8*, 69–73.

[4] Wicki, R., Sauter, P., Mettler, C., Natsch, A. et al., *Eur. J. Clin. Microbiol. Infect. Dis.* 2000, *19*, 427–432.

[5] Olsufjev, N. G., Emelyanova, O. S., Dunayeva, T. N., *J. Hyg. Epidemiol. Microbiol. Immunol.* 1959, *3*, 139–149.

[6] Sandström, G., Sjöstedt, A., Forsman, M., Pavlovich, N. V., Mishankin, B. N., *J. Clin. Microbiol.* 1992, *30*, 172–175.

[7] Gurycova, D., *Eur. J. Epidemiol.* 1998, *14*, 797–802.

[8] Forsman, M., Sandström, G., Sjöstedt, A., *Int. J. Syst. Bacteriol.* 1994, *44*, 38–46.

[9] Hollis, D. G., Weaver, R. E., Steigerwalt, A. G., Wenger, J. D. et al., *J. Clin. Microbiol.* 1989, *27*, 1601–1608.

[10] Broekhuijsen, M., Larsson, P., Johansson, A., Byström, M. et al., *J. Clin. Microbiol.* 2003, *41*, 2924–2931.

[11] Görg, A., Obermaier, C., Boguth, G., Csordas, A. et al., *Electrophoresis* 1997, *18*, 328–337.

[12] Brown, R. E., Jarvis, K. L., Hyland, K. J., *Anal. Biochem.* 1989, *180*, 136–139.

[13] Havlasova, J., Hernychova, L., Halada, P., Pellantova, V. et al., *Proteomics* 2002, *2*, 857–867.

[14] Prior, R. G., Klasson, L., Larsson, P., Williams, K. et al., *J. Appl. Microbiol.* 2001, *91*, 614–620.

[15] Marchler-Bauer, A., Anderson, J. B., DeWeese-Scott, C., Fedorova, N. D., Geer, L. Y. et al., *Nucleic Acids Res.* 2003, *31*, 383–387.

[16] Tatusov, R. L., Natale, D. A., Garkavtsev, I. V., Tatusova, T. A. et al., *Nucleic Acids Res.* 2001, *29*, 22–28.

[17] Karlin, S., Brocchieri, L., Trent, J., Blaisdell, B. E., Mrazek, J., *Theor. Popul. Biol.* 2002, *61*, 367–390.

[18] Mattow, J., Jungblut, P. R., Schaible, U. E., Mollenkopf, H. J. et al., *Electrophoresis* 2001, *22*, 2936–2946.

[19] Chirwa, N. T., Herrington, M. B., *Microbiology* 2003, *149*, 525–535.

[20] Sjobring, U., Pohl, G., Olsen, A., *Mol. Microbiol.* 1994, *14*, 443–452.

[21] Busiello, I., Acquaviva, R., Di Popolo, A., Blanchard, T. G. et al., *Cell Microbiol.* 2004, *6*, 255–267.

[22] Shibayama, K., Kamachi, K., Nagata, N., Yagi, T. et al., *Mol. Microbiol.* 2003, *47*, 443–451.

[23] Zaborina, O., Dhiman, N., Ling Chen, M., Kostal, J. et al., *Microbiology* 2000, *146*, 2521–2530.

[24] Gray, C. G., Cowley, S. C., Cheung, K. K., Nano, F. E., *FEMS Microbiol. Lett.* 2002, *24*, 53–56.

[25] Fuller, T. E., Thacker, B. J., Mulks, M. H., *Infect. Immun.* 1996, *64*, 4659–4664.

[26] Bandyopadhyay, P., Byrne, B., Chan, Y., Swanson, M. S., Steinman, H. M., *Infect. Immun.* 2003, *71*, 4526–4535.

[27] Haile, Y., Bjune, G., Wiker, H. G., *Microbiology* 2002, *148*, 3881–3886.

[28] Rasmussen, B. A., Bush, K., Keeney, D., Yang, Y. et al., *Antimicrob. Agents Chemother.* 1996, *40*, 2080–2086.

[29] Kok, M., Bühlmann, E., Pechere, J. C., *Microbiology* 2001, *147*, 727–733.

[30] Chavez-Rios, R., Arias-Romero, L. E., Almaraz-Barrera Mde, J., Hernandez-Rivas, R. et al., *Mol. Biochem. Parasitol.* 2003, *127*, 151–160.

[31] Hintz, M., Reichenberg, A., Altincicek, B., Bahr, U. et al., *FEBS Lett.* 2001, *509*, 317–322.

[32] Poquet, Y., Kroca, M., Halary, F., Stenmark, S. et al., *Infect. Immun.* 1998, *66*, 2107–2114.

[33] Drake, S. L., Sandstedt, S. A., Koomey, M., *Mol. Microbiol.* 1997, *23*, 657–658.

[34] Kang, Y., Liu, H., Genin, S., Schell, M. A., Denny, T. P., *Mol. Microbiol.* 2002, *46*, 427–437.

[35] Sakai, D., Komano, T., *J. Bacteriol.* 2002, *184*, 444–451.

[36] Golovliov, I., Sjostedt, A., Mokrievich, A., Pavlov, V., *FEMS Microbiol Lett.* 2003, *222*, 273–280.

14
Proteome comparison of *Vibrio cholerae* cultured in aerobic and anaerobic conditions*

Biao Kan, Hajar Habibi, Monika Schmid, Weili Liang, Ruibai Wang, Duochun Wang and Peter R. Jungblut

The pathogen *Vibrio cholerae* causes severe diarrheal disease in humans. This environmental inhabitant has two distinct life cycles, in the environment and in the human small intestine, in which it differs in its multiplication behavior and virulence expression. Anaerobiosis, limitation of some nutrient elements, and excess burden from host metabolism reactants are the major stresses for *V. cholerae* living in intestine, in comparison to conditions in the environment and laboratory medium. For an insight into the response of *V. cholerae* to different microenvironments, we cultured the bacteria in aerobic and anaerobic conditions, and compared the whole cell proteome by two-dimensional electrophoresis. Among the protein spots identified, some protein species involved in aerobic respiration and the nutrient carbohydrate transporters were found to be more abundant in aerobic conditions, and some enzymes for anaerobic respiration and some stress response proteins were found more abundant in anaerobic culture. One spot corresponding to flagellin B subunit was decreased in anaerobic conditions, which suggests correlation with the meticulous regulation of bacterial motility during infection in the host intestine. This proteome analysis is the starting point for in-depth understanding of *V. cholerae* behavior in different environments.

14.1
Introduction

Cholera is a severe diarrheal disease caused by *Vibrio cholerae*, a Gram-negative genus of *Vibrionacea*. There are more than 200 serogroups in this genus, yet to date only toxigenic serogroup O1 and O139 [1, 2] are the known etiologic agents of epidemic cholera. Human cholera symptoms result from the colonization of the small

---

* Originally published in Proteomics 2004, 10, 3061–3067

Proteomics of Microbial Pathogens. Edited by Peter R. Jungblut and Michael Hecker
Copyright © 2007 WILEY-VCH Verlag GmbH & Co. KGaA, Weinheim
ISBN: 978-3-527-31759-2

intestine and the secretion of cholera toxin. The severity of disease ranges from asymptomatic to severe, the so-called "cholera gravis" [3]. *V. cholerae* serogroup O1 has two biotypes, classical and El Tor. The seventh pandemic, which began in 1961, was triggered by El Tor biotype, whereas previously classical vibrios were prevalent. In the 1990s, the most reported cases *per* year were more than 590 000 in 1991; the minimum was still near 150 000 in 1996 [4].

V. cholerae is a microbial inhabitant of coastal and estuarine water, associated with zooplankton, phytoplankton, and other aquatic flora and fauna (reviewed in [5, 6]). Some studies show the relationship between cholera epidemics with the bacteria environmental survival and persistence [7, 8]. When *V. cholerae* infects humans, large numbers of bacteria are excreted with the stool. Therefore there are two major living environments for the toxigenic *V. cholerae*: persistence in the aquatic environment and explosive multiplication in the human intestine. Different environmental factors, such as temperature, pH, oxygen level and nutrition, evoke different physiological responses of the microorganism. Bacteria alter their metabolism and expression of virulent factors through signal transduction pathways. Environmental temperature and pH affect the expression of cholera toxin and the major colonization factor, toxin-coregulated pilus (TCP), in *V. cholerae*. The survival of *V. cholerae* can alternatively assume a viable but nonculturable state and a 'rugose' survival form in the infertile environment. Growth of *V. cholerae* under low-iron conditions induces the expression of several new outer membrane proteins (OMPs) compared with cells grown in iron-rich media [5, 6].

Learning about the physiological states of *V. cholerae* in the human intestinal environment, namely investigating *in vivo* induced genes, may facilitate our understanding of its pathogenesis. The whole genome of an El Tor biotype strain has been sequenced [9], which is beneficial to the research of global gene regulation under various conditions. Using a microarray technique, Merrell *et al.* [10] analyzed the transcriptional profiling of *V. cholerae* harvested directly from stool samples, compared with the cultured bacteria. Xu *et al.* [11] compared the global transcriptional pattern of *V. cholerae* cells grown in the rabbit small intestinal environment to those grown in rich medium under aerobic conditions. Different expression of physiological and behavioral states of the bacteria were found, especially the virulence-related genes. The analysis suggests that nutrient limitation (particularly iron) and anaerobiosis are major stresses experienced by *V. cholerae* during their growth in the rabbit upper intestine.

Proteomics can help us directly discern the biology of bacteria at the protein species level, and protein composition change under different conditions [12, 13]. 2-DE combined with mass spectrometry provides a high-throughput technique to analyze the actual protein composition of bacteria at a defined point in time. It cannot be predicted that the level of transcriptional mRNA matches the actual protein amount of the primary translation products, because these may undergo different post-translational modifications. Here we present the global proteome profile of the genome-sequenced *V. cholerae* strain N16961 under common *in vitro* culture conditions. Since the human intestine is an anaerobic environment, we further analyzed the proteome profile of the culture under anaerobic conditions, to approximate the *in vivo* situation.

14.2
Materials and methods

14.2.1
Strains and culture

N16961 and Wujiang-2, strains of Serogroup O1, El Tor biotype of *V. cholerae*, were used in this study. The whole genome of N16961 has been sequenced [9].

To prepare the whole cell protein samples used for 2-DE analysis, a single fresh cultured colony of N16961 was picked from Luria-Bertani (LB) agar, transferred into 5 mL LB liquid media, and incubated for 8 h at 37°C at 200 rpm; 0.5 mL of culture was transferred into fresh 5 mL LB, and shaken for 3 h at 250 rpm. For the aerobic incubation, 1 mL of culture was transferred to a flask with 150 mL LB, and shaken for 7 h in aerobic conditions. For anaerobic incubation, another parallel culture flask was placed into a sealed anaerobic jar (Oxoid, Basingstoke, Hampshire, England), a bag of gas generating kit (Oxoid) was placed in the jar to produce anaerobic conditions, and a piece of anaerobic indicator (Oxoid) was used to monitor the anaerobic condition. Each culture condition was repeated three times and gels were produced for each of these preperations.

14.2.2
Sample preparation for 2-DE

For each flask, the culture was harvested after centrifugation for 10 min at 4°C and 5000 rpm. The supernatant was removed and the pellet resuspended with precooled 50 mL PBS, then centrifuged for 10 min at 5000 rpm and 4°C. The supernatant was removed and the pellet resuspended with precooled 50 mL PBS containing one Complete Protease Inhibitor Cocktail Tablet (Roche, Penzberg, Germany), and centrifuged as before. The pellet was weighed, then stored at $-86°C$.

14.2.3
2-DE

The pellet weight in mg was multiplied by 1.08, 0.1, 0.1, 0.1 and 0.5 to obtain the volumes in µL to be added to the pellet for urea, CHAPS (40%), Servalyte (Serva, Heidelberg, Germany) pI 2–4 (40%), DTT (1.4 M) and double-distilled water, respectively, to obtain final concentrations of 9 M urea, 70 mM DTT, 2% Servalyte pI 2–4 (Serva, Heidelberg, Germany) and 2% CHAPS [14]. After addition of these constituents to the pellet, the mixture was left at room temperature and shaken for 20 min and then centrifuged at $100\,000 \times g$ for 30 min. The supernatant was collected and the protein concentrations were 22.0 ± 3 and 13.8 ± 2 mg/mL for aerobic grown and anaerobic grown cultures, respectively. IEF and SDS-PAGE were run as previously described [14, 15]. A 23×30 cm 2-DE gel system was used. Sample loading was 50 µg for the 0.75 mm thick silver-stained gel, and 300 µg for the 1.5 mm thick Coomassie Brilliant Blue G-250-stained gel. The gels were scanned, and then the spots were numbered using TopSpot (available free from: http://www.mpiib-berlin.mpg.de/2D-PAGE/). Global profiles of the gels were compared visually to assign clear differences in intensity. For 24 spots picked out randomly from the spots with different intensity between aerobic and anaerobic conditions, the intensity of each spot, from four gels of each of aerobic culture and anaerobic culture, was measured with TopSpot and the significance of differences was analyzed by *t*-test within Excel (Microsoft).

14.2.4
MS

The spots in Coomassie-stained gels were excised, digested with Promega (Madison, WI, USA) sequencing grade modified trypsin and used to do MS with a MALDI-mass spectrometer (Voyager Elite; Perseptive, Framingham, MA, USA) [15]. The peptide mass fingerprint obtained for each protein digest was analyzed using the program MS-Fit (UCSF Mass Spectrometry Facility at http://prospector.ucsf.edu) or MASCOT (Matrix Science, London, UK, at http://www.matrixscience.com) in the NCBInr or Swiss-Prot database. The searches were performed with a mass tolerance of 100 ppm.

14.2.5
Electron microscope

The strains N16961 and Wujiang-2 were cultured as described in Section 2.1, then centrifuged and resuspended with PBS. The samples were adhered to the copper grids used for electron microscopy, and stained with 1% uranyl acetate before microscopy.

14.3
Results and discussion

The proteomes of three independent cultures of *V. cholerae* in aerobic and anaerobic conditions were separated on large 2-DE gels. The database of 2-DE of *V. cholerae* is available at http://www.mpiib-berlin.mpg.de/2D-PAGE/. Most proteins were located in the range of pI 3–6, especially the proteins of high and middle M_r. We compared *V. cholerae* 2-DE gels of anaerobic and aerobic culture from three independently prepared samples each (Fig. 1). Within one pair of gels about 133 spots were found visually to be more abundant in the aerobically cultured silver-stained gel, compared to the anaerobic gel, and about 111 spots displayed higher intensities in the anaerobic gel. Ten of the more abundant spots in the aerobic and 14 spots more abundant in the anaerobic culture were checked for reproducibility and found to be reproducibly different in a dataset of four gels of aerobic and four gels of anaerobic condition. All these eight gels were from different sample preparations. These spots were identified by MS analysis (Tables 1 and 2).

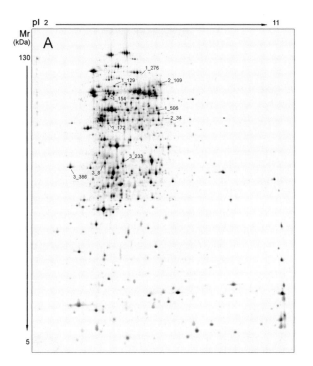

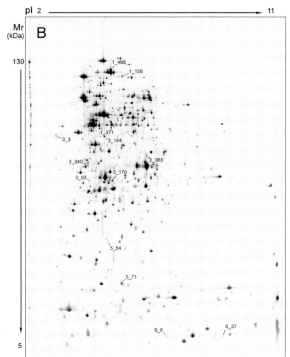

Fig. 1 Overview of 2-DE profile of *V. cholerae* strain N16961 cultured in aerobic (A) and anaerobic (B) conditions in LB medium, by silver staining. Identified spots with comparatively higher densities (see Tables 1 and 2) are marked in the corresponding gels.

Tab. 1 Identified protein species more abundant in aerobic culture

Spot no.	Intensity comparison[a]	SC%[b]	Score MASCOT	Proteins	Locus tag	Mass kDa	pI	Biological role
1_276	26.9 ± 15.7/0	26	152	conserved hypothetical protein	VC1872	74.1	5.56	
1_154	132.0 ± 17.1/43.7 ± 19.2	47	167	conserved hypothetical protein	VC2507	51.3	5.09	
1_506	203.0 ± 30.3/0	31	180	tryptophanase	VCA0161	52.9	5.80	Energy metabolism
2_34	58.1 ± 23.6/4.9 ± 3.4	36	177	glucose-1-phosphate adenylyltransferase	VC1727	45.5	5.90	Energy metabolism
3_233	654.1 ± 111.1/86.2 ± 18.6	33	96	iron(III) ABC transporter, periplasmic iron-compound-binding protein	VCA0685	37.7	5.91	Transport and binding proteins
1_129	280.6 ± 54.3/61.0 ± 12.9	28	91	phosphoribosylaminoimidazole carboxamide formyltransferase/IMP cyclohydrolase	VC0276	57.3	5.21	Nucleotides
2_109	408.6 ± 48.3/72.8 ± 26.9	41	174	succinate dehydrogenase, flavoprotein subunit	VC2089	64.5	6.09	Energy metabolism
3_386	570.1 ± 97.9/68.9 ± 21.5	72	264	galactoside ABC transporter, periplasmic D-galactose/D-glucose-binding protein	VC1325	34.8	4.68	Transport and binding proteins
3_8	636.4 ± 190.0/114.0 ± 41.8	42	125	amino acid ABC transporter, periplasmic amino acid-binding protein	VC1362	36.8	5.19	Transport and binding proteins
1_172	315.2 ± 73.2/146.7 ± 28.6	64	262	flagellin FlaB	VC2142	39.5	5.05	Cellular processes

a) The comparison of spots in aerobic to those in anaerobic gels. The intensity value for each spot, measured using TopSpot, is average ± SD. For all comparisons the p values are < 0.05.
b) Sequence coverage (percentage)

Tab. 2 Identified protein species more abundant in anaerobic culture

Spot no.	Intensity comparison[a]	SC%[b]	Score MASCOT	Proteins	Locus tag	Mass kDa	pI	Biological role
1_64	2.38 ± 0.8/30.7 ± 9.7	18	67	methyl-accepting chemotaxis protein	VC1403	68.0	4.85	Cellular processes
1_156	0/121.5 ± 27.9	42	230	lysine decarboxylase, inducible	VC0281	82.8	5.43	Energy metabolism
3_365	21.3 ± 12.1/85.9 ± 44.3	49	135	glyceraldehyde-3-phosphate dehydrogenase	VC2000	35.3	5.80	Energy metabolism
3_93	15.4 ± 3.9/72.2 ± 62.3	36	167	outer membrane protein OmpK	VC2305	33.4	5.96	Cell envelope
5_54	14.8 ± 6.7/44.1 ± 22.2	38	118	autoinducer-2 production protein	VC0557	19.1	5.30	Cellular processes
6_6	13.7 ± 10.4/43.1 ± 31.5	48	56	phosphocarrier protein HPr	VC0966	91.0	6.26	Transport and binding proteins
6_37	13.7 ± 16.2/93.4 ± 30.5	77	68	cold shock domain family protein	VCA0933	76.2	8.09	Regulatory functions
1_468	442.8 ± 42.2/843.5 ± 115.3	54	271	phosphate acetyltransferase	VC1097	76.8	5.28	Energy metabolism
3_3	79.8 ± 46.8/225.4 ± 122.3	48	179	maltoporin	VCA1028	45.0	4.44	Transport and binding proteins
3_144	65.1 ± 17.7/153.4 ± 18.6	70	297	peptidase, M20A family	VC1343	39.3	5.17	Protein fate/protein synthesis
3_340	49.0 ± 19.6/105.7 ± 25.4	58	109	cytidine deaminase	VC1231	31.9	4.84	Nucleotides
3_170	144.7 ± 20.5/244.5 ± 55.0	64	175	conserved hypothetical protein	VC0134	30.8	5.23	
5_71	84.3 ± 26.7/452.7 ± 122.6	69	174	16 kDa heat shock protein A	VC0018	16.8	5.28	Protein fate/protein synthesis
1_371	290.2 ± 37.0/423.8 ± 52.6	50	170	malate oxidoreductase, putative	VC2681	46.1	5.14	Energy metabolism

a) Comparison of spots in aerobic to those in anaerobic gels. The intensity value for each spot, measured by using TopSpot, is average ± SD. For all comparisons the p values are < 0.05, except for spots "**3_93**" (0.06) and "**6_6**" (0.06).
b) Sequence coverage (percentage)

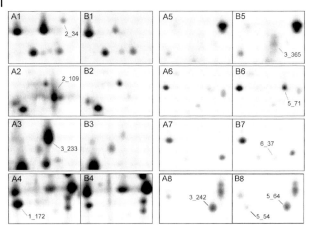

Fig. 2 Part view of selected spots with different abundance in aerobic and anaerobic cultures. The markers containing 'A' represent culture in aerobic conditions, and 'B' culture in anaerobic conditions.

14.3.1
Identified protein species more abundant in aerobic culture

Among the spots we analyzed, some protein species involved in energy metabolism were found with higher staining intensity on gels of the aerobic culture, these included enzymes such as glucose-1-phosphate adenylytransferase (spot no. 2_34 in the aerobic gel, A1 and B1 in Fig. 2) and succinate dehydrogenase (spot no. 2_109, parts A2 and B2 in Fig. 2). Glucose-1-phosphate adenylytransferase catalyzes the biosynthesis of glycogen at the first step of gluconeogenesis. This indicates that under aerobic and nutrient rich conditions, the gluconeogenesis pathway by which redundant glucose is stored as glycogen was enhanced. Succinate dehydrogenase is utilized during the aerobic respiration of the bacteria and catalyzes the reaction of succinate to fumarate in the TcA cycle. This may suggest active glycolysis and anabolism of carbohydrate of the culture in the nutrient media and under aerobic conditions. Further proteins of the TCA cycle are to be expected on the 2-DE patterns and it will be interesting to show if they behave in the same way. Another reductase which is involved in the interconversion of fumarate and succinate, fumarate reductase (Frd), was also found to be highly expressed in anaerobic growth in the rabbit ileal loop model [11].

Some spots corresponding to sugar, amino acid and iron transporters were found to have higher abundance compared with anaerobic culture. These include periplasmic binding proteins of galactoside ABC transporter (VC1325), amino acid ABC transporter (VC1362), and iron(III) ABC transporter (VCA0685, spot no. 3_233 in the gel, A3 and B3 in Fig. 2). This indicates active transport and metabolism in the aerobic culture and nutrient media. We consider the increase of the periplasmic iron-compound-binding protein (VCA0685) species may result from a different mechanism than *in vivo*; it showed a 2.7-fold induction in the rabbit ileal loop compared with culture in LB in the microarray-based study [11, supporting information]. In aerobic culture, the increase of this protein species may be due to other mechanisms compared with those in the small intestine, which is an iron-starved environment and the bacteria need to actively acquire iron [11, 16].

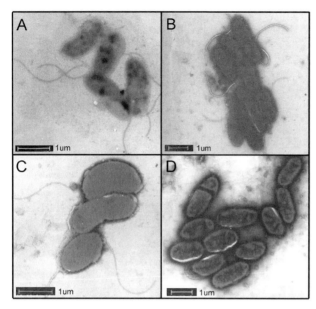

Fig. 3 The electron micrographs of *V. cholerae* strains cultured in aerobic and anaerobic LB media. A, strain N16961 in aerobic conditions; B, strain N16961 in anaerobic conditions; C, strain Wujiang-2 in aerobic conditions; D, strain Wujiang-2 in anaerobic conditions.

We found one spot ascribed to the flagellin subunit B (FlaB) decreased in abundance in anaerobic culture, compared with aerobic conditions (spot No. 1_172, A4 and B4 in Fig. 2). Although motility is important for full virulence of *V. cholerae* in the small intestine, the turnoff of flagellar synthesis might be advantageous to avoid immune recognition and clearance [17]. In *V. cholerae*, it was shown that motility phenotypes and the expression of some virulence genes are inversely related [18]. Whereas FlaB is not the essential subunit of flagella [17], its definitive role in formation or motility of flagella is uncertain. In our electron microscope analysis, we did not find obvious differences of flagella between the cultures of strain N16961 in the aerobic and anaerobic conditions, but for another El Tor strain, Wujiang-2, distinct differences were observed and less cells cultured in anaerobic condition had flagella (Fig. 3). The reported data from microarray-based analysis showed *flaB* with *flaE* and *flaC* transcribed in the rabbit ileal loop was 2.6 to 3.8 times higher than in LB [11]. In summary of the above we consider the regulation of flagella formation and bacterial motility in different environments is meticulous, especially during different periods of propagation in host intestine. Further identification needs to be performed to provide detailed information.

14.3.2
Identified protein species which are more abundant in anaerobic culture

In anaerobic conditions, some spots attributed to carbohydrate transport proteins showed higher abundance compared with aerobic culture. Phosphocarrier protein HPr (VC0966) is a component of the phosphoenolpyruvate-dependent sugar phosphotransferase system (PTS), a major carbohydrate active-transport system.

This suggests the enhanced utility of phosphoenolpyruvate-dependent sugar in anaerobic respiration of *V. cholerae*. Another two proteins were identified as belonging to the same group, maltoporin (VCA1028) for the transport of maltose and maltodextrins, and OmpK, which is suggested by sequence similarity to belong to the channel Tsx. In *Escherichia coli*, Tsx functions as a substrate-specific channel for deoxynucleosides and the antibiotic albicidin, and as a receptor for bacteriophages and colicins [19].

Some spots with higher abundance are related to carbohydrate metabolism, including inducible lysine decarboxylase (VC0281), phosphate acetyltransferase (VC1097), peptidase M20A family (VC1343), cytidine deaminase (VC1231), glyceraldehyde-3-phosphate dehydrogenase (VC2000, spot no. 3_365 in the anaerobic gel, A5 and B5 in Fig. 2) and putative malate oxidoreductase (VC2681). Glyceraldehyde-3-phosphate dehydrogenase plays an important role in glycolysis. This enzyme catalyzes the production of 1,3-bisphosphoglycerate, which is the first compound with high-energy bond during glycolysis. Malate oxidoreductase catalyzes malate to pyruvate and CO_2 in oxidation, its increased amount indicates the anaerobic respiration of *V. cholerae*.

We also found some protein species related to stress response having higher abundance in anaerobic culture. These include 16 kDa heat shock protein A (VC0018, spot no. 5_71, A6 and B6 in Fig. 2) and a cold shock domain family protein (VCA0933, spot no. 6_37, A7 and B7 in Fig. 2). This suggests that these regulons or chaperones might have a common response to some environmental stresses for bacterial survival. We found two spots (no. 3_242 in aerobic culture and no. 5_64 in anaerobic culture, A8 and B8 in Fig. 2) corresponding to autoinducer-2 production protein (VC0557) similar in abundance, whereas another spot in anaerobic culture, no. 5_54 which was also identified as VC0557, was more abundant. These isomers might play different roles in signal activation relating to different conditions. In quorum sensing system of *V. cholerae* [19], high expression of autoinducer-2 may activate the expression of hemagglutinin-protease (HAP), and decrease the expression of TCP. The result is that adhesion of bacterial cells is inhibited and may facilitate establishment of new colonization sites in the host intestine and escape from the host body.

14.4
Concluding remarks

Human intestine is anaerobic, compared with the natural environment and *in vitro* culture media for *V. cholerae*. Here we cultured the bacteria *in vitro* in aerobic and anaerobic conditions, to try to observe the protein amount difference involved in growth metabolism and virulence, which may be related to the change of this single factor. Most proteins we analyzed belong to the aerobic or anaerobic metabolism of carbohydrates. In aerobic cultures, the increased abundance of some protein species involved in substrate transport, amino acid metabolism and energy metabolism indicated active growth. In anaerobic cultures, the amount of proteins involved in glycolysis and other anaerobic respiration was enhanced. Certain sugar transporters and some stress response protein species also increased, compared with those in aerobic culture, which need further analysis on their effects on metabolism adaptability in the anaerobic environment. Some protein species directly or indirectly involved in motility were also found to be more abundant. This may suggest complex and aborative regulation during bacterial colonization, multiplication and spread. This proteome analysis resulted in the basic proteome of the cellular proteins of *V. cholerae* N16961 strain. It is a starting point for detection of antigens by immunoproteomics for the understanding of its pathogenesis and vaccine development. It is also helpful for further experiments to understand physiological adaptation to different microenvironments of *V. cholerae*.

This work was part supported by the National Basic Research Priorities Programme (Grant G1999054102 to B.K.), Ministry of Science and Technology, P.R.China. We thank Anne Meyer-Scholl for reading the manuscript and R. Stein for building up the proteome database.

14.5
References

[1] Ramamurthy, T., Garg, S., Sharma, R., Bhattacharya, S. K. et al., *Lancet* 1993, *341*, 703–704.

[2] Cholera Working Group, *Lancet* 1993, *342*, 387–390.

[3] Kaper, J. B., Morris, J. G., Levine, M. M., *Clin. Microbiol. Rev.* 1995, *8*, 48–86.

[4] WHO, *Weekly Epidemiol. Rec.* 2002, *77*, 257–264.

[5] Miller, M. B., Skorupski, K., Lenz, D. H., Taylor, R. K., Bassler, B. L., *Cell* 2002, *110*, 303–314.

[6] Faruque, S. M., Albert, M. J., Mekalanos, J. J., *Microbiol. Mol. Bio. Rev.* 1998, *62*, 1301–1314.

[7] Pascual, M., Rodo, X., Ellner, S. P., Colwell, R., Bouma, M. J., *Science* 2000, *289*, 1766–1769.

[8] Lobitz, B., Beck, L., Huq, A., Wood, B. et al., *Proc. Natl. Acad. Sci. USA* 2000, *97*, 1438–1443.

[9] Heidelberg, J. F., Eisen, J. A., Nelson, W. C. et al., *Nature* 2000, *406*, 477–484.

[10] Merrell, D. S., Butler, S. M., Qadri, F., Dolganov, N. A. et al., *Nature* 2002, *417*, 642–645.

[11] Xu, Q., Dziejman, M., Mekalanos, J. J., *Proc. Natl. Acad. Sci. USA* 2003, *100*, 1286–1291.

[12] Jungblut, P. R., *Microbes Infect.* 2001, *3*, 831–840.

[13] Washburn, M. P., Yates, III. J. R., *Cur. Opin. Microbiol.* 2000, *3*, 292–297.

[14] Jungblut, P. R., Bumann, D., Haas, G., Zimny-Arndt, U. et al., *Mol. Microbiol.* 2000, *36*, 710–725.

[15] Jungblut, P., Otto, A., Zeindl-Eberhart, E., Pleissner, K. P. et al., *Electrophoresis* 1994, *15*, 685–707.

[16] Wyckoff, E. E., Stoebner, J. A., Reed, K. E., Payne, S. M., *J. Bacteriol.* 1997, *179*, 7055–7062.

[17] Klose, K. E., Mekalanos, J. J., *J. Bacteriol.* 1998, *180*, 303–316.

[18] Gardel, C. L., Mekalanos, J. J., *Infect. Immun.* 1996, *64*, 2246–2255.

[19] Nieweg, A., Bremer, E., *Microbiology* 1997, *143*, 603–615.

15
Induction of *Mycobacterium avium* proteins upon infection of human macrophages*

Lara Brunori, Federico Giannoni, Luca Bini, Sabrina Liberatori, Cristiane Frota, Peter Jenner, Ove Fredrik Thoresen, Graziella Orefici and Lanfranco Fattorini

Induction of *Mycobacterium avium* proteins labelled with [^{35}S]methionine and mRNAs upon infection of the human macrophage cell line THP-1 was investigated by two-dimensional gel electrophoresis-mass spectrometry and reverse transcriptase-polymerase chain reaction (RT-PCR), respectively. *M. avium* overexpressed proteins within the macrophages that are involved in fatty acids metabolism (FadE2, FixA), cell wall synthesis (KasA), and protein synthesis (EF-tu). The correlation of differential protein and mRNA expression varied between good and no correlation. Overall, these four proteins may be involved in the adaptation and survival of *M. avium* within human macrophages.

Mycobacterium avium complex organisms are nontuberculosis mycobacteria relatively avirulent for healthy subjects, although they can cause disseminated infections in patients with AIDS [1]. *M. avium* is phagocytosed by macrophages and multiply within the phagosomes, which do not fuse with lysosomes and do not acidify due at least in part to the exclusion of the vacuolar proton-ATPase [2–4]. The macrophage exerts a selective pressure on the mycobacteria residing within it influencing the expression of gene products important for the survival of the bacteria in the phagosomal compartment. Various approaches have been used to identify genes associated with *M. avium* virulence in human cells including DNA selective capture of transcribed sequences and subtractive hybridization techniques [5, 6]; the protein profile of intracellular *M. avium* in mouse macrophages was also reported [7]. Gene expression is controlled at mRNA and protein levels, then the study of the interactions between *M. avium* and the macrophage cannot be limited to the analysis of transcriptional products and a comparative evaluation of protein and gene expression is necessary. To this purpose, we investigated protein and gene expression profiles upon infection of the human macrophage cell line THP-1 with *M. avium*. This was achieved by 2-D gel electrophoresis of [^{35}S]methionine-labelled proteins and MALDI-

---

* Originally published in Proteomics 2004, 10, 3078–3083

Proteomics of Microbial Pathogens. Edited by Peter R. Jungblut and Michael Hecker
Copyright © 2007 WILEY-VCH Verlag GmbH & Co. KGaA, Weinheim
ISBN: 978-3-527-31759-2

TOF MS; in addition, RT-PCR analysis was used to examine mRNA steady-state levels of specific genes.

M. avium strain 905 (virulent transparent colonies) [8] was grown at 37°C for 3 days under agitation in Middlebrook 7H9 liquid medium (Difco Laboratories, Detroit, MI, USA) to an optical density of 0.3 at 500 nm, corresponding to approximately 5×10^8 colony forming units (CFU)/mL. Microscopic examination showed that at this stage of growth most cells were in the form of coccobacilli. In preliminary experiments we observed that in Middlebrook 7H9 bacteria elongated to form filaments that fragmented into coccobacilli at the late-exponential phase; the coccobacilli were efficiently internalized by the human monocyte-derived cell line THP-1 (data not shown) and were used throughout the study. The THP-1 cells were grown in tissue culture plates in RPMI 1640 medium (Gibco Laboratories, Grand Island, NY, USA) supplemented with 2 mM L-glutamine, 10% heat-inactivated foetal calf serum (FCS), 1 mM sodium pyruvate, 1% nonessential amino acids and 100 µg/mL kanamycin (RPMI) at 37°C in a 5% CO_2-humidified incubator. For infection experiments THP-1 cells were passaged three times in kanamycin-free RPMI and incubated with 50 ng/mL phorbol myristate acetate for 24 h at 2×10^6 cells/mL for differentiation in macrophages. Nonadherent cells were removed by washing. M. avium organisms were incubated at 37°C for 30 min in kanamycin-free RPMI supplemented with 10% heat-inactivated human serum; opsonised bacteria were added to THP-1 cells at a ratio of 200 M. avium:1 THP-1. After 5 h, THP-1 cells were extensively washed and incubated for an additional 16–18 h. In metabolic labelling experiments, infected cells were incubated for 16–18 h in methionine-free kanamycin-free RPMI containing 50 µCi (1.8×10^6 Bq) [^{35}S]methionine/mL (Amersham Biosciences, Uppsala, Sweden). THP-1 cell viability was assessed by Trypan blue exclusion. M. avium intracellular growth was determined by microscopic examination of infected cells by Kinyoun staining (bioMerieux, Marcy-l'Etoile, France) and by determination of the numbers of the CFU in the THP-1 cell lysates as described by Sato and Tomioka [9]. Ninety-three percent THP-1 cells resulted to be infected and contained 21 ± 3 M. avium/cell, whereas the number of extracellular bacteria was negligible; the numbers of CFU per THP-1 cell were 24 ± 7 and 30 ± 10 at 0 and 18 h post-infection, respectively. Two different media were chosen as extracellular controls: 7H9 and RPMI. A growth curve was determined for these two media at 0, 24, 48, and 72 h. In 7H9, the mean CFU values were 1.3×10^7/mL and 5.5×10^8/mL at 0 and 72 h, respectively, whereas no CFU increase was observed in RPMI. THP-1 cells were lysed in PBS containing 0.25% w/v SDS; the intracellular bacteria were centrifuged at $14\,000 \times g$ for 20 min and the pellet was extensively washed in PBS containing 1% v/v Tween 80. The pellet was suspended in a solution containing 4% w/v CHAPS, 8 M urea, 65 mM DTT, 35 mM Tris, and bacteria were lysed by vigourous vortexing in the presence of 100 µm diameter glass beads for 10 min. M. avium organisms were incubated in kanamycin-free RPMI or Middlebrook 7H9 (1×10^9 CFU) and uninfected THP-1 cells were processed in the same manner of infected THP-1 cells and used as controls. No protein spots were detected in 2-D gels of uninfected THP-1 cells (data not shown).

M. avium lysates were diluted to a final volume of 350 µL in rehydration solution (8 M urea, 2% w/v CHAPS with traces of bromophenol blue), additioned with 2 µL IPG buffer pH 3–10 and loaded on nonlinear (NL) 3–10 IPG Strips (Amersham Biosciences), according to the manufacturer's instructions. The IEF was run in a IPGPhor apparatus (Amersham Biosciences), with a final voltage of 70 000 Vh. The second-dimensional separation was performed in a 9–16% gradient SDS-PAGE. Radioactive gels were transferred on nitrocellulose as described by Towbin et al. [10] and exposed using a phosphorimager (Amersham Biosciences). Molecular mass and isoelectric point were determined in comigration studies with human sera [11]. Preparative 2-D gels were performed for MS by running proteins obtained from large-scale M. avium-infected THP-1 cells (3×10^8 cells) without [^{35}S]methionine, and stained with colloidal Coomassie blue. Scanned 2-D gels were analyzed with the Melanie 3 software (GeneBio, Geneva, Switzerland). Quantitative variations of proteins were determined as volumes of spots normalized to the volume of all spots in each gel [12]. Protein spots of interest were cut out of the preparative gels using a pipette tip and processed for MS as described by Courchesne et al. [13] with minor modifications.

  The gel was destained and dehydrated in 50 mM NH_4HCO_3 and acetonitrile. The gel pieces were rehydrated with 50 mM NH_4HCO_3, and the proteins reduced with 20 mM DTT in 50 mM NH_4HCO_3 for 1 h at 37°C and washed with 50 mM NH_4HCO_3. The proteins were alkylated with 25 mM iodoacetamide in 50 mM NH_4HCO_3, incubated at room temperature in the dark for 30 min, washed with 50 mM NH_4HCO_3, and dehydrated with acetonitrile. The proteins were digested overnight at 32°C with ice-cold 5 mM NH_4HCO_3 containing 2 µg/µL trypsin (Promega, Madison, WI, USA). The peptide samples were acidified with 2% trifluoroacetic acid prior to MS analysis. Thin layer matrix surfaces of α-cyano-4-hydroxycinnamic acid mixed with nitrocellulose were prepared as described by Shevchenko et al. [14]. The acidified digest was deposited onto the thin layer and air-dried prior to rinsing with water. A Reflex III MALDI-TOF mass spectrometer (Bruker Daltonik, Bremen, Germany) equipped with the Scout-384 probe was used to obtain positive ion mass spectra of digested proteins with pulsed ion extraction in reflectron mode. Data were obtained using the following parameters: 26 kV accelerating voltage, 30 kV reflectron voltage, 70% grid voltage, 75 ns pulse delay time, low mass gate of 700 Da, and a deflection voltage of 2 kV. Calibration was achieved using internal mass calibration of the trypsin (Promega; modified, sequencing grade) autodigestion peptides. Peptide mass fingerprints were searched against the nonredundant databases available for M. avium 104 and M. tuberculosis H37Rv from the National Centre for Biotechnology Information. Partial enzymatic cleavages leaving two cleavage sites, oxidation of methionine, pyroglutamic acid formation at N-terminal glutamine, and modification of cysteine by acrylamide were considered in these searches. As to RT-PCR experiments, intracellular and extracellular bacteria were pelleted and resuspended in chloroform-methanol (3:1) and vigourously vortexed before addition of 4 volumes of Trizol (Gibco). 2 to 5 µg total RNA was treated with 50 units DNase I for 1 h at 37°C, extracted with phenol-chloroform, and precipitated with ethanol. Approximately 1 µg of RNA was reverse-

Tab. 1 Oligonucleotides used in RT-PCR analysis

Primer	Sequence
FadE2-S	GCCAAAGCCAGCGACTACCACA
FadE2-AS	TGGCGAACAGGTGCAGCGTCT
KasA-S	GAGCTACGACCTGATGAACGAGG
KasA-AS	TGACATCCGGCCACCCGGAT
EF-Tu-S	CAAGTACCCGGACCTGAACGAGT
EF-Tu-AS	CTTGATACCACGCAGCAGCAGAC
FixA-S	CGACGCGGTGCTGGACGAGAT
FixA-AS	GCCATGATGCCCTTGAACGAG
SigA-S	GCCACCCAGCTGATGTCGGAGAT
SigA-AS	CGTCGCGTAGGTGGAGAACTTGT
23S-S	TTTGGGGTGTAGTGGCGTGTTCTG
23S-AS	CCGGCGATCTGGGCTGTTTC

transcribed in ImProm-II reaction buffer, 3 mM MgCl$_2$, 0.5 mM each dNTP, 500 ng random hexamers, and 1 µL ImProm-II Reverse Transcriptase (Promega) for 1 h at 42°C. The exponential range of amplification for control genes *sigA* (28 cycles) and 23S (18 cycles) was established using as template serial dilutions of *M. avium* DNA (not shown). To amplify the selected genes, cDNA preparations from intracellular and extracellular bacteria were used in equal amounts, as determined using sub-plateau amplifications of the reference genes *sigA* and 23S. Reactions were performed in standard solute concentrations, and PCR conditions for all primer pairs were as follows: 3 min at 95°C, denaturation at 95°C for 40 s, and a single annealing-extension step at 72°C for 2 min. The sense (S) and antisense (AS) oligonucleotides used are shown in Tab. 1.

Fig. 1 shows the 2-D gel electrophoresis profiles of proteins from extra- (Fig. 1A) and intracellularly (Fig. 1B) grown mycobacteria labelled with [^{35}S]methionine; in Fig. 1C, enlargements of areas from *in vitro* (RPMI and Middlebrook 7H9 media) and intracellular (Intra) *M. avium* are shown. Three major spots (A2, A3, A4) were upregulated within THP-1 cells (Intra) compared to RPMI or 7H9 media; spot A1 was upregulated intracellularly compared to RPMI but not 7H9 medium. These proteins may play a role in adaptation and survival of *M. avium* within the macrophages since they were induced intracellularly at an extent higher than in the medium used for THP-1 infection, RPMI 1640, or even in a standard mycobacterial culture medium, Middlebrook 7H9. Spots induction may derive from a macrophage specific effect or from the physiological state of bacteria in relation to the growth conditions. Growth curves indicated that our bacteria can either rapidly grow (in 7H9) or not (in RPMI and THP-1 cells). These observations suggest that, up to 18 h, intracellular organisms are physiologically more similar to those recovered from RPMI than 7H9 and that induction of specific spots inside THP-1 cells is not an indirect effect of growth but is closely related to the intracellular life. Spots A2, A3, A4 were identified by MALDI-TOF MS as elongation factor Tu (EF-Tu), electron transfer flavoprotein β-subunit (FixA), acyl-CoA dehydrogenase

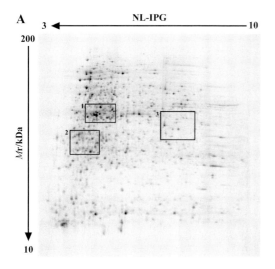
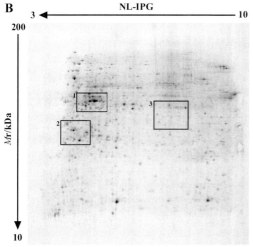
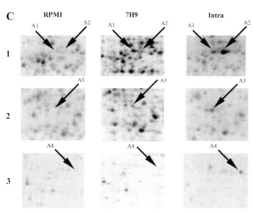

Fig. 1 Representative 2-D gel electrophoresis transferred on nitrocellulose membrane of [^{35}S]methionine-labelled proteins from *M. avium* 905 recovered (A) after incubation in RPMI 1640 medium or (B) after phagocytosis by THP-1 cells; squares 1, 2, and 3 indicate areas with differential spot intensities. Proteins were separated in the first dimension by isoelectric focusing using nonlinear (NL) 3–10 IPG strips and in the second dimension by a 9–16% gradient SDS-PAGE. The set of 2-D gels shown here is representative of three independent experiments. (C) Panel inserts of enlarged squared areas (1, 2, 3) where spots A1, A2, A3, and A4 were upregulated inside THP-1 cells (Intra), compared to incubation in RPMI 1640 medium and Middlebrook 7H9 medium.

Tab. 2 Identification by MALDI-MS of *M. avium* 905 proteins differentially expressed inside THP-1 cells

Spot	Mass (kDa)	p*I*	No. of peptides matched	Sequence coverage	Protein[a]	Rv No.	% of identity between *M. avium* and *M. tuberculosis*
A1	42.0	5.14	9/37	54/416	KasA	Rv2245	96
A2	42.0	5.22	9/16	121/396	EF-Tu	Rv0685	98
A3	32.1	4.83	14/18	134/263	FixA	Rv3029c	91
A4	37.8	6.66	28/63	211/406	FadE2	Rv0154c	85

a) KasA: [β]-ketoacyl-ACP synthase; EF-Tu: elongation factor EF-Tu; FixA: electron transfer flavoprotein (β)-subunit; FadE2: acyl-CoA dehydrogenase

Protein names and Rv No. are deduced from http://www.sanger.ac.uk/Projects/M_tuberculosis/Gene_list/. Percentages of amino acid identities between *M. avium* 104 and *M. tuberculosis* are shown.

(FadE2), respectively, corresponding to *Mycobacterium tuberculosis* Rv0685, Rv3029c, and Rv0154c, respectively. Spot A1 was identified as β-ketoacyl-ACP-synthase (KasA), corresponding to *M. tuberculosis* Rv2245 (Tab. 2).

The strong increase of EF-Tu inside THP-1 cells, but not in RPMI, seems to be related to the macrophage environment. EF-Tu was reported to be upregulated also in *M. bovis* BCG grown within THP-1 cells [15]. *M. avium* arrest of phagosome maturation is essential for growth inside the macrophage and requires accessibility to iron [16]. It is known that *M. tuberculosis* EF-Tu is upregulated by high iron concentrations [17], then overexpression of *M. avium* EF-Tu may be related to iron uptake by the mycobacterium inside the phagosome. Some support to this hypothesis comes from the knowledge that intraphagosomal *M. tuberculosis* acquires iron from extracellular transferrin and intracellular iron pools [18]. Also, *M. avium* EF-Tu presence in 7H9-grown bacteria may be related to the medium iron content.

FadE2 and FixA can be sequentially involved in the β-oxidation of fatty acids, which are a major source of carbon and energy for mycobacteria [19]. *Eschericha coli* FixA protein resembles the β-subunit of an electron transfer flavoprotein typically transferring electrons by FadE dehydrogenases to a ubiquinone oxidoreductase during fatty acid oxidation [20–22]. *E. coli* FixA has a 32% amino acid identity to *M. avium* FixA, so it is possible that, in *M. avium*, FadE2 is coupled with FixA in utilizing fatty acids inside the macrophage. This is in keeping with the knowledge that isocitrate lyase, an enzyme using fatty acids *via* the glyoxylate shunt, is upregulated during infection of macrophages by *M. avium* [23]. It has been reported that FixA of *M. tuberculosis* represents a new T cell antigen, as determined by measuring the interferon-γ (IFN-γ) response in mice [24].

KasA is an enzyme involved in the synthesis of mycolic acids, which are major components of the mycobacterial cell wall. The corresponding spot (spot A1) was almost absent in RPMI, whereas in 7H9 its intensity was comparable to that seen in

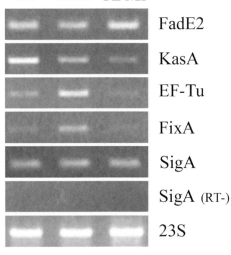

Fig. 2 Representative RT-PCR analysis of transcripts relative to the proteins synthesized by M. avium 905 in Middlebrook 7H9 medium, inside THP-1 cells (Intra), or in RPMI 1640 medium. To amplify kasA, EF-Tu, and fixA cDNAs, 28 cycles were used, whereas 32 cycles were necessary for fadE2 cDNA and 18 cycles for 23S. An RT mix without reverse transcriptase was amplified with sigA primers (28 cycles) to detect genomic DNA contamination (RT-). Three separate experiments were performed, all showing constant sigA/23S ratios in different samples, validating the use of either of these two transcripts as housekeeping genes.

THP-1-infected cells. This indicates that KasA may be not specifically induced within the macrophages, but the reasons for the discrepancy observed between the two media are unknown. Interestingly, InhA, another enzyme of the mycolic acid biosynthetic chain was found to be upregulated in BCG grown within THP-1 cells [15].

Next, we verified whether the protein inductions observed in 2-D gels were paralleled at the mRNA level by RT-PCR (Fig. 2). To verify equal loading of cDNA in each PCR, we amplified sigA mRNA, a transcription factor used as housekeeping control gene in M. tuberculosis studies [25] and 23S mRNA, amplified in a reduced number of cycles, given its high abundance. Overall results indicate that protein increases observed in 2-D gels did not always parallel upregulation of the corresponding mRNAs. For instance, while FixA mRNA steadily increased inside THP-1 cells in three consecutive experiments compared to both RPMI and 7H9, EF-Tu was induced more intracellularly than in RPMI (in two out of three experiments) and 7H9 (in one out of three experiments). As to KasA, transcription within THP-1 cells was repeatedly higher than in RPMI but lower than in 7H9. A major difference between protein and mRNA signal was seen for FadE2, with slightly higher transcription in RPMI or 7H9 than inside THP-1 cells. Overall, the correlation of protein and mRNA levels varied between good (FixA), moderate (EF-Tu) and no correlation (FadE2 and KasA). Discrepancies between mRNA and protein profiles can be attributed to mRNA stability, translational regulation, post-translational processing or a combination of these [26]; furthermore, a gene may give rise to more than one spot on the 2-D gels [27–29]. So far, in our knowledge, only another study compared mycobacterial differential gene and protein expression [29].

In conclusion, our data show that M. avium responds to macrophage phagocytosis by upregulating genes involved in biosynthetic and metabolic activities such as EF-Tu, FadE2, FixA and, possibly, KasA, that may play a role in survival and adaptation to the intracellular life. It is likely that additional genes are regulated in a less dramatic fashion, yet difficult to identify, at different time points after phagocytosis. The mechanisms by which mycobacteria survive the hostile environment of the macrophage are fundamental to our understanding of their virulence and pathogenicity and have been the objective of various studies on the transcriptional response [29–31]. In the present investigation, intracellular and extracellular proteomes of *M. avium* were compared and used to identify proteins whose induction was examined by mRNA semiquantitative analysis; our approach may represent a model to further characterize temporal expression of proteins and genes in response to different environmental stimuli.

This work was supported in part by grant No. 2017/RI of the Istituto Superiore di Sanità and grants No. 0AL/F and 1AF/F2 from the Ministero della Salute, Istituto Superiore di Sanità. The work at the National Institute of Medical Research was supported by the Medical Research Council (UK). We thank Dr. Steven A. Howell for help with the mass spectrometry.

15.1
References

[1] Inderlied, C. B., Kemper, C. A., Bermudez, L. E., *Clin. Microbiol. Rev.* 1993, *6*, 266–310.

[2] de Chastellier, C., Frehel, C., Offredo, C., Skamene, E., *Infect. Immun.* 1993, *61*, 3775–3784.

[3] Crowle, A. J., Dahl, R., Ross, E., May, M. H., *Infect. Immun.* 1991, *59*, 1823–1831.

[4] Sturgill-Koszycki, S., Schlesinger, P. H., Chakraborty, P., Haddix, P. L. et al., *Science* 1994, *263*, 678–681.

[5] Hou, J. Y., Graham, J. E., Clark-Curtiss, J. E., *Infect. Immun.* 2002, *70*, 3714–3726.

[6] Plum, G., Brenden, M., Clark-Curtiss, J. E., Pulverer, G., *Infect. Immun.* 1997, *65*, 4548–4557.

[7] Sturgill-Koszycki, S., Haddix, P. L., Russell, D. G., *Electrophoresis* 1997, *18*, 2558–2565.

[8] Fattorini, L., Li, B., Piersimoni, C., Tortoli, E. et al., *Antimicrob. Agents Chemother.* 1995, *39*, 680–685.

[9] Sato, K., Tomioka, H., *J. Antimicrob. Chemother.* 1999, *43*, 351–357.

[10] Towbin, H., Staehelin, T., Gordon, J., *Proc. Natl. Acad. Sci. USA* 1979, *76*, 4350–4354.

[11] Bjellqvist, B., Hughes, G. J., Pasquali, C., Paquet, et al., *Electrophoresis* 1993, *14*, 1023–1031.

[12] Zucchi, I., Bini, L., Valaperta, R., Ginestra, A. et al., *Proc. Natl. Acad. Sci. USA* 2001, *98*, 5608–5613.

[13] Courchesne, P. L., Patterson, S. D., in: Link, A. J. (Ed.), *2-D Proteome Analysis Protocols*, Humana Press, Totowa, NJ 1999, pp. 487–511.

[14] Shevchenko, A., Wilm, M., Vorm, O., Mann, M., *Anal. Chem.* 1996, *68*, 850–858.

[15] Monahan, I. M., Betts, J., Banerjee, D. K., Butcher, P. D., *Microbiology* 2001, *147*, 459–471.

[16] Kelley, V. A., Schorey, J. S., *Mol. Biol. Cell.* 2003, *14*, 3366–3377.

[17] Wong, D. K., Lee, B. Y., Horwitz, M. A., Gibson, B. W., *Infect. Immun.* 1999, *67*, 327–336.

[18] Olakanmi, O., Schlesinger, L. S., Ahmed, A., Britigan, B. E., *J. Biol. Chem.* 2002, *277*, 49727–49734.

[19] McKinney, J. D., Honer zu Bentrup, K., Munoz-Elias, E. J., Miczak, A. et al., *Nature* 2000, *406*, 735–738.

[20] Nunn, W. D., in: Neidhart, F. C. (Ed.), *Escherichia coli and Salmonella typhimurium. Cellular and Molecular Biology*, American Society for Microbiology, Washington, DC, USA 1987, pp. 285–301.

[21] Walt, A., Kahn, M. L., *J. Bacteriol.* 2002, *184*, 4044–4047.

[22] Tsai, M. H., Saier, M. H. Jr., *Res. Microbiol.* 1995, *146*, 397–404.

[23] Honer Zu Bentrup, K., Miczak, A., Swenson, D. L., Russell, D. G., *J. Bacteriol.* 1999, *181*, 7161–7167.

[24] Covert, B. A., Spencer, J. S., Orme, I. M., Belisle, J. T., *Proteomics* 2001, *1*, 574–586.

[25] Manganelli, R., Voskuil, M. I., Schoolnik, G. K., Smith, I., *Mol. Microbiol.* 2001, *41*, 423–437.

[26] Anderson, L., Seilhamer, J., *Electrophoresis* 1997, *18*, 533–537.

[27] Mattow, J., Jungblut, P. R., Muller, E. C., Kaufmann, S. H., *Proteomics* 2001, *1*, 494–507.

[28] Mattow, J., Schaible, U. E., Schmidt, F., Hagens, K. *et al., Electrophoresis* 2003, *24*, 3405–3420.

[29] Betts, J. C., Lukey, P. T., Robb, L. C., McAdam, R. A., Duncan, K., *Mol. Microbiol.* 2002, *43*, 717–731.

[30] Schnappinger, D., Ehrt, S., Voskuil, M. I., Liu, Y., Mangan, J. A. *et al., J. Exp. Med.* 2003, *198*, 693–704.

[31] Cappelli, G., Volpe, P., Sanduzzi, A., Sacchi, A., Colizzi, V. *et al., Infect. Immun.* 2001, *69*, 7262–7270.

16
Proteomics-based identification of novel *Candida albicans* antigens for diagnosis of systemic candidiasis in patients with underlying hematological malignancies*

Aida Pitarch, Joaquín Abian, Montserrat Carrascal, Miguel Sánchez,, César Nombela and Concha Gil

Systemic candidiasis remains a major cause of disease and death, particularly among patients suffering from hematological malignancies. In an attempt to contribute to the discovery of useful biomarkers for its diagnosis and therapeutic monitoring, we embarked on a mapping of *Candida albicans* immunogenic proteins specifically recognized by antibodies produced during the natural course of systemic *Candida* infection in this high-risk population. About 85 immunoreactive protein species were detected with systemic candidiasis patients' serum specimens by using immunoproteomics (*i.e.*, two-dimensional electrophoresis followed by Western blotting), and identified through a combination of peptide mass fingerprinting by matrix-assisted laser desorption/ionization-time of flight-mass spectrometry (MALDI-TOF-MS), *de novo* peptide sequencing using nano-electrospray ionization-ion trap (ESI-IT) MS, and genomic database searches. This proteomic approach has led to the characterization of 42 different housekeeping enzymes as *C. albicans* antigens. Their biological significance is also discussed. Furthermore, this study is the first to report that 26 of them exhibit antigenic properties in *C. albicans*, and 35 of them become targets of the human antibody response to systemic candidiasis. Our findings suggest that the production of antibodies to *C. albicans* phosphoglycerate kinase and alcohol dehydrogenase during systemic candidiasis could be associated with a differentiation of the human immune response. We also highlight the relationship between changes in maintenance of circulating levels of specific anti-*Candida* antibodies and patients' outcome. Some of these variations, especially the rise of high anti-enolase antibody concentrations, appear to be related to recovery from systemic candidiasis in these patients, which might serve as markers for predicting their outcome. This approach could therefore provide new challenges for diagnosis and clinical follow-up of these fungal infections, and even for antifungal drug or vaccine design.

* Originally published in Proteomics 2004, 10, 3084–3106.

16.1 Introduction

The last three decades have witnessed an alarming increase in the incidence of systemic candidiasis among patients suffering from hematological malignancies, especially leukemias and lymphomas [1–4]. This rise results from a host defense impairment coupled with barrier disruption due, in turn, to changes in the therapeutic management of the underlying malignancy (*i.e.*, more aggressive cytotoxic chemotherapy, total-body irradiation, bone marrow transplantation (BMT), development of graft-*versus*-host disease (GVHD) and a protracted granulocytopenic state, central intravenous catheterization, continuous corticosteroid therapy, and prolonged use of multiple broad-spectrum antibiotics, to name but a few [4–7]), leading to *Candida* colonization and subsequent bloodstream and deep-non-mucosal-tissue invasion. Autopsy data show that approximately 20–50% of patients with blood cell malignancies have histological evidence of invasive fungal infections [3, 8, 9]. Currently, leukemic patients receiving allogeneic BMT represent one of the most relevant populations at risk of systemic *Candida* infections [4, 10, 11].

Lamentably, the mortality associated with systemic candidiasis is dreadfully high in these patients [4, 11, 12]. In BMT recipients with *Candida* infections, for instance, this mortality rate can reach 73% [12]. The poor outcome of these high-risk patients with systemic candidiasis is directly related to (i) persistence of immunodeficiency, (ii) delay in diagnosis of this systemic fungal infection, because of nonspecific clinical signs and symptoms of early infection, unfeasibility to perform invasive diagnostic procedures given the underlying conditions of these patients, and lack of rapid, specific, and sensitive diagnostic tests [13, 14], and (iii) limitations in the treatment of candidiasis due to reduced efficacy and side-effects of currently available drugs, and appearance of antifungal resistance [15–17]. In addition, it is also noteworthy that on account of the inherent difficulties in the diagnosis of systemic candidiasis, the antifungal treatment is often given empirically [6, 13, 18, 19], and consequently many patients are overtreated with unnecessary highly nephrotoxic agents.

For all the aforementioned reasons, there is a pressing need to devise reliable diagnostic strategies for systemic candidiasis in this high-risk population. Although diverse innovative approaches are gaining an eminent interest in recent years [20, 21], none of them, however, has yet achieved widespread clinical use. Until recently, standard serological tests to detect anti-*Candida* antibodies were considered to have little diagnostic accuracy since these assays displayed a low specificity to discriminate between systemic candidiasis and colonization (owing to the opportunistic character of *Candida albicans*) [22, 23], and a low sensitivity as the patients had been immunocompromised by underlying disease and/or its treatment (*e.g.*, antitumor agents or steroids) [22]. Nevertheless, several promising findings have recently revealed that these two drawbacks may be avoided when antibody detection tests are carried out using (i) specific antigens and/or antigenic epitopes, instead of crude antigens, selected appropriately and correlated significantly with systemic infection, to reduce the false-positive rate of the assay [21, 24], and (ii) highly sensitive techniques such as immunoproteomics (*i.e.*, the combination of two-dimensional gel electrophoresis along with immunoblot analysis [25]) to minimize the source of the false-negative results [26, 27]. So far few *C. albicans* immunogenic proteins that induce antibody response to systemic candidiasis have been characterized [24, 26–29]. In the light of this, an important issue is to define the comprehensive pattern of immunorelevant antigens specifically recognized by antibodies produced in the course of systemic *Candida* infection in order to appraise useful combinations for an efficient diagnosis and therapeutic monitoring of these infections and/or for further vaccine or antifungal agent designs in future research. In doing so, recent studies [30, 31] have demonstrated that antibody-mediated immunity may confer protection against disseminated candidiasis in experimental models.

To address these matters, we embarked on an immunoreactive *C. albicans* protein mapping by using immunoproteomics. We report the identification of a repertoire of novel *C. albicans* antigens that elicit a specific immune response in systemic candidiasis patients undergoing malignant hematological disorders. Furthermore, we take advantage of the Pasteur Institute's *C. albicans* genomic database (CandidaDB; France) of the paradigm strain SC5314, which is recently in the public domain for database searching. We also show preliminary results revealing differences in antigen recognition patterns correlated with infection progression. These data are currently compiled in our public 2-DE database (at http://babbage.csc.ucm.es/2d/2d.html [32]). This pilot work thus represents the keystone to construct a *C. albicans* antigen database that could contribute to the development of reliable serological tests for detecting and monitoring systemic candidiasis.

16.2
Materials and methods

16.2.1
Human serum samples

Four cases of laboratory-confirmed systemic candidiasis (3 males and 1 female; age range, 38–67 years; median age, 53 years) were identified and recruited prospectively among patients undergoing hematological malignancies between June 1998 and May 1999 in the Hematology and Oncology Department at the Salamanca University Hospital, Spain, and followed longitudinally. Paired serum specimens were collected from these patients, after informed consent, both at the time of diagnosis (before treatment) and at one month after starting systemic *Candida* infection for antibody testing. Previous antifungal agent prophylaxis was not administered in any patients. Sera from 14 subjects without clinical or microbiological evidence of systemic candidiasis, including 8 underlying malignancy-matched (1:2 ratio) patients (5 males and 3 females; age range, 28–73 years; median age, 55 years) and 6 healthy individuals (4 males and 2 females; age range, 35–69 years; median age, 52 years), were randomly selected and used as controls. All serum samples were kept frozen at $-80°C$, and thawed just before analysis. Every enrolled patient with systemic candidiasis manifested clinical signs of infection or sepsis and yielded the same *Candida* spp. in two or more blood cultures and/or in culture from at least three noncontiguous sites with an inadequate response to broad-spectrum antibiotics. This group consisted of myelodysplastic syndrome (case 1), lymphoma (case 2), and leukemia (case 3 and 4) patients. Data on demographic and baseline clinical characteristics (age, gender, underlying diseases, predisposing factors for systemic candidiasis, antifungal drug therapy, and outcome) of the study patients suffering from systemic *Candida* infection were collected by reviewing their medical records, and are summarized in Tab. 1. Both patients and clinical information were treated at the physician's discretion.

16.2.2
Preparation of *C. albicans* protoplast lysates

Protoplast lysates of *C. albicans* strain SC5314 (a clinical isolate [33]) were exploited as antigen source, and prepared essentially as reported [26]. Briefly, yeast cells were grown in YED medium (1% Difco yeast extract, 2% peptone, and 2% glucose), and incubated at 28°C and 80 rpm first in a pretreatment solution (10 mM Tris-HCl, pH 9, 5 mM EDTA, 1% 2-mercaptoethanol) for 30 min and then in a solution containing 30 µg/mL Glusulase (Du Pont, Boston, MA, USA) and 1 M sorbitol (up to 5×10^8 cells/mL) until obtaining more than 90% protoplasts. After three gentle washes with 1 M sorbitol, protoplasts were resuspended in 200 µL cold lysis buffer (50 mM Tris-HCl, pH 7.5, 150 mM NaCl, 1 mM DTT, 0.5 mM PMSF, and 5 µg/mL of leupeptin, pepstatin, and antipain (Sigma, St. Louis, MO, USA)), vortexed for 1 min, and cooled on ice for 1 min. The clarified supernatant was stored at −80°C. The protein concentration was measured with the Bradford assay (Bio-Rad, Hercules, CA, USA), using bovine serum albumin (Sigma) as calibrator [34].

16.2.3
Two-dimensional polyacrylamide gel electrophoresis (2-D PAGE)

Either 500 µg (analytical runs) or 5 mg (preparative runs) of protein diluted in 350 µL lysis buffer (7 M urea, 2 M thiourea, 2% CHAPS, 65 mM dithioerythritol (DTE), 0.5% IPG buffer 3–10, bromophenol blue) were adsorbed onto 18 cm immobilized pH 3–10 nonlinear gradient (IPG) strips (Amersham Biosciences, Uppsala, Sweden) at 15°C for 16 h, and then focused on an IPGphor IEF unit (Amersham Biosciences) by using a step-wise increasing voltage (*i.e.*, (i) for analytical runs: 500 V for 1 h, 500–2000 V for 1 h, and 8000 V for 5.5 h; (ii) for preparative runs: 500 V for 1 h, 1000 V for 1 h, 2000 V for 1 h, 2000–5000 V for 3 h, and 8000 V for 11 h). After equilibrating the IPG strips as reported elsewhere [35], proteins were separated on SDS-polyacrylamide gels (10% T; 1.6% C) at 40 mA per gel in a Protean II cell (Bio-Rad). Analytical or preparative gels were subsequently visualized by silver staining as described [36] or according to Shevchenko *et al.* [37], respectively. 2-D images were captured by scanning the 2-DE gels using a GS-690 imaging densitometer (Bio-Rad) and digitalized with Multi-Analyst software pro-

Tab. 1 Demographic and clinical characteristics of the four cases of systemic candidiasis with underlying hematological malignancies recruited in this study

Patient	Case 1	Case 2	Case 3	Case 4
Sex	Female	Male	Male	Male
Age (years)	56	67	38	50
Underlying diseases				
Primary condition	Myelodysplastic syndrome (MDS)	Non-Hodgkin's lymphoma (NHL)	Acute myelogenous leukemia (AML)	Chronic lymphoblastic leukemia (CLL)
Other associated conditions	None	Chronic renal insufficiency	Graft-versus-host disease (GVHD)[a]	Graft-versus-host disease (GVHD)[a]
		Steroid diabetes	Tuberculosis	Chronic obstructive pulmonary disease (COPD)
Predisposing factors				
Cytotoxic chemotherapy	No	Yes	Yes	Yes
Total-body irradiation	No	No	No	No
Transplants	No	Yes[b]	Yes[c]	Yes[c]
Granulocytopenia	Yes	Yes	Yes	Yes
Adrenal corticosteroids	No	Yes	Yes	Yes
Central venous catheters	No	Yes	Yes	Yes
Broad-spectrum antibiotics	No	Yes	Yes	Yes
Treatment (antifungal drugs)[d]	Amphotericin B	Amphotericin B	Amphotericin B	Amphotericin B plus fluconazole
Patients' outcome	Alive[e]	Alive[e]	Alive[e]	Dead

a) GVHD is a well-known complication of patients treated with allogeneic BMT, which is caused by allogeneic lymphocytes T from graft.
b) Patient who received autologous peripheral blood progenitor cell transplantation. In addition, this patient was a renal transplant recipient two years ago.
c) Patient who received allogeneic BMT.
d) After the onset of systemic candidiasis. Previous antifungal drug prophylaxis was not administered in any patients.
e) Without evidence of fungal infection.

gram (Bio-Rad). M_r and pI values were estimated after calibration of gels with internal 2-D PAGE standards (Bio-Rad) by using the Melanie 3.0 software program (GenBio, Geneva, Switzerland).

16.2.4
Immunoblot analysis

Proteins were transferred from analytical 2-DE gels to nitrocellulose membranes (HyBond ECL; Amersham Biosciences) at 50 mA overnight by electroblotting following standard protocols [38]. Blots were blocked with 5% nonfat dry milk in TBS for 2 h, and then rinsed with washing buffer (TBS containing 0.01% Tween 20) and incubated with human serum at a 1:100 dilution for 1 h. After three washes, blots were incubated again with a horseradish peroxidase-labeled antihuman immunoglobulin G (IgG) antibody (Amersham Biosciences) at a 1:1000 dilution for 2 h, and then rinsed twice. Immunoblots were revealed by the ECL detection system (Amersham Biosciences) using high-performance films (Hyper-film ECL; Amersham Biosciences). Each blot was stripped (up to three times) with a solution containing 130 mM glycine-HCl, pH 2.2, 1% SDS, and 0.05% NP-40 for 2 h, and immediately used again with a different serum, and developed as described above. Furthermore, each serum specimen was hybridized onto at least two different blots, one of them being used for the first time with each sample. Films were digitalized as indicated in Section 2.3. The densitometry data on the Western blot intensity signals were subsequently evaluated by using the Melanie 3.0 program. Immunoblots and silver-stained gels were aligned to pI and M_r and then matched by Melanie 3.0.

16.2.5
Peptide sample preparation for MS and MS/MS analyses

16.2.5.1 In-gel digestion
Selected spots were manually excised from preparative 2-DE gels, destained with 100 mM sodium thiosulfate and 30 mM potassium ferricyanide [39], and then rinsed twice in 25 mM NH_4HCO_3 and once in water, shrunk with 100% ACN, and vacuum-dried in a SpeedVac (Thermo Savant, Holbrook, NY, USA). Gel pieces were subsequently reduced with 10 mM DTE in 25 mM NH_4HCO_3 at 56°C for 30 min, and S-alkylated with 55 mM iodoacetamide in 25 mM NH_4HCO_3 in the dark for 20 min. After five washes with 25 mM NH_4HCO_3 and ACN alternately, gel pieces were vacuum-dried and then incubated with 12.5 ng/µL sequencing-grade trypsin (Roche, Mannheim, Germany) in 25 mM NH_4HCO_3 at 37°C overnight. Thereafter, the supernatant was removed and stored. Tryptic peptides were also recovered from the gel by sequential extraction with 50% ACN in 1% TFA and 100% ACN. Finally, all extracts were pooled and vacuum-concentrated.

16.2.5.2 Desalting
Prior to nanoelectrospray MS analysis, the tryptic peptide samples were individually desalted and concentrated using in-tip reverse-phase resins (ZipTip$_{C18}$; Millipore, Bedford, MA, USA), according to the manufacturer's recommendations with some modifications. In brief, each peptide extract was vacuum-dried, and then reconstituted in 10 µL 0.1% TFA. In parallel, a ZipTip$_{C18}$ microcolumn was equilibrated for peptide binding with 50% ACN, and washed with 0.1% TFA. The

reconstituted sample was subsequently loaded onto the tip. The adsorbed peptides were washed ten times first with 10 µL 0.1% TFA and then with 10 µL of Milli-Q grade water (Millipore) to remove any contaminants. The tryptic peptides were eluted from the microcolumn with 5 µL of 60% methanol in 1% acetic acid.

16.2.6
Mass spectrometric analysis (MS and MS/MS)

16.2.6.1 Peptide mapping by MALDI-TOF-MS
MS analyses were carried out on a MALDI-TOF Voyager-DE STR mass spectrometer (PerSeptive Biosystems, Framingham, MA, USA), equipped with delayed extraction, and operated in positive ion reflector mode. One µL of each peptide mixture was crystallized with 0.4 µL of freshly prepared α-cyano-4-hydroxycinnamic acid matrix solution (3 mg/mL; Sigma) in 50% ACN containing 0.01% TFA onto a MALDI target plate. Spectra were acquired over the m/z range of 700–4500 Da. All mass spectra were first calibrated externally with the Sequazyme peptide mass standards kit (PerSeptive Biosystems) and then recalibrated internally using three peptide fragments from trypsin autolysis ($[M+H]^+$, m/z 805.42, 2163.06 and 2273.16). Tryptic, monoisotopic peptide mass lists were generated and exploited for database searching (see Section 2.7 for further details).

16.2.6.2 Peptide fragmentation and sequencing by nanoESI-IT-MS
MS/MS analyses were performed on an LCQ ion trap mass spectrometer (ThermoFinnigan, San Jose, CA, USA), fitted with a nanoelectrospray ionization (nanoESI) source (Protana, Odense, Denmark), and operated in the positive ion mode. Four µL of the eluate of each peptide sample was introduced into a gold-coated borosilicate capillary (Protana). Transfer of the sample to the capillary was assisted by a microcentrifuge. A new needle was used for each analysis to avoid the risk of cross-contamination between different peptide digests. The spray needle was placed approximately 4 mm from the LCQ inlet orifice. Both this position and nanoESI phenomena were visualized and checked under video-microscopy. Nanoelectrospray was initiated by applying a 0.85 kV spray voltage, and the temperature capillary was set at 120°C. MS/MS spectra were acquired using data-dependent scanning in "triple-play" mode, which comprises three sequential scans: (i) a full-range MS scan in which ions were collected under a total of 100–200 microscans with a maximum ion injection time of 400 ms, covering the mass range from m/z 175–2000 Da; (ii) a narrow-range, high resolution zoom scan on a selected ion from the first MS scan to resolve its isotopic distribution and determine its charge state, and (iii) an MS/MS scan on this ion, using an isolation width of 3.0 m/z units and a normalized collision energy ranging from 25 to 45%, depending on the charge of the precursor ion, which was adjusted for optimum CID fragmentation. The two last scan events were performed again with several further precursor ions. Additionally, high-order MS^3 experiments were also explored to obtain more accurate sequence information. All spectra were acquired and processed using the

Xcalibur software program (ThermoFinnigan). MS/MS spectra were interpreted manually. To this end, amino acid sequences were deduced by the mass differences between y- or b-ion "ladder" series resulting from the CID fragmentation spectra of the selected tryptic peptides.

16.2.7
Database searching

The peptide mass fingerprints produced by MALDI-TOF-MS were searched against the Swiss-Prot/TrEMBL nonredundant protein database (http://www.expasy.ch/sprot) and the freely available *C. albicans* genomic database, CandidaDB (downloaded in FASTA format from http://genolist.pasteur.fr/CandidaDB for MS applications) to obtain protein candidates, using MS-Fit (http://prospector.ucsf.edu), ProFound (http://prowl.rockefeller.edu) and/or Mascot (http://www.matrixscience.com) software programs. Peptide sequences were identified from their CID product ion spectra both (i) by manual interpretation, and then used to search Swiss-Prot/TrEMBL database and/or CandidaDB for protein identity with NCBI's Basic Local Alignment Search Tool (BLAST; at http://www.ncbi.nlm.nih.gov/entrez) and/or MS-Pattern (http://prospector.ucsf.edu) algorithm, and (ii) by retrieving from such databases using MS-Tag (http://prospector.ucsf.edu) and/or Mascot software programs. Initial search parameters were as follows: all M_r and pI ranges, carboxyamidomethylation of cysteine, oxidation of methionine, acetylation of N-terminus, one missed cleavage site, monoisotopic molecular masses, peptide mass tolerance of $\pm$ 50 ppm and MS/MS ion mass tolerance of $\pm$ 0.5 Da. The identified proteins were named using CandidaDB accession numbers. Nucleotide sequence data for *C. albicans* were obtained from the Stanford Genome Technology Center website at http://www.sequence.stanford.edu/group/candida. Sequencing of *C. albicans* was accomplished with support of the NIDR and the Burroughs Wellcome Fund. Information about coding sequences and proteins were obtained from CandidaDB available at http://www.pasteur.fr/Galar_Fungail/Candida DB/, which has been developed by the Galar Fungail European Consortium (QLK2-2000-00795).

16.3
Results

16.3.1
Mapping of *C. albicans* immunogenic proteins

16.3.1.1 **Overall 2-D *C. albicans* antigen recognition pattern**
Serum specimens from four cases of systemic candidiasis patients with underlying hematological malignancies (Tab. 1) were initially used to detect specific antigens on blots of 2-DE-separated *C. albicans* cytoplasmic proteins from the strain SC5314. Similar 2-D antigen recognition profiles were yielded with serum samples assayed

on two independently prepared sets of membranes, confirming their reproducibility. In parallel, a silver-stained 2-DE *C. albicans* protein pattern was also performed to match the immunostained 2-D patterns and estimate the pI and M_r of immunoreactive spots using the Melanie 3.0 program (Tab. 2). This 2-DE pattern was exploited as a reference map of *C. albicans* immunogenic proteins (Fig. 1) offering reference points (pI, M_r) for spots detected on the immunoblots. Unfortunately, three immunostained rod-like spots with pI range of 5.0–7.5 and M_r of 50–52 kDa (mp50, mp51, and mp52) could neither be correlated to visible proteins from this reference 2-D pattern nor be excised from a preparative gel for further identification. Approximately 85 (~6.5%) of the 1300 protein spots visualized on a silver-stained 2-DE gel of *C. albicans* cytoplasmic extracts reacted with systemic candidiasis patients' sera. On the whole, immunoreactive spots were uniformly spread within the gel. Nevertheless, about one-third of them (almost 30 spots) were found within the pI range from 5.0 to 7.5 and the M_r range from 35 to 52 kDa (Fig. 1 and Tab. 2) which, in turn, corresponded both to major immunodominant proteins developed on the films following ECL detection and to the most prominent spots revealed in the silver-stained pattern. Interestingly, several hybridizing spots showed analogous reactivity and nearly identical M_r, but different pI, suggesting that these series of spots could represent different isoforms of a single protein.

16.3.1.2 Identification of *C. albicans* immunoreactive proteins by peptide mass fingerprinting (PMF)

The seroreactive protein spots detected in this way were excised from a single silver-stained preparative 2-DE gel and subjected to in-gel tryptic digestion. The resulting tryptic peptides of each destained, reduced, and carbamidomethylated protein spots were firstly analyzed by MALDI-TOF-MS to determine their molecular masses, the spectra being manually acquired and processed. Unambiguous matches to proteins from the Swiss-Prot data bank, and/or to the products of ORFs predicted in the recently public *C. albicans* genomic database of the strain SC5314 (CandidaDB), were achieved for 80 of the 82 gel-excised spots with 10–65% coverage of their amino acid sequences. However, these 80 *C. albicans* protein identities represented only a total of 40 different proteins – due to the presence of isoforms as expected. Antigenic properties for 24 of them are described for the first time in this study (see Tab. 2). The PMF-identified proteins along with their sequence coverage and their number of matching peptides are summarized in Tab. 2, and their position in the reference 2-DE map is illustrated in Fig. 1. In general, their pI and M_r determined in this map were found to coincide with the theoretical values of the matched proteins (data not shown). Unsurprisingly, the identified proteins proved to be members of diverse groups with housekeeping functions, including chaperones, metabolic enzymes, translation-involved proteins, porins and redox enzymes, among others (Tab. 2). These proteins were clustered in four functional categories. PMF data also revealed one of the identified antigens to be a protein of unknown function.

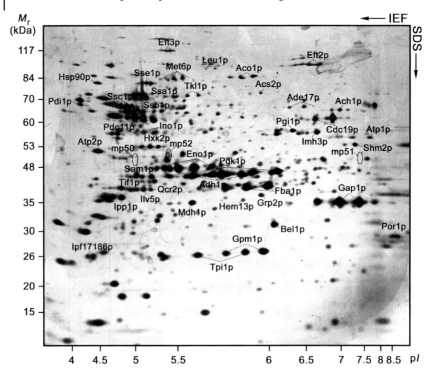

Fig. 1 Silver-stained 2-DE map of soluble *C. albicans* cytoplasmic proteins. Labeled spots represent immunogenic proteins that stimulate the human immune system during systemic candidiasis. Spot names refer to those in Tab. 2. The broken circles depict the relative 2-D position of three rod-like spots (mp50, mp51, and mp52) immunodetected on 2-D blots but not visualized in silver-stained 2-DE gels.

16.3.1.3 Peptide sequencing of *C. albicans* immunoreactive proteins

In an endeavor to obtain *de novo* peptide sequence information for those immunogenic proteins either that rendered unsuccessful PMF results (*i.e.*, two spots with (pI, M_r) coordinates: (6.3, 58) and (5.2, 39, Tab. 2) or that have hitherto been unreported as *C. albicans* antigens (24 proteins) and/or unannotated in the Swiss-Prot/TrEMBL knowledgebase (23 proteins; see Tab. 2), nanoESI – coupled with a single desalting and concentration step – IT-MS was accomplished on their remaining tryptic peptide digests using data-dependent switching from MS to MS/MS mode (Fig. 2). For each chosen protein spot, several positively charged precursor ions were specifically trapped and fragmented by CID generating their own MS/MS spectra. From one to five good tandem mass spectra, which could thus supply structural information, were attained for 19 of the 31 spots analyzed (Tab. 2). A total of 38 amino acid sequences with a length ranging from 7 to 19 amino acids were deduced *de novo* from the series of y- or b-ions deriving from these CID ion-product spectra both by manual interpretation and with computer assistance (Tab. 3). Subsequent database searching as detailed in Section 2.7 returned 19 protein identities,

Tab. 2 Clustering of C. albicans immunogenic proteins identified by MALDI-TOF-MS according to their main biological function

Name[a]	Full name	Accession number		M_r[b] (kDa)	pI[b]	Matching peptides (No.)	Sequence coverage (%)	co-MS/MS[c]	Antibody response reported in[d]	
		Candida DB	Swiss-Prot database						humans	animal models

I. Chaperones and heat shock proteins
I.1. Members of the HSP70 multigene family

Ssa1p	70-kDa heat shock protein	CA2857	P41797	70	5.07	14	37		–	+[27]
				70	5.13	8	15			
Ssb1p	70-kDa heat shock protein	CA3534	P87222	66	4.95	14	24		–	+[27]
				66	5.03	12	27			
Ssc1p	70-kDa heat shock protein	CA4474	P83784[e]	68	4.81	7	13	✓	–	–
				68	4.89	11	25			
				67	4.96	8	16			
Sse1p	70-kDa heat shock protein	CA1911	P83785[e]	80	5.13	25	47	✓	–	–
				80	5.18	16	31			

I.2. Others

Hsp90p	90-kDa heat shock protein	CA4959	P46598	82	4.48	16	23		–[f]	–
Pdi1p	Protein-disulfide isomerase	CA1755	n.a.[g]	70	4.20	5	17	×	–	–
				68	4.20	12	22			

II. Metabolic enzymes
II.1. Carbohydrate metabolism-involved enzymes
II.1.a. Glycolytic enzymes

Hxk2p	Hexokinase II	CA0127	P83776[e]	53	5.17	5	18	✓	–	–
Pgi1p	Glucose-6-phosphate isomerase	CA3559	P83780[e]	58	6.33	–[h]	–[h]	✓	–	–
Fba1p	Fructose biphosphate aldolase	CA5180	Q9URB4	39	5.85	10	40		–	+[27]
				40	6.00	15	65			

Tab. 2 Continued

Name[a]	Full name	Accession number		M_r[b] (kDa)	pI[b]	Matching peptides (No.)	Sequence coverage (%)	co-MS/MS[c]	Antibody response reported in[d]	
		Candida DB	Swiss-Prot database						humans	animal models
Tpi1p	Triose phosphate isomerase	CA5950	Q9P940	26	5.59	4	22	✓[i]	–	+[27]
				27	5.85	5	23			
Gap1p	Glyceraldehyde-3-phosphate dehydrogenase	CA5892	Q92211	35	7.00	10	41		+[24, 26]	+[24, 27]
Pgk1p	Phosphoglycerate kinase	CA1691	P46273	35	7.38	11	42		+[24, 26]	+[27]
				44	5.40	9	22			
				44	5.53	8	20			
				45	5.62	7	19			
				45	5.75	10	34			
				46[i]	5.89	8	20			
				46[i]	6.12	5	16			
Gpm1p	Phosphoglycerate mutase	CA4671	P82612	27	5.73	9	49	✓[i]	+[28]	–
				28	5.96	16	56			
Eno1p	Enolase	CA3874	P30575	48[i]	5.25	9	24		+[24, 26]	+[24, 27]
				48[i]	5.33	11	28			
				48[i]	5.49	13	37			
				48[i]	5.57	20	52			
				49[i]	5.65	6	18			
Cdc19p	Pyruvate kinase	CA3483	P46614	62	6.64	7	25	✓[i]	+[28]	+[27]
				62	6.89	14	50			
II.1.b. Fermentative enzymes										
Pdc11p	Pyruvate decarboxylase	CA2474	P83779[e]	63[i]	5.00	14	37	✓	–	+[29]

Tab. 2 Continued

Name[a]	Full name	Accession number		M_r[b] (kDa)	pI[b]	Matching peptides (No.)	Sequence coverage (%)	co-MS/MS[c]	Antibody response reported in[d]	
		Candida DB	Swiss-Prot database						humans	animal models
Adh1p	Alcohol dehydrogenase	CA4765	P43067	63[i]	5.09	19	44		–	+[27]
				60[i]	5.22	24	51			
				44	5.56	5	15			
				45	5.66	7	23			
				45[i]	5.70	18	46			
				45[i]	5.83	15	38			
II.1.c. Tricarboxylic acid cycle enzymes										
Aco1p	Aconitate hydratase (Aconitase)	CA3546	P82611	84	5.82	17	15		+[28]	+[27]
				85	5.90	11	16			
Mdh1p	Mitochondrial malate dehydrogenase	CA5164	P83778[e]	35	5.56	19	74	✓	–	–[k]
II.1.d. Miscellaneous										
Tkl1p	Transketolase 1	CA3924	n.a.	73	5.54	5	16	×	–	+[29]
				73	5.59	10	28			
Ino1p	Myo-inositol-1-phosphate synthase	CA5986	P42800	57	5.02	6	15	×	–	–
				57	5.11	9	25			
				57	5.21	12	24			
Acs2p	Acetyl-coenzyme-A synthetase	CA2858	n.a.	74	5.95	13	18	×	–	–[k]
II.2. Fatty acid metabolism-involved enzymes										
Ach1p	Acetyl-coenzyme-A hydrolase	CA0345	P83773[e]	66	7.00	9	31	✓	–	–

Tab. 2 Continued

Name[a]	Full name	Accession number		M_r[b] (kDa)	pI[b]	Matching peptides (No.)	Sequence coverage (%)	co-MS/ MS[c]	Antibody response reported in[d]	
		Candida DB	Swiss-Prot database						humans	animal models

II.3. Amino acid metabolism-involved enzymes

Met6p	Methionine synthase	CA0653	P82610	84	5.27	15	12		+[28]	+[27]
				84	5.35	12	18			
				84	5.44	20	17			
				85	5.52	26	32			
				85	5.55	14	18			
Sah1p	S-Adenosyl-L-homocysteine hydrolase	CA3018	P83783[e]	50	5.37	11	22	✓	–	–
Ilv5p	Ketol-acid reducto-isomerase	CA1983	n.a.	39	5.10	10	24	×	–	–[k]
Shm2p	Serine hydroxymethyltransferase	CA0895	O13426	53	7.29	4	10	✓	–	–
Leu1p	3-isopropylmalate dehydratase	CA5842	n.a.	91	5.63	11	16	×	–	–

II.4. Nucleotide metabolism-involved enzymes

Imh3p	Inosine-5′-monophosphate dehydrogenase	CA1245	O00086	57	6.67	9	24		–	+[27]
Ade17p	5-aminoimidazole-4-carboxamide ribotide transformylase	CA4513	n.a.	67	6.28	8	23	×	–	–
				67	6.49	10	29			
				67	6.70	6	15			

II.5. Energetic central intermediary metabolism-involved enzymes

Qcr2p	Ubiquinol-cytochrome c reductase 40 kDa chain II	CA2065	P83782[e]	39	5.18	–[h]	–[h]	✓	–	–
Atp1p	F1F0-ATPase complex, F1 α-subunit	CA4457	n.a.	55	7.65	11	21	×	–	–

Tab. 2 Continued

Name[a]	Full name	Accession number		M_r[b] (kDa)	pI[b]	Matching peptides (No.)	Sequence coverage (%)	co-MS/MS[c]	Antibody response reported in[d]	
		Candida DB	Swiss-Prot database						humans	animal models
Atp2p	F1F0-ATPase complex, F1 β-subunit	CA4362	n.a.	52	4.48	14	34	×	–	–
Ipp1p	Inorganic pyrophosphatase	CA0870	P83777[e]	36	4.83	5	23	✓	–	–
III. Translation apparatus-involved proteins										
Eft3p	Translation elongation factor 3	CA3081	P25997	117 117	5.39 5.45	14 31	13 28	✓	–	–
Eft2p	Translation elongation factor 2	CA2810	O13430	93 93	6.55 6.72	12 15	18 25	✓	–	–
Tif1p	Translation initiation factor 4A	CA2939	P87206	44	5.09	11	30	×	–	–
Bel1p	40S small subunit ribosomal protein	CA4588 CA4589[l]	P83774[e]	31	6.07	11[m] 6[n]	71[m] 57[n]	✓	–	+[29]
IV. Miscellaneous										
Por1p	Mitochondrial outer membrane porin	CA0919	P83781[e]	29 29	8.55 8.78	5 6	23 31	✓	–	–
Grp2p	Reductase (similary to plant di-hydroflavonol-4-reductases)	CA2644	P83775[e]	37	5.95	8	36	✓	–	–
				37	6.31	11	39			
Hem13p	Coproporphyrinogen III oxidase	CA0517	n.a.	36	5.83	6	16	×	–	–
Ipf17186p	Protein of unknown function	CA0828	n.a.	26	4.32	6	25	×	–	–
				26	4.40	6	42			
mp50	50-kDa mannoproteins[o]	n.a.	n.a.	50	5.05	n.a.	n.a.		–	–

Tab. 2 Continued

Name[a]	Full name	Accession number		M_r[b] (kDa)	pI[b]	Matching peptides (No.)	Sequence coverage (%)	co-MS/ MS[c]	Antibody response reported in[d]	
		Candida DB	Swiss-Prot database						humans	animal models
mp51	51-kDa mannoproteins[o]	n.a.	n.a.	51	7.45	n.a.	n.a		–	–
mp52	52-kDa mannoproteins[o]	n.a.	n.a.	52	5.36	n.a.	n.a		–	–

a) Protein names according to the *C. albicans* genomic database (CandidaDB) from the Pasteur Institute (at http://genolist.pasteur.fr/CandidaDB).
b) Experimental M_r, and pI values (calculated by using the Melanie 3.0 program).
c) Proteins analyzed by nanoESI-IT MS. ✓ denotes proteins that were either identified or confirmed unambiguously by peptide sequencing (refer to Tab. 3), and × those that did not yield good MS/MS spectra or successful structural information.
d) Antibody responses to *C. albicans* proteins reported previously in systemic candidiasis patients and/or animal models upon artificial immunization are labeled with "+". References are shown in brackets. The dashes in the column named as "humans" indicate that antigenic properties for such proteins were described for the first time in this study.
e) Swiss-Prot accession number assigned for this protein sequenced *de novo* in the present study (refer to Tab. 3) and annotated in the Swiss-Prot and TrEMBL knowledgebase. So far, this protein had not been registered in the Swiss-Prot/TrEMBL data bank.
f) Antigenic traits were reported for the 47-kDa heat-stable breakdown product of Hsp90p in patients with systemic candidiasis [70], but not for the 82-kDa form of this protein (see Section 4.1.1).
g) n.a., not applicable
h) No conclusive identification was achieved by PMF.
i) The most acidic isoform, hitherto uncharacterized as a *C. albicans* antigen, was confirmed by MS/MS.
j) Isoform recognized by controls' serum specimens
k) Immunogenic properties for this protein in a murine model have been submitted for publication by our research group [49].
l) The Bel1p PMF matched to two CandidaDB entries corresponding to the two predicted exons of the *BEL1* gen (refer to Fig. 4A).
m) Number of tryptic peptide masses matching the encoding product of the exon 1 of the *C. albicans BEL1* gene (CandidaDB entry, CA4588), or percentage of amino acid sequence coverage for the product of this predicted ORF. See Fig. 4A.
n) Number of tryptic peptide masses matching the encoding product of the exon 2 of the *C. albicans BEL1* gene (CandidaDB entry, CA4589), or percentage of amino acid sequence coverage for the product of this predicted ORF. Refer to Fig. 4A.
o) Rod-like spots immunodetected on 2-D blots but not visualized in silver-stained 2-DE gels.

which permitted the characterization of a further two novel *C. albicans* antigens (*i.e.*, glucose-6-phosphate isomerase (Pgi1p) and ubiquinol-cytochrome-*c* reductase (Qcr2p)), and the confirmation of PMF-obtained results for the rest of them. In all, 42 different *C. albicans* antigens were identified unequivocally by MALDI-TOF-MS and nano-ESI-IT MS. Of these, 26 had not been characterized as *C. albicans* antigens previously. However, prior to publication of the *C. albicans* genomic database, only 8 of these 26 novel antigens could be cross-species identified.

In most cases, MS/MS analyses resulted in only internal sequence information on the different *C. albicans* antigens identified. Nonetheless, the *N*-terminal sequence of the mature protein Shm2p (serine hydroxymethyltransferase) was also determined *de novo* as Ac-SAYALSQSHR, this being yielded by the doubly charged precursor ion at *m/z* 582.2 (Fig. 3 and Tab. 3). This finding enabled two typical cotranslational modifications – taking place on most of the eukaryotic proteins – to be uncovered. These comprised (i) removal of the *N*-terminal methionine, given that this preceded an active amino acid with a small side chain (*i.e.*, serine), whereupon the action of *N*-terminal methionylaminopeptidases can be facilitated, and (ii) *N*-terminal acetylation.

The deduced sequences were in agreement with those expected. Amazingly, nanoESI-IT-MS data disclosed that the *C. albicans* 40S small subunit ribosomal protein (Bel1p) – identified previously by PMF – contained peptide sequences matching the encoding products of two ORFs from the *C. albicans* genome. These corresponded in turn to the two computationally predicted exons of its respective *BEL1* gene, which obviously provided a single protein identity (Tab. 3 and Fig. 4). A more exhaustive analysis of its related MALDI-TOF spectrum revealed additional peptide peaks that matched these two CandidaDB entries, vouching for such results (Fig. 4A). Accordingly, the *BEL1* gene model that predicted a genomic region containing two exons was consistent with our MS data.

16.3.1.4 Reference 2-DE *C. albicans* antigen pattern display on the Net

The data referred to here were compiled and organized in our dynamic yeast 2-DE database, namely COMPLUYEAST-2DPAGE, which is freely available on the ExPASy Web server at URL address http://www.expasy.ch/ch2d/2d-index.html or http://babbage.csc.ucm.es/2d/2d.html [32]. This electronic 2-DE database includes cross-references to the universal Swiss-Prot protein knowledgebase. However, so far, only 19 of the 42 immunogenic proteins identified in this research are registered in the Swiss-Prot data bank, as indicated in Tab. 2. In view of this, the peptide sequences obtained *de novo* for 13 of the remaining proteins (*i.e.*, Ssc1p, Sse1p, Hxk2p, Pgi1p, Pdc11p, Mdh1p, Ach1p, Sah1p, Bel1p, Por1p, Grp2p, Qcr2p, Ipp1p; see Tab. 2 for their full names) were annotated in this database and assigned Swiss-Prot accession numbers. As a result, the protein sequence data reported in this paper for these 13 proteins will appear in the Swiss-Prot and TrEMBL knowledgebase under the accession numbers listed in Tab. 3. Through these Swiss-Prot accession numbers, which have been used as active hypertext links, each immu-

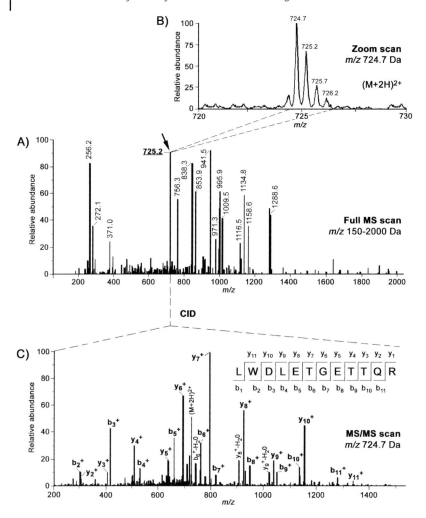

Fig. 2 Schematic representation of data-dependent switching from MS to MS/MS mode performed on a LCQ ion trap mass spectrometer. The example protein illustrated in this figure is *C. albicans* Bel1p (refer to Tab. 3 for further details). (A) Tryptic peptides are detected by using a full-range MS scan (m/z 150–2000 Da), leading to the peptide mass mapping of the tryptic digest of the protein spot of interest. (B) The isotopic distribution of one of the most abundant ions (marked by an arrow in the full mass spectrum) is resolved by a narrow-range, high-resolution zoom scan (m/z 720–730 Da in this illustration) revealing its charge state. In the example, two isotope peaks are shown for a doubly charged precursor ion at m/z 724.7. (C) The fragmentation by low-energy CID of the ion selected for the zoom-scan results in sequence information through a product-ion scan (MS/MS analysis). The peptide sequence (LWDLETGETTQR) is deduced *de novo* from the labeled b- and y- ions in the CID fragmentation spectrum, which correspond N- and C-terminal fragments of the peptide generated by breakage at its peptide bonds, respectively (inset).

Tab. 3 Summary of *de novo* peptide sequence data obtained for the *C. albicans* immunogenic proteins identified unambiguously by nanoESI-IT MS

Cluster[a]	Protein name	m/z observed	Charge	(M+H)+ matched	Peptide sequence[b]	Modification[c]	Residue[d] Start	Residue[d] End	Missed cleavage sites	New Swiss-Prot entries
I.1	Ssc1p	564.6	2	1127.6	VQDQIQQLR	–	589	597	0	P83784[e]
	Sse1p	658.0	2	1314.7	GIDIVVNEVSNR	–	23	34	0	P83785[e]
II.1.a	Hxk2p	531.6	2	1061.5	GWSIDGIEGK	–	122	131	0	P83776[e]
		667.5	2	1332.7	LILLEFAEEKK	–	261	271	1	
	Pgi1p	617.4	2	1232.6	DTAQEVDDVLK	–	121	131	0	P83780[e]
		847.1	2	1692.6	DAMFAGDHINTTEDR	–	86	100	0	
	Tpi1p	524.4	2	1047.5	GGVTLDVC$_c$AR	C$_c$:143	136	145	0	–
		716.2	2	1431.7	VILC$_c$IGETLEER	C$_c$:126	123	134	0	
	Gpm1p	619.7	2	1237.6	AQTXEAYGQEK	–	99	109	0	–
	Cdc19p	516.2	2	1030.5	TANDVLEIR	–	221	229	0	–
II.1.b	Pdc11p	498.2	2	996.4	IYEVEGMR	–	37	44	0	P83779[e]
		677.3	2	1352.6	NIVEFHSDYTK	–	307	317	0	
		784.5	2	1566.8	FGGVYVGSLSKPEVK	–	261	275	1	
		828.0	2	1655.8	SINPNYTPVPVPETK	–	342	356	0	
		992.8	2	1983.5	WAGNANELNAGYAADGYAR	–	45	63	0	
II.1.c	Mdh1p	590.2	2	1178.4	VTDLALYDIR	–	43	52	0	P83778[e]
		683.1	2	1364.9	DDLFNTNASIVR	–	103	114	0	
		732.7	2	1463.6	DGAGSATLSMAQAGAR	–	233	248	0	
		904.3	2	1806.9	GAPGVAADVSHVPTNSTVK	–	53	71	0	
II.2	Ach1p	624.9	2	1247.6	WAENNMILTR	–	82	91	0	P83773[e]
		655.3	2	1328.7	SQVVSNSPEIIR	–	348	359	0	
II.3	Sah1p	629.1	2	1256.6	VPAINVNDSVTK	–	178	189	0	P83783[e]
		630.6	2	1259.6	SKFDNLYGC$_c$R	C$_c$:198	190	199	1	
	Shm2p	582.2	2	1161.2	S$_a$AYALSQSHR[f]	S$_a$:1	1	11	0	–

Tab. 3 Continued

Cluster[a]	Protein name	m/z observed	Charge	(M+H)⁺ matched	Peptide sequence[b]	Modification[c]	Residue[d] Start	Residue[d] End	Missed cleavage sites	New Swiss-Prot entries
II.5	Qcr2p	598.9	2	1196.2	VLVAGTSAYC$_c$R	–	189	199	0	P83782[e]
	Ipp1p	703.7	2	1405.7	ALPTNTFTGQEAR	–	219	231	0	P83777[e]
		462.6	2	923.4	ATNEWFR	–	186	192	0	–
III.	Eft3p	675.9	2	1349.5	NNFANMADPEAR	–	299	310	0	–
	Eft2p	509.6	2	1017.5	ISPPVVSYR	–	556	564	0	–
		590.0	2	1178.5	EGPIFGENC$_c$R	C$_c$:681	673	682	0	–
		852.8	1	852.8	SPNKHNR	–	579	585	1	–
	Bel1p	656.2	2	1311.6	AEVHALAFSPNR[g]	–	59	70	0	P83774[e]
		696.2	2	1391.6	LTGGEDNQYGIPK[h]	–	47	59	0	–
		724.7	2	1448.7	LWDLETGETTQR[h]	–	92	103	0	–
		904.5	2	1807.4	ISPSDQSSTVISASWDK[h]	–	157	173	0	–
IV.	Por1p	658.4	2	1314.6	GELDTSVVPNGAR	–	96	108	0	P83781[e]
	Grp2p	601.9	2	1202.5	GDPSQADAWKK	–	297	307	1	P83775[e]
		605.2	2	1208.4	VVVTSSYAAVGR	–	128	139	0	–

a) According to Tab. 2
b) Amino acid sequences deduced manually from MS/MS data are depicted in bold. I/L and Q/K as the *C. albicans* genomic database (CandidaDB, at http://www.genolist.Pasteur.fr/CandidaDB).
c) C$_c$ denotes carbamidomethylated cysteine and S$_a$ acetylated serine (protein with acetylated N-terminus). Cysteines were carbamidomethylated during equilibration (reduction/alkylation with iodoacetamide) of IPG gels.
d) Residue denotes the position of the identified sequence in the protein.
e) The peptide sequences reported on this protein were annotated in the Swiss-Prot and TrEMBL knowledgebase, and will appear in this data bank under this accession number.
f) Observed N-terminal sequence of mature protein
g) Peptide sequence located in the encoding product of the exon 2 of the *C. albicans* BEL1 gene
h) Peptide sequence located in the encoding product of the exon 1 of the *C. albicans* BEL1 gene

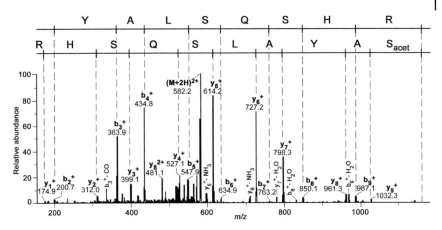

Fig. 3 MS/MS product ion spectrum acquired from a doubly charged precursor ion at m/z 582.2 for the N-terminal tryptic peptide of Shm2p. Series of b- and y-ions are labeled. The peptide sequence deduced de novo, from both series derived from this spectrum, is given at the top of this panel in N- to C-terminal (above) and C- to N-terminal (below) direction. I/L and Q/K as the CandidaDB. The generated amino acid sequence disclosed two cotranslational processes, i.e., cleavage of N-terminal methionine and N-terminal acetylation (Sacet).

nogenic protein annotated in our database can be connected bidirectionally with its specific entry in the Swiss-Prot database, allowing comprehensive information about the protein in question to be retrieved.

16.3.2
Comparison of 2-D *C. albicans* antigen recognition patterns obtained with serum samples from patients with and without systemic candidiasis

With the aim of excluding the occurrence of nonspecific reactions, the blotted 2-D *C. albicans* protein maps were also hybridized individually with serum samples from subjects without systemic *Candida* infection (see Section 2.1 for further details) using the same dilutions as for positive sera. Only four of the most abundant *C. albicans* proteins (Eno1p (enolase), Pgk1p (phosphoglycerate kinase), Adh1p (alcohol dehydrogenase), and Pdc11p (pyruvate decarboxylase)) were immunodetected using control serum specimens (Tab. 2), which, however, showed a considerably lower level of reactivity with these proteins than systemic candidiasis patients' samples. The rest of the antigens identified in this study seemed to be specifically immunorecognized by systemic candidiasis patients' sera. Intriguingly, the most prominent isoforms of Pgk1p and Adh1p (i.e., spots whose (pI, M_r) coordinates were (5.89, 46) and (6.12, 46) for Pgk1p, and (5.70, 45) and (5.83, 45) for Adh1p) reacted with both patients' and controls' serum samples, whereas the remaining protein species, corresponding to the most acidic isoforms, were only recognized by systemic candidiasis patients' sera (Fig. 5).

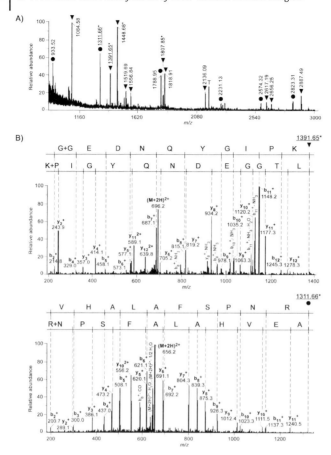

Fig. 4 Characterization of *C. albicans* Bel1p by MS. (A) MALDI-TOF spectrum of the tryptic digest of Bel1p. Its peptide masses were used for the CandidaDB search and retrieved the encoding product of *C. albicans BEL1* gene's exon 1 (CA4588). Further peptide peaks were retrospectively found to match to the encoding product of the *BEL1* gene's exon 2 (CA4589) after a detailed visual analysis of its PMF and database searches. Filled triangles and circles designate peptide peaks matching the products encoded by *BEL1* exon 1 and exon 2, respectively. The peptides analyzed by nanoESI-IT MS are depicted with asterisks. Peaks of trypsin autolysis products are labeled with "T". (B) Fragmentation patterns of two doubly charged precursor ions at m/z 696.2 (top) and 656.2 (bottom) for two Bel1p tryptic peptides (underlined in A) obtained by nanoESI-IT MS. The y series of ions and those from the b series are given. I/L and Q/K as the CandidaDB. The *de novo* peptide sequences are shown at the top of each panel in *N*- to *C*-terminal (above) and *C*- to *N*-terminal (below) direction. These mapped to the protein sequences encoded by the products of *C. albicans BEL1* gene's exon 1 (top panel) and exon 2 (bottom panel). The presence of peptide sequences (derived from a single spot) matching two CandidaDB entries promoted its confirmation by corresponding peptide signals in the PMF as indicated in (A). These MS data support the predicted *BEL1* gene model containing two exons.

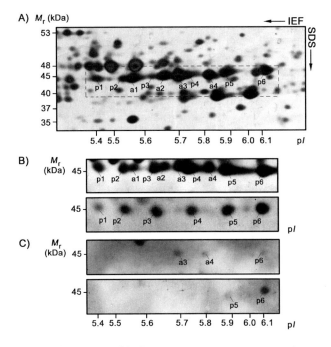

Fig. 5 Differentiation of the human immune response against the different *C. albicans* Pgk1p and Adh1p isoforms. (A) Enlargement of a section of the silver-stained 2-DE map illustrated in Fig 1. The relative positions of the different reactive isoforms of *C. albicans* Pgk1p (spots p1–6) and Adh1p (spots a1–4) immunodetected on 2-D blots are labeled. Refer to Tab. 2 for detailed data on their pI and M_r. The broken rectangle shows the close-up section of the 2-D Western blots exhibited in (B) and (C). (B) Expanded sections of 2-D immunoblots obtained using serum specimens from two cases of systemic candidiasis recruited in this study (cases 1 and 2; top and bottom panels, respectively). The different isoforms of Pgk1p (both panels) and Adh1p (top panel) detected as two trains of six and four hybridizing spots, respectively, could be ubiquitously expressed during systemic candidiasis. (C) Close-up of representative 2-D Western blots obtained using controls' serum samples. No reactivity was observed against the acidic isoforms of Pgk1p and Adh1p using these samples. In doing so, post-translational modifications may mediate changes in their epitope specificities, triggering the production of specific antibodies during systemic *Candida* infection.

16.3.3
Differences in the 2-D *C. albicans* antigen recognition profile associated with infection progression

We have also conducted a preliminary study to establish the differences in antigen recognition occurring during the course of the systemic *Candida* infection. To do this, the 2-D *C. albicans* protein blots were probed with systemic candidiasis patients' serum specimens collected at two time points: (i) at the time of diagnosis, in which previous antifungal prophylaxis had not been administered in any patients, and (ii) at one month after the fungal infection had started, when all patients had received amphotericin B as antifungal treatment. Additionally, each blot was stripped and then hybridized with other specimens in an attempt to ensure the close variations in their antigenic repertoire. These serum samples were taken from patients who suffered from hematological malignancies, including myelodysplastic syndrome (case 1), lymphoma (case 2), and leukemia (cases 3 and 4 receiving allogeneic BMT, who also developed graft-*versus*-host disease). Most of these cases had several risk factors that predisposed them to systemic candidiasis (see Tab. 1). Fortunately, three cases (cases 1–3) underwent an improvement of their clinical status with disappearance of fungal infection, whereas the remaining case (case 4), for whom the antifungal therapy (*i.e.*, amphotericin B plus fluconazole) was not effective, died within 45 days from systemic candidiasis.

Not unexpectedly, the 2-D antigen recognition profile associated with infection progression for the case with a fatal outcome did not correlate with the counterparts for patients recovering from systemic candidiasis who, on the whole, had similar 2-D antigen recognition profiles (Tab. 4). The overall immunoreactivity pattern found with serum samples from patients who survived was as follows. In parallel with clinical improvement, circulating serum levels of antibodies to *C. albicans* (i) Eno1p (high levels) were maintained or increased, (ii) Pgk1p (high levels) and 52-kDa mannoprotein (mp52) remained unaffected or declined, (iii) methionine synthase (Met6p) and Pdc11p decreased or even disappeared, (iv) transketolase (Tkl1p) was maintained or vanished, (v) reductase (Grp2p) disappeared, (vi) triose phosphate isomerase (Tpi1p), glyceraldehyde-3-phosphate dehydrogenase (Gap1p), and different members of heat shock proteins appeared or remained almost unchanged, (vii) fructose biphosphate aldolase (Fba1p) continued unaltered, and (viii) remaining immunogenic proteins were found to be ambiguous, since their tendencies could not be well defined with the cases investigated. Fig. 6A shows differences in a characteristic immunoreactivity profile associated with infection progression among patients who recovered from systemic *Candida* infection. Intriguingly, the number of *C. albicans* proteins recognized by these serum specimens enhanced in parallel with infection evolution, *i.e.*, from 12–21 to 18–28 immunogenic proteins (Tab. 4). In contrast, in the patient case who did not survive, serum concentrations of antibodies to *C. albicans* (i) Eno1p (low levels) declined, (ii) Pgk1p, Gap1p, Fba1p and Adh1p increased, (iii) Met6p, Grp2p and Cdc9p (pyruvate kinase) were maintained, (iv) some members of Hsp70p family appeared, and (v) Tpi1p, Pdc11p, Tkl1p, mp52 and other antigens (see Tab. 4) were not detected. This profile is illustrated in Fig. 6B.

16.4
Discussion

The purpose of this study was to launch into the expression profiling of *C. albicans* immunogenic proteins that induce a specific antibody response during the course of systemic *Candida* infection in a well-known high-risk population, *i.e.*, in patients

Tab. 4 Summary of the differences in the *C. albicans* antigen recognition profile along the course of systemic candidiasis among the cases recruited in this study

Immunogenic proteins		Cases recovering from systemic candidiasis						Fatal case		Frequency of spots (isoforms)
		Case 1 (MDS)		Case 2 (NHL)		Case 3 (AML)		Case 4 (CLL)		
Cluster[a]	Name	Before[b]	After[c]	Before	After	Before	After	Before	After	
I.1	Ssa1p	–	–	–	+	–	+	–	–	2
	Sb1p	–	–	–	+	–	+	–	+	2
	Ssc1p	–	–	+	++	–	+	–	+	3
	Sse1p	+	–	–	–	–	+	–	+	2
I.2	Hsp90p	–	–	–	++	–	+	–	–	1
	Pdi1p	–	–	–	+	–	+	–	–	2
II.1.a	Hkx2p	–	–	–	+	–	–	–	–	1
	Pgi1p	–	–	+	+	+	+	–	+	1
	Fba1p	–	–	+	+	–	+	+	++	2
	Tpi1p	–	+++	+	+	–	+	–	–	2
	Gap1p	–	+++	++	++	–	++	++	+++	2
	Pgk1p	+++++	+++++	+++	+	++++	+++	++	+++	6
	Gpm1p	–	–	–	–	–	+	–	+	2
	Eno1p	+++++	+++++	+++	++++	+++	+++++	++	+	5
	Cdc19p	–	++	+	+	–	–	+	+	2
II.1.b	Pdc11p	++++	+	+	–	+	–	–	–	3
	Adh1p	++	++++	–	–	–	–	+	++++	4
II.1.c	Aco1p	–	+	–	–	–	+	+	+	2
	Mdh1p	–	–	–	–	++	+	–	–	1
II.1.d	Tkl1p	+	+	+	+	++	–	–	–	2
	Ino1p	++	–	–	–	++	–	–	–	3
	Acs2p	–	+	–	+	–	–	–	–	1
II.2	Ach1p	–	+	+	+	–	–	–	–	1

Tab. 4 Continued

Immunogenic proteins		Cases recovering from systemic candidiasis						Fatal case		Frequency of spots (isoforms)
		Case 1 (MDS)		Case 2 (NHL)		Case 3 (AML)		Case 4 (CLL)		
Cluster[a]	Name	Before[b]	After[c]	Before	After	Before	After	Before	After	
II.3	Met6p	++++	+++	+	–	+++	–	+	+	5
	Sah1p	–	–	+	+	+	–	–	–	1
	Ilv5p	–	+	–	+	+	+	–	–	1
	Shm2p	–	+	–	–	–	+	–	–	1
	Leu1p	–	+	–	–	–	–	–	–	1
II.4	Imh3p	–	+	–	+	–	–	–	–	1
	Ade17p	–	+	+	++	–	–	–	–	3
II.5	Qcr2p	–	–	+	+	–	–	–	–	1
	Atp1p	–	–	+	–	+	–	–	–	1
	Atp2p	–	–	–	–	+	–	–	–	1
	Ipp1p	–	–	+	–	+	–	–	–	1
III.	Eft3p	–	–	–	–	+	+	–	–	2
	Eft2p	–	+	–	+	–	–	–	–	2
	Tif1p	–	–	–	+	–	+	–	–	1
	Bel1p	–	–	–	–	+	–	–	–	1
IV.	Por1p	–	–	+	–	+	–	+	–	2
	Grp2p	+	–	+	–	–	–	–	+	2
	Hem13p	+	–	–	+	–	+	–	–	1
	Ipf17186p	–	–	–	–	+++	+	–	–	2
	mp50	+++	–	+++	+++	–	–	–	–	1
	mp51	–	–	+++	++++	–	–	–	–	1
	mp52	++++	+	+++	+++	+	+	–	–	1
Overall[d]		12/45 (26.7%)	18/45 (40.0%)	21/45 (46.7%)	28/45 (62.2%)	16/45 (35.6%)	19/45 (42.2%)	8/45 (17.8%)	14/45 (31.1%)	85

Tab. 4
a) According to Tab. 2
b) At the time of diagnosis of systemic candidiasis (before antifungal treatment)
c) At one month after starting systemic candidiasis
d) Number of different *C. albicans* immunogenic proteins recognized by each serum specimen.

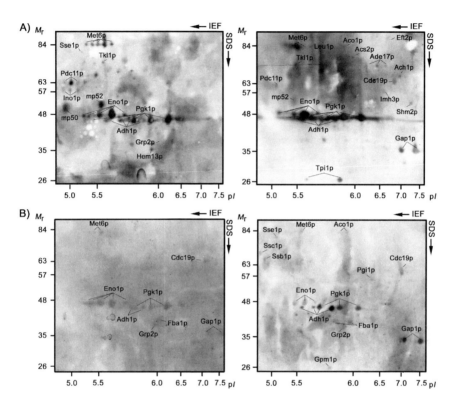

Fig. 6 Differences in the 2-D *C. albicans* antigen recognition profile occurring during the course of the systemic *Candida* infection, using hematological malignancy patients' serum specimens collected at the time of its diagnosis (left panels) and at one month after starting the fungal infection (right panels). See Tab. 4 for further details. Spot names refer to those in Tab. 2. (A) Representative 2-D immunoreactivity pattern found with serum samples from patients recovering from systemic candidiasis (case 1, with underlying myelodysplastic syndrome). (B) 2-D immunoreactivity pattern obtained using the collected samples from the patient who died witin 45 days from systemic candidiasis (case 4, suffering from chronic lymphoblastic leukemia).

with underlying hematological malignancies, in an attempt to characterize potential candidates for improving the current diagnosis and follow-up of systemic candidiasis. Despite the fact that these immunocompromised patients may present a delayed, reduced, or absent antibody response [22], the combination of the high

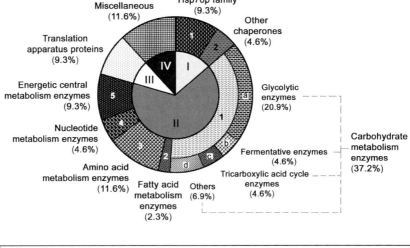

Fig. 7 Pie charting summarizing the distribution of main functional classes (inner circle) and subclasses (outer circle) of the *C. albicans* housekeeping enzymes identified in this study as targets of human immune response to systemic candidiasis. Roman and Arabic numerals, and letters refer to those in Tables 2–4. See Tab. 2 for further details.

resolving power of 2-DE gels, the specificity of antigen-antibody reaction (immunoblotting), and the extreme sensitivity of ECL detection, however, has afforded the establishment of a reference 2-DE *C. albicans* antigen recognition pattern for future large-scale screenings of systemic candidiasis patients' serum specimens. Recently, this proteomic approach has been successfully applied to identify antigens both of *C. albicans* [26–29, 40] and of other human pathogens [25, 41–46].

16.4.1
C. albicans housekeeping enzymes can stimulate the human immune system during systemic candidiasis

The entire genome sequence of a free-living organism is found to be crucial for the high-throughput characterization of its proteome [47]. In the light of this, the recent availability of the CandidaDB – a genomic database for the *C. albicans* strain SC5314 in the public domain – in conjunction with MS analyses have allowed the unambiguous identification of 42 different *C. albicans* protein targets of the human immune response to systemic candidiasis, 24 of which displayed several isoforms on 2-DE profiles. The presence of these protein species may be due either to posttranslational modifications, such as phosphorylation, sulfation, or acetylation (regarded as essential components for protein structure and/or enzyme activity), or to artificial chemical modifications occurring during sample preparation [48]. To our knowledge, this report is the first to prove (i) antigenic properties for 26 of them in *C. albicans* (3 of which have been in turn submitted for publication in a murine model by our research group [49]), and (ii) the ability for 35 of them to stimulate the human immune system during natural systemic *Candida* infection (see Tab. 2 for further details). It is worth mentioning that notwithstanding the fact that some of these identified antigens were only found in individual samples, the prevalence of all these proteins appears to be high when serum specimens from patients with systemic candidiasis belonging to different risk groups were similarly assayed only at the time of its diagnosis (A. Pitarch *et al.*, unpublished observations), strengthening their potential for diagnosis and/or follow-up of these fungal infections.

Our results evidenced at least three main functional categories of housekeeping proteins as *C. albicans* antigens (Fig. 7 and Tab. 2). The role that certain housekeeping enzymes play as elicitors of antibody production during candidiasis is not unprecedented [24, 27, 28]. Several putative explanations could account for this finding. First, it is conceivable that a large amount of these housekeeping enzymes may be released in the human body as a result of damage or lysis inflicted on *C. albicans* cells by the host during invasive infection, and subsequently double as antigens. Second, since most of these proteins are also known to be present on the cell surface of *C. albicans* [24, 36, 50, 51] and/or other species [28, 52–56], their immunogenic nature may also be attributed to this extracellular location, which enables these proteins to be exposed naturally even in absence of damage to *C. albicans* cells [24]. Third, regardless of their location, the immunodominant trait of these proteins could, however, be due to their relative abundance in the *C. albicans* proteome, this facilitating its visibility and targeting by the host immune effector cells. Fourth, several groups have suggested that some of these housekeeping enzymes may behave as moonlighting proteins [57–59] implicated in multiple functions relying on their subcellular location, cell type, oligomeric state, cellular concentration of ligand or substrate, binding sites, complex formation, *etc.* [57]. For instance, the bacterial elongation factor EF-Tu can take part in peptide elongation, and in parallel be found moonlighting as a chaperone involved in protection from stress [60] and in immunogenicity (*i.e.*, in eliciting a specific immune response) [25, 42]. The existence of these multifunctional proteins might save large amounts of energy in protein biosynthesis and DNA replication, and in turn in growth and reproduction [57]. Finally, the presence of these ubiquitous and highly conserved housekeeping enzymes, whose epitopes can be shared by several infectious agents, could supply the host immune system with a universal signal for infection and allow the subsequent development of a natural resistance to infection [24, 61].

16.4.1.1 Heat shock proteins (Hsps)

Hsps are often accepted as immunodominant antigens in a widespread range of infections [24, 62, 63], and are sometimes associated with protective immunity [64, 65]. Our findings disclosed that in addition to Ssa1p [27, 66], Ssa2p [67], and Ssb1p

[27] of the *C. albicans* Hsp70p family identified previously as antigens, another further two members of this family, *i.e.*, Ssc1p (associated with *C. albicans* cell surface [51]) and Sse1p (with calmodulin-binding activity in the baker's yeast [68]) could also act as targets of the human antibody response to systemic candidiasis. Intriguingly, *C. albicans* Hsp90p may exist both as an 82-kDa form in the cell wall and cytoplasm and as 76 and 47-kDa fragments in the cytoplasm [69]. Nonetheless, only antibodies directed against the 47-kDa [70] and 82-kDa forms (in this study) have been detected in systemic candidiasis, indicating that the epitopes exposed naturally by these forms are not shared by the 76-kDa peptide. *C. albicans* Hsp90p may result in more resistant to host defense mechanisms, serving as a virulence factor [71, 72].

16.4.1.2 Metabolic proteins

We identified some proteins belonging to main pathways of carbohydrate, fatty acid, amino acid, nucleotide and energy metabolism as inducers of human immune response to systemic *Candida* infection (listed in Tab. 2). Among them, glycolytic and fermentative enzymes, which are also present on the *C. albicans* cell surface [36, 50, 51], are major antigenic determinants during candidiasis [24, 28], some of which can play a putative protective role (discussed below) [29, 73]. Surprisingly, all enzymes of the highly conserved glycolytic pathway proved to be immunogens in patients with systemic candidiasis, including those hitherto uncharacterized as *C. albicans* antigenic proteins, *i.e.*, Hxk2p (hexokinase) and Pgi1p (in this research), and 6-phosphofructokinase (A. Pitarch *et al.*, unpublished observations). Supporting these findings, diverse groups have reported that some *C. albicans* glycolytic and fermentative enzymes seem to interact specifically with human extracellular matrix proteins, such as plasminogen [74], plasmin [75], fibronectin [76, 77], laminin [76], and integrin-like vitronectin [77].

It is also to note that some identified antigens with metabolic functions may also trigger antibody responses to other fungal infections. For instance, malate dehydrogenase was found as a *Paracoccidioides brasiliensis* immunogenic protein [78] as well as a *Malassezia furfur* allergen [79]. Interestingly, the abundance of *C. albicans* Eno1p, Fba1p, Tkl1p, Ino1p, Met6p, Sah1p, Ilv5p, Ade17p, and Qcr2p (see Tab. 2 for their full name) appeared to be increased in response to antifungal treatment with mulundocandin [80].

16.4.1.3 Elongation factors and ribosomal proteins

There is growing evidence that these proteins, colocalized at the cell envelope of different microorganisms [36, 51, 53, 56], also play an important role in the immune response to a variety of infections, given that some of them have already been identified as antigens and proposed as potential vaccines [81–84]. The *C. albicans* elongation factor 1 was found as a plasminogen-binding protein [74]. Furthermore, the *C. albicans* elongation factor 2 (Eft2p) was predicted to be target for sordarin (an antifungal drug) [85], raising the question as to whether it might also be considered as a candidate vaccine target against systemic candidiasis mimicking their counterparts in other species. Amazingly, the ribosomal protein Bel1p contains four WD-40 repeats in its sequence, which may mediate protein-protein interactions. In this sense, these WD-domains might be implicated in the antigen-antibody interaction.

16.4.1.4 Miscellaneous proteins

Although no immunogenic role has been described for the *C. albicans* porin Por1p to date, several previous studies have revealed that some antigenically conserved outer membrane porins can be exploited as a vaccine target to protect humans against infection by *Pseudomonas aeruginosa* [86], *Neisseria meningitidis* [87], *Haemophilus influenzae* [88], *Helicobacter pylori* [89], or *Leptospira* spp. [90], to name but a few.

16.4.2
Natural anti-*Candida* antibodies might be correlated with differentiation of the human immune response

Our results revealed that low levels of circulating antibodies directed against some abundant *C. albicans* glycolytic and fermentative enzymes (Eno1p, Pgk1p, Adh1p, and Pdc11p) were often present in subjects with no evidence of candidiasis. The presence of these natural anti-*Candida* antibodies may be attributable to continuous exposure of these antigens during harmless colonization of *C. albicans*, i.e., a commensal inhabitant of the human microflora [22, 23]. Alternatively, considering the ubiquitous nature, the great abundance and the high degree of sequence homology of these enzymes across species [91], they could, therefore, cross-react with antibodies elicited by other human commensal or infectious agents.

A relevant finding of our study is that some acidic isoforms of these enzymes or, more exactly, of Pgk1p and Adh1p proved to be specifically recognized by systemic candidiasis patients' serum specimens. These data suggest that their epitopes – elicitors of a specific immune response – may be altered as a result of the potential post-translational modifications undergone by these protein species. Consequently, these epitopes could either not induce an antibody response during *C. albicans* colonization or not be shared by other commensal or infectious agents. Strengthening this idea, differentiation of the host immune response can, therefore, take place against these post-translationally modified forms of Pgk1p or Adh1p, as reported previously for ribosomal proteins from *H. pylori* [25] and *Salmonella* spp. [92]. Although *N*-terminal acetylation has recently been proposed for *C. albicans* Pgk1p and Adh1p on the basis of prediction methods of *Saccharomyces cerevisiae* protein sequences [74], post-translational modifications of these *C. albicans* enzymes, however, have not yet been characterized. Accordingly, further studies are now required to shed light on the type of post-translational modification(s) for these *C. albicans* immunogenic proteins and its/their relationship with discrimination of the immune response to these isoforms. Interestingly, specific epitopes of these acidic isoforms might be exploited as candidates for the development of future diagnostic tests for systemic candidiasis, owing to the precise epitopic specificity of these anti-Pgk1p and anti-Adh1p antibodies mounted by patients with systemic candidiasis.

16.4.3
Serum levels of specific anti-*Candida* antibodies could be useful for the clinical follow-up of systemic candidiasis

In a previous study, different serologic responses to systemic candidiasis along the course of such infections were outlined between systemic candidiasis-resistant and -susceptible mouse strains immunized with sublethal and lethal doses, respectively [27]. Nonetheless, since antigens and/or epitopes recognized in animal models upon artificial immunization may diverge from those detected in humans, the antigen recognition profiling during natural infection and disease progression using human serum specimens is a prerequisite to search for reliable serological markers for the clinical follow-up of systemic candidiasis in infected patients and/ or for monitoring the efficacy of antifungal treatment regimens. Indeed, little is known about the progression and clinical implications of the human immune response before and after complete eradication of systemic candidiasis as a result of the antifungal therapy administered. It has been proposed that detection of antibodies against *C. albicans* germ tubes might be useful for therapeutic monitoring of invasive candidiasis in patients undergoing hematological malignancies [93].

Although, so far only a small number of paired patient sera could be assayed, some promising observations in the antibody response monitoring were, however, achieved. Detection of specific antibodies directed against Hsp90p, members of Hsp70p family, Fba1p, Tpi1p, Gap1p, Pgk1p, Eno1p, Pcd11p, Tkl1p, Met6p, Grp2p, and mp52 might be of prognostic value for systemic candidiasis. High serum levels of anti-enolase antibodies could be associated with recovery from systemic candidiasis in humans. This is in agreement with the fact that antibodies against *C. albicans* enolase seem to be protective in experimental infection models [27, 29, 73], in which it was postulated that elevated anti-Eno1p antibody titers could help to extend the survival rate of mice with systemic candidiasis. Likewise, the maintenance or development of an antibody response against Hsp90p, Tpi1p or a 52-kDa mannoprotein in patients who survived the infection could also be related to the protective capacity reported for antibodies to *C. albicans* Hsp90p [94], Tpi1p [29], or mannans [31, 95], respectively. The falling or even disappearance of serum anti-Met6p or anti-Pgk1p antibody levels might indicate that the antifungal therapy cleared the fungal infection, consistent with the detection of high levels of these antibodies in lethal *C. albicans* infections [27]. We are currently working to confirm these preliminary findings using a large number of 2-D antigen recognition patterns.

In conclusion, the delineation of *C. albicans* protein antigens that are naturally exposed during systemic candidiasis, especially those that become targets of the human antibody response, may certainly offer a new perspective on the detection and therapeutic monitoring of these infections. More than 40 candidate *C. albicans* antigens for its diagnosis were identified in this preliminary study and some of these revealed differences in their recognition patterns associated with infection progression (*i.e.*, with the efficacy of antifungal therapy). However, the full potential of this set of antigens for diagnosis and/or follow-up of systemic candidiasis is currently being addressed in our laboratory by a large-scale screening of patients' and controls' serum specimens. Furthermore, our study is also being extended to the characterization of *C. albicans* yeast and hyphal cell wall proteins [36] that elicit a specific immune response in patients with systemic candidiasis. All these identifications are deposited in our public Web-based proteomic database [32], and can be searched and retrieved in a user-friendly query interface using Swiss-Prot/ TrEMBL accession numbers or entry names. It is, therefore, our hope that these data serve as a basis for further studies focused on improving serological tests for diagnosis and/or prognosis of systemic candidiasis, and even on the design of vaccines or antifungal drugs for these infections.

We are indebted to Dr. A. Jiménez (from the Department of Internal Medicine, Clinic Hospital from Salamanca, Spain) for supplying human serum samples. We also thank M. L. Hernáez (from Genomics and Proteomics Center, Complutense University, Madrid, Spain, accessible through the Net at our URL address http://www.ucm.es/info/ gyp/proteomica) for her assistance with the MALDI-TOF spectra acquisition. This work was supported by grants SAF 2000-0108 and BIO-2003-00030 from Comisión Interministerial de Ciencia y Tecnología (CICYT) and CPGE 1010/2000 Strategic Groups from Comunidad Autónoma de Madrid (CAM), Spain. A. Pitarch was recipient of a fellowship from the Merck, Sharp & Dohme (MSD) Special Chair in Genomics and Proteomics, Spain.

16.5
References

[1] Anaissie, E., Pinczowski, H., *Recent Results Cancer Res.* 1993, *132*, 137–145.

[2] Anttila, V. J., Elonen, E., Nordling, S., Sivonen, A. *et al.*, *Clin. Infect. Dis.* 1997, *24*, 375–380.

[3] Bodey, G., Bueltmann, B., Duguid, W., Gibbs, D. *et al.*, *Eur. J. Clin. Microbiol. Infect. Dis.* 1992, *11*, 99–109.

[4] Goodrich, J. M., Reed, E. C., Mori, M., Fisher, L. D. *et al.*, *J. Infect. Dis.* 1991, *164*, 731–740.

[5] Karabinis, A., Hill, C., Leclercq, B., Tancrede, C. *et al.*, *J. Clin. Microbiol.* 1988, *26*, 429–432.

[6] Guiot, H. F., Fibbe, W. E., 't Wout, J. W., *Clin. Infect. Dis.* 1994, *18*, 525–532.

[7] Bow, E. J., Loewen, R., Cheang, M. S., Schacter, B., *Clin. Infect. Dis.* 1995, *21*, 361–369.

[8] Iwasaki, H., Misaki, H., Nakamura, T., Ueda, T., Arisawa, M., *Int. J. Hematol.* 2000, *71*, 266–272.

[9] Yamazaki, T., Kume, H., Murase, S., Yamashita, E., Arisawa, M., *J. Clin. Microbiol.* 1999, *37*, 1732–1738.

[10] Bow, E. J., *Br. J. Haematol.* 1998, *101* Suppl 1, 1–4.

[11] Verfaillie, C., Weisdorf, D., Haake, R., Hostetter, M. *et al.*, *Bone Marrow Transplant.* 1991, *8*, 177–184.

[12] Meyers, J. D., *Semin. Oncol.* 1990, *17*, 10–13.

[13] Jones, J. M., *Clin. Microbiol. Rev.* 1990, *3*, 32–45.

[14] Pfaller, M. A., *Mycopathologia* 1992, *120*, 65–72.
[15] Edwards, J. E., Jr., *N. Engl. J. Med.* 1991, *324*, 1060–1062.
[16] Kontoyiannis, D. P., Lewis, R. E., *Lancet* 2002, *359*, 1135–1144.
[17] Sanglard, D., Odds, F. C., *Lancet Infect. Dis.* 2002, *2*, 73–85.
[18] D'Antonio, D., Iacone, A., Schioppa, F. S., Bonfini, T., Romano, F., *Curr. Microbiol.* 1996, *33*, 118–122.
[19] Philpott-Howard, J., *Curr. Opin. Infect. Dis.* 1996, *9*, 218–222.
[20] Reiss, E., Morrison, C. J., *Clin. Microbiol. Rev.* 1993, *6*, 311–323.
[21] Ponton, J., Moragues, M. D., Quindos, G., in: Calderone, R. A. (Ed.), *Candida and Candidiasis*, ASM Press, Washington, DC 2002, pp. 395–425.
[22] Buckley, H. R., Richardson, M. D., Evans, E. G., Wheat, L. J., *J. Med. Vet. Mycol.* 1992, *30* Suppl 1, 249–260.
[23] De Repentigny, L., Kaufman, L., Cole, G. T., Kruse, D. et al., *J. Med. Vet. Mycol.* 1994, *32* Suppl 1, 239–252.
[24] Martinez, J. P., Gil, M. L., Lopez-Ribot, J. L., Chaffin, W. L., *Clin. Microbiol. Rev.* 1998, *11*, 121–141.
[25] Haas, G., Karaali, G., Ebermayer, K., Metzger, W. G. et al., *Proteomics* 2002, *2*, 313–324.
[26] Pitarch, A., Pardo, M., Jimenez, A., Pla, J. et al., *Electrophoresis* 1999, *20*, 1001–1010.
[27] Pitarch, A., Diez-Orejas, R., Molero, G., Pardo, M. et al., *Proteomics* 2001, *1*, 550–559.
[28] Pardo, M., Ward, M., Pitarch, A., Sanchez, M. et al., *Electrophoresis* 2000, *21*, 2651–2659.
[29] Fernandez-Arenas, E., Molero, G., Nombela, C., Diez-Orejas, R., Gil, C., *Proteomics* 2004, *4*, 1204–1215.
[30] Han, Y., Cutler, J. E., *Infect. Immun.* 1995, *63*, 2714–2719.
[31] Han, Y., Ulrich, M. A., Cutler, J. E., *J. Infect. Dis.* 1999, *179*, 1477–1484.
[32] Pitarch, A., Sanchez, M., Nombela, C., Gil, C., *J. Chromatogr. B* 2003, *787*, 129–148.
[33] Gillum, A. M., Tsay, E. Y., Kirsch, D. R., *Mol. Gen. Genet.* 1984, *198*, 179–182.
[34] Bradford, M. M., *Anal. Biochem.* 1976, *72*, 248–254.
[35] Valdes, I., Pitarch, A., Gil, C., Bermudez, A. et al., *J. Mass Spectrom.* 2000, *35*, 672–682.
[36] Pitarch, A., Sanchez, M., Nombela, C., Gil, C., *Mol. Cell Proteomics* 2002, *1*, 967–982.
[37] Shevchenko, A., Wilm, M., Vorm, O., Mann, M., *Anal. Chem.* 1996, *68*, 850–858.
[38] Gallager, S., Winston, S. E., Fuller, S. A., Hurrell, J. G. R., in: Ausubel, F. M., Brent, R., Kingston, R. E., Moore, D. D. et al. (Eds.), *Current Protocols in Molecular Biology*, Greene Publishing Associates and Wiley Interscience, New York 1993, pp. 10.8.1–10.8.21.
[39] Gharahdaghi, F., Weinberg, C. R., Meagher, D. A., Imai, B. S., Mische, S. M., *Electrophoresis* 1999, *20*, 601–605.
[40] Pitarch, A., Sanchez, M., Nombela, C., Gil, C., *J. Chromatogr. B* 2003, *787*, 101–128.
[41] Jungblut, P. R., Grabher, G., Stoffler, G., *Electrophoresis* 1999, *20*, 3611–3622.
[42] Sanchez-Campillo, M., Bini, L., Comanducci, M., Raggiaschi, R. et al., *Electrophoresis* 1999, *20*, 2269–2279.
[43] Jungblut, P. R., Schaible, U. E., Mollenkopf, H. J., Zimny-Arndt, U. et al., *Mol. Microbiol.* 1999, *33*, 1103–1117.
[44] Kimmel, B., Bosserhoff, A., Frank, R., Gross, R. et al., *Infect. Immun.* 2000, *68*, 915–920.
[45] Dea-Ayuela, M. A., Ubeira, F. M., Pitarch, A., Gil, C. et al., *Parasite* 2001, *8*, S117–S119.
[46] Havlasova, J., Hernychova, L., Halada, P., Pellantova, V. et al., *Proteomics* 2002, *2*, 857–867.
[47] Liska, A. J., Shevchenko, A., *Proteomics* 2003, *3*, 19–28.
[48] Sickmann, A., Marcus, K., Schafer, H., Butt-Dorje, E. et al., *Electrophoresis* 2001, *22*, 1669–1676.
[49] Fernandez-Arenas, E., Molero, G., Nombela, C., Diez-Orejas, R., Gil, C. *Proteomics* 2004, in press. DOI 10.1002/pmic.200400292.
[50] Chaffin, W. L., Lopez-Ribot, J. L., Casanova, M., Gozalbo, D., Martinez, J. P., *Microbiol. Mol. Biol. Rev.* 1998, *62*, 130–180.

[51] Urban, C., Sohn, K., Lottspeich, F., Brunner, H., Rupp, S., *FEBS Lett.* 2003, *544*, 228–235.
[52] Hughes, M. J., Moore, J. C., Lane, J. D., Wilson, R. et al., *Infect. Immun.* 2002, *70*, 1254–1259.
[53] Lim, D., Hains, P., Walsh, B., Bergquist, P., Nevalainen, H., *Proteomics* 2001, *1*, 899–909.
[54] Singh, V. K., Jayaswal, R. K., Wilkinson, B. J., *FEMS Microbiol. Lett.* 2001, *199*, 79–84.
[55] Chivasa, S., Ndimba, B. K., Simon, W. J., Robertson, D. et al., *Electrophoresis* 2002, *23*, 1754–1765.
[56] Marques, M. A., Chitale, S., Brennan, P. J., Pessolani, M. C., *Infect. Immun.* 1998, *66*, 2625–2631.
[57] Jeffery, C. J., *Trends Biochem. Sci.* 1999, *24*, 8–11.
[58] Pancholi, V., *Cell Mol. Life Sci.* 2001, *58*, 902–920.
[59] Ejiri, S., *Biosci. Biotechnol. Biochem.* 2002, *66*, 1–21.
[60] Caldas, T. D., El Yaagoubi, A., Richarme, G., *J. Biol. Chem.* 1998, *273*, 11478–11482.
[61] Kauffmann, S. H. E., *Immunol. Today* 1990, *11*, 129–136.
[62] Moseley, P., *Immunopharmacology* 2000, *48*, 299–302.
[63] Maresca, B., Kobayashi, G. S., *Experientia* 1994, *50*, 1067–1074.
[64] Pockley, A. G., *Lancet* 2003, *362*, 469–476.
[65] Matthews, R., Burnie, J., *Trends Microbiol.* 1996, *4*, 354–358.
[66] La Valle, R., Bromuro, C., Ranucci, L., Muller, H. M. et al., *Infect. Immun.* 1995, *63*, 4039–4045.
[67] Lopez-Ribot, J. L., Alloush, H. M., Masten, B. J., Chaffin, W. L., *Infect. Immun.* 1996, *64*, 3333–3340.
[68] Mukai, H., Kuno, T., Tanaka, H., Hirata, D. et al., *Gene* 1993, *132*, 57–66.
[69] Burt, E. T., Daly, R., Hoganson, D., Tsirulnikov, Y. et al., *Ann. Clin. Lab. Sci.* 2003, *33*, 86–93.
[70] Matthews, R., Burnie, J., Tabaqchali, S., *J. Clin. Microbiol.* 1987, *25*, 230–237.
[71] Zhang, X., Essmann, M., Burt, E. T., Larsen, B., *J. Infect. Dis.* 2000, *181*, 1441–1446.
[72] Hodgetts, S., Matthews, R., Morrissey, G., Mitsutake, K. et al., *FEMS Immunol. Med. Microbiol.* 1996, *16*, 229–234.
[73] van Deventer, H. J., Goessens, W. H., van Vliet, A. J., Verbrugh, H. A., *Clin. Microbiol. Infect.* 1996, *2*, 36–43.
[74] Crowe, J. D., Sievwright, I. K., Auld, G. C., Moore, N. R. et al., *Mol. Microbiol.* 2003, *47*, 1637–1651.
[75] Jong, A. Y., Chen, S. H., Stins, M. F., Kim, K. S. et al., *J. Med. Microbiol.* 2003, *52*, 615–622.
[76] Gozalbo, D., Gil-Navarro, I., Azorin, I., Renau-Piqueras, J. et al., *Infect. Immun.* 1998, *66*, 2052–2059.
[77] Klotz, S. A., Pendrak, M. L., Hein, R. C., *Microbiology* 2001, *147*, 3159–3164.
[78] da Fonseca, C. A., Jesuino, R. S., Felipe, M. S., Cunha, D. A. et al., *Microbes Infect.* 2001, *3*, 535–542.
[79] Onishi, Y., Kuroda, M., Yasueda, H., Saito, A. et al., *Eur. J. Biochem.* 1999, *261*, 148–154.
[80] Bruneau, J. M., Maillet, I., Tagat, E., Legrand, R. et al., *Proteomics* 2003, *3*, 325–336.
[81] Ribeiro, L. A., Azevedo, V., Le Loir, Y., Oliveira, S. C. et al., *Appl. Environ. Microbiol.* 2002, *68*, 910–916.
[82] Oliveira, S. C., Splitter, G. A., *Vaccine* 1996, *14*, 959–962.
[83] Iborra, S., Soto, M., Carrion, J., Nieto, A. et al., *Infect. Immun.* 2003, *71*, 6562–6572.
[84] Coler, R. N., Skeiky, Y. A., Bernards, K., Greeson, K. et al., *Infect. Immun.* 2002, *70*, 4215–4225.
[85] Dominguez, J. M., Martin, J. J., *Antimicrob. Agents Chemother.* 1998, *42*, 2279–2283.
[86] Larbig, M., Mansouri, E., Freihorst, J., Tummler, B. et al., *Vaccine* 2001, *19*, 2291–2297.
[87] Wright, J. C., Williams, J. N., Christodoulides, M., Heckels, J. E., *Infect. Immun.* 2002, *70*, 4028–4034.
[88] Neary, J. M., Yi, K., Karalus, R. J., Murphy, T. F., *Infect. Immun.* 2001, *69*, 773–778.
[89] Peck, B., Ortkamp, M., Nau, U., Niederweis, M. et al., *Microbes. Infect.* 2001, *3*, 171–179.
[90] Haake, D. A., Mazel, M. K., McCoy, A. M., Milward, F. et al., *Infect. Immun.* 1999, *67*, 6572–6582.
[91] Canback, B., Andersson, S. G., Kurland, C. G., *Proc. Natl. Acad. Sci. USA* 2002, *99*, 6097–6102.

[92] Adams, P., Fowler, R., Howell, G., Kinsella, N. *et al.*, *Electrophoresis* 1999, *20*, 2241–2247.

[93] Garcia-Ruiz, J. C., del Carmen, A. M., Regulez, P., Quindos, G. *et al.*, *J. Clin. Microbiol.* 1997, *35*, 3284–3287.

[94] Matthews, R., Hodgetts, S., Burnie, J., *J. Infect. Dis.* 1995, *171*, 1668–1671.

[95] Han, Y., Kanbe, T., Cherniak, R., Cutler, J. E., *Infect. Immun.* 1997, *65*, 4100–4107.

Index

a
Accessory gene regulator 225
Active immunisation 181
Aerobic culture 267
Anaerobic culture 267
Antibodies 181
Antigens 181, 289
Antigen selection 21

b
Bacillus subtilis 31
Bacterial mixtures 75
Bacterial pathogenesis 203
Bartonella henselae 203
Basic proteins 121
Bioinformatics 63

c
Candida albicans 181
Cell wall 153
Chlamydia 1
Chlamydophila 1
Cold shock proteins 75

d
Databases 63

f
Francisella tularensis ssp. 249

g
Genome 1

h
Hematological malignancy 289

i
Immunoproteome 21
ionization-time of flight mass spectrometry 75

l
Listeria monocytogenes 153

m
Mass spectrometry 203
Matrix assisted laser desorption/ionization-time of flight-mass spectrometry 75, 97, 279
Metabolism 267
Microorganisms 63
Minimal protein identifier 97
Moonlighting proteins 153
Mycobacterium avium 279
Mycobacterium leprae 121
Mycobacterium tuberculosis 97

n
Narrow range pH gradient 141

o
On-probe digestion 75
Outer membrane protein 203

p
Peptide mass fingerprinting 97
Physiological proteomics 31
Plasminogen 153
Post-translational modification 141
Protection 181
Proteome 1, 121, 267
Proteome analysis 249
Proteomics 63

r

Rapid species identification 75
Reverse transcriptase-polymerase chain reaction 279
Review 31

s

Secretome 21
Selective solubilization 75
SigB 31, 225
Staphylococcal accessory regular 225
Staphylococcus aureus 225
Stress adaptation reactions 31
Subcellular fractions 121
Subunit vaccine 21
Surface plasmon resonance 153
Systemic candidiasis 279

t

Therapeutic monitoring 289
THP-1 cells 279
Two-dimensional blot overlay 141
Two-dimensional gel electrophoresis 63, 97, 121, 203, 279
Two-dimensional polyacrylamide gel electrophoresis 1, 289

u

Unimolecular decomposition analysis 75

v

Vaccine 181
Vegetive Bacillus 75
Vibrio cholerae 267
Virulence factors 225, 249

Related Titles

Omenn, G. S. (Ed.)

Human Plasma Proteomics

2006
ISBN-13: 978–3-527–31757–8
ISBN-10: 3–527–31757–0

Lion, N., Rossier, J. S., Girault, H. (Eds.)

Microfluidic Applications in Biology

From Technologies to Systems Biology

2006
ISBN-13: 978–3-527–31761–5
ISBN-10: 3–527–31761–9

Humphery-Smith, I., Hecker, M. (Eds.)

Microbial Proteomics

Functional Biology of Whole Organisms

2006
ISBN-13: 978–0-471–69975–0
ISBN-10: 0–471–69975–6

Hacker, J., Dobrindt, U. (Eds.)

Pathogenomics

Genome Analysis of Pathogenic Microbes

2006
ISBN-13: 978–3-527–31265–8
ISBN-10: 3–527–31265-X

Sanchez, J.-C., Corthals, G. L., Hochstrasser, D. F. (Eds.)

Biomedical Applications of Proteomics

2004
ISBN-13: 978–3-527–30807–1
ISBN-10: 3–527–30807–5

Proteomics of Microbial Pathogens. Edited by Peter R. Jungblut and Michael Hecker
Copyright © 2007 WILEY-VCH Verlag GmbH & Co. KGaA, Weinheim
ISBN: 978–3-527–31759–2